TRACE ELEMENTS IN COAL

Volume I

Author

Vlado Valković

Institute Ruder Bošković
Zagreb, Yugoslavia

CRC Press, Inc.
Boca Raton, Florida

Library of Congress Cataloging in Publication Data
Valković, V.
Trace elements in coal

Includes bibliographical references.
1. Coal — Analysis. 2. Trace elements—
Analysis. I. Title.
TP325.V34 662.6′22 82-4386
ISBN 0-8493-5491-9 (v. 1) AACR2
ISBN 0-8493-5492-7 (v. 2)

Direct all inquiries to CRC Press, Inc., 2000 Corporate Blvd., N.W., Boca Raton, Florida, 33431.

International Standard Book Number 0-8493 (Volume I) 5491-9
International Standard Book Number 0-8493 (Volume II) 5492-7
Library of Congress Card Number 82-4386

Printed in the United States.

PREFACE

Coal has been widely used as a principal energy source since the 18th century. It is believed that coal was the main driving force behind the Industrial Revolution; for the first time in history, a truly cheap metal, iron, was made available. By the end of the 19th century coal surpassed wood as the main energy source in the U.S. (see Figure 1, Chapter 1, showing projections of the U.S. Energy Information Administration). Since then it has been used as the principal fuel for heating of buildings, manufacturing processes, and electrical power generation. For a long time coal was king.

Then, at the beginning of the 1950s it was dethroned by petroleum with the help of attacks by environmentalists. The problems of energy and environment were brought together. In order to maintain or increase the standard of living, an increase in the development of energy sources is required. So, to preserve the quality of life, the production of energy in an environmentally acceptable manner is required. So far, energy has been extracted from coal by direct combustion in steam boilers and generation of electricity. This results in damage to the environment, with the worst offender probably being sulfur in the coal.

With the predictable exhaustion of oil reserves and the finite amounts of deposits of some elements, attention is again being focused on coal. There have been some very important developments in coal utilization technology in the areas of coal liquefaction, coal gasification, and advanced methods of coal combustion to generate electric power. It may turn out to be equally important that coal combustion produces an ash which may offer an economical raw material for resource recovery. It has been generally accepted that coal technologies (unlike nuclear technologies, for example) can proceed in a largely empirical way. However, recently the need to accelerate the transition from experience-based technology to science-based technology has been recognized. This new approach can produce more efficient and cleaner technologies.

Coal has been described as a "bridge to the future". It is the most plentiful fossil fuel in the world, and it has a potential for filling a growing proportion of the demand for energy. It is found around the globe, but three countries (the U.S. the U.S.S.R., and China) own nearly two thirds of all known coal reserves. At present rates of consumption, these reserves would last the world more than 200 years, according to conservative estimates. Furthermore, geologists think the world probably has 15 times this much coal.

Developments in the U.S. are of great interest. The goal of the U.S. government to reduce dependency on imported fossil fuels, as well as some other factors, is moving the U.S. toward an economy based on coal as the primary fossil fuel. The use of coal as fuel is expected to triple in the period 1975 to 2000, from approximately 10 quads (1 quad = 10^{15} Btu) to 30 quads (see Figure 1, Chapter 1). Electric power utilities are the major users of coal, and most of the projected increase will also be for the generation of power. Minor increases in coal use as a chemical feedstock to replace petroleum sources is anticipated, but not in the next few years. The use of coal in coking will probably remain at about the present level.

Because of the way it was formed, coal may contain every naturally occurring element. Early chemical analyses of coal and its ashes were limited to the major elements (C, H, O, N, S, Si, Fe, Al, K, Ca, and Mg). Trace elements is often defined as any element whose concentration in coal is 1000 ppm (0.1%) or less. However, elemental composition is very variable, and for the sake of completeness, major elements should be included in discussions on elemental composition of coals.

The first trace element analyses of coal were performed in the 1930s. In the U.S., an ongoing program for the sampling and analysis of coal for trace elements began in 1948. At that time the lack of trace element analysis resulted from the difficulty of applying the classical wet chemistry methods of analysis to trace constituents. However, the advent of newer techniques of instrumental analysis has made analysis of trace elements more feasible. In recent years, general interest in chemical analysis of coal and coal related materials has significantly increased to the state where adequate analytic procedures and suitable standards are currently available for many trace elements. The number of complete modern analyses for trace elements in coal has increased greatly in recent years. As of 1979, the largest accumulation of these data, on 3700 samples of U.S. coal, is publicly available from the data bank of the U.S. Geological Survey in its National Coal Resources Data System.

The main emphasis of this book is on the occurrence and distribution of trace elements in coal. Nature and mode of occurrence of trace elements, variations in distributions, and concentration levels in ash are discussed in detail. The accumulated knowledge about all elements detected in coal is summarized. Special attention is paid to rare and uncommon elements and to radionuclides. Trace elements distribution in different phases of coal utilization processes is discussed next. Possibilities of trace element recovery from coal and coal ash are elaborated.

Environmental considerations are of great importance for the acceptance of coal-utilization techniques. Trace element contaminants from coal-fired power plants are studied by many researchers. The effects of radionuclides released in the environment are critically summarized.

Methods for the measurements of trace elements concentration in coal and its ash are described in the last chapter of Volume II. Optical methods, X-ray analysis, and nuclear methods are presented, in addition to classical chemical methods.

THE AUTHOR

Vlado Valković, Ph.D., is presently Scientific Advisor at the Institute Ruder Bošković, Zagreb, Croatia, Yugoslavia.

Dr. Valković graduated in 1961 from the University of Zagreb, with a B.A. degree in experimental physics and obtained his M.A. degree in nuclear physics in 1963 and his Ph.D. degree in 1964. The thesis title was "Nuclear Reactions with 14.4 MeV Neutrons on Light Elements".

Dr. Valković has been associated with two institutions during all his professional career — one in the U.S., one in Yugoslavia. The institutions are the Institute Ruder Bŏsković in Zagreb and the T. W. Bonner Nuclear Laboratories at Rice University, Houston, Texas.

Dr. Valković is a fellow of the American Physical Society, a member of the European Physical Society, Yugoslav Physical Society (Croatian section), Society for Environmental Geochemistry and Health, and other professional and honorary organizations.

His research has been done on both sides of the Atlantic: in the U.S. as well as in a number of laboratories in Europe.

Dr. Valković has published more than 100 research papers on nuclear physics and applications of nuclear techniques to the problems in biology, medicine, environmental research, and trace element analysis. He is the author of five additional books: *Trace Element Analysis* (Taylor and Francis, London, 1975), *Nuclear Microanalysis* (Garland, New York, 1977), *Trace Elements in Human Hair* (Garland, New York, 1977), *Trace Elements in Petroleum* (Petroleum Publishing, Tulsa, 1978),and *Analysis of Biological Material for Trace Elements Using X-ray Spectroscopy* (CRC Press, Boca Raton, 1980). His current major research interests include the study of the role and movements of the elements in nature.

To Georgia

TABLE OF CONTENTS

Volume I

Volume II

Chapter 1

COAL — ORIGIN, CLASSIFICATION, PHYSICAL AND CHEMICAL PROPERTIES

I. COAL ORIGIN

Coal is one of the fossil fuels found on our planet. It represents the accumulation of organic materials in sedimentary strata. The basic difference between coal and other fossil fuels (oil and gas) is in the fact that coal does not migrate but it undergoes *in situ* compaction with time, to form various ranks of coal. In many cases coal still contains recognizable source material and is composed chiefly of compressed and altered remains of terrestrial plant material such as wood bark, roots, leaves, spores, and seeds.

The results of accumulation of *in situ* residues and important debris in swamps leads to the formation of peat. After peat has accumulated for a period, it must be buried under mineral sediment, generally clay, silt, and sand. In time, the peat forms beds of coal that range from a few centimeters to many meters in thickness. The coal is often interbedded with shale, sandstone, and other sedimentary rocks. A single stratigraphic sequence may include several coal beds. Coal-bearing strata may include alternating marine and nonmarine beds. The coal beds are in the nonmarine parts of the section and are of brackish water or freshwater origin, though some peat swamps received occasional marine incursions.

In terms of biogeochemical cycles, coal represents the fixation of carbon from atmospheric carbon dioxide by plant growth over extensive geologic time. The release of all this accumulated fixed carbon into the atmosphere as carbon dioxide may result in an unbalanced biogeochemical carbon cycle. In the long term, the consequences of this imbalance may be the most significant impact of coal utilization, and these are currently under study by many scientists. Other elements will be mobilized to a far less dramatic level but in sufficient quantities to warrant concern (see References 1 to 27).

Van Krevelen[27] gives the following account of the processes involved in coal bed formation: during the period of coal formation lasting from the lower Carboniferous to the Permian, uninterrupted swamps covered large areas of the Northern Hemisphere. Due to the great depth of Carboniferous coal formations, it seems likely that these swamps underwent a process of subsidence that was largely balanced by sedimentation. Such sunken areas, called geosynclines, are usually formed by strong lateral, compressive forces acting on the crust of the Earth. The geosynclines probably did not sink at a constant rate, but alternated between periods of rapid and slow subsidence. During the periods of slow subsidence, shallow lagoons formed and sedimentation occurred at a more rapid rate than did the sinking, resulting in luxuriant growth of aquatic plants. The plant debris settled to the bottom and was acted upon by microorganisms to form peat deposits, which grew until the rate of subsidence increased again. Eventually, the swamp was submerged and covered by sediment, and a coal seam was formed. The pattern was probably the one most often followed in formation of coal deposits; however, some coal seams were probably formed by the build-up of silt in deeper waters.

It is generally accepted that there were at least two stages of coal formation from plant material: (1) the biochemical period of accumulation and preservation of plant material as peat and (2) the geochemical period of conversion of peat to coal. Although most coal researchers accepted this theory, agreement on the details of the actual chemical and physical changes that occurred in the process is not so widespread.[28]

Spackman[29] states that coal seams are laminar, sedimentary rock bodies ranging in thickness from a centimeter to several hundred meters. Most of the economically significant seams of the U.S. are 6 cm to 7 m thick, a few being as thick as 25 m. A seam may consist of a number of layers (lithobodies), and these layers may contain different mixtures of macerals and minerals, giving them different characteristics. Although the classification of coal components, macerals and minerals, will be discussed later, here we shall summarize van Krevelen's[27] description of the formation of major lithotypes.

Fusain was formed under very dry conditions. Many researchers feel that, because of its similarity to charcoal, fusain may have formed as the result of forest fires caused by lightning. Another theory suggests that it was produced by some exothermic microbial process.

Vitrain was formed more slowly under semidry conditions, as in a swamp where the groundwater level was just under the surface. The dead plant material sunk into the wet humus soil and was preserved. Because the stagnant water contained little oxygen, aerobic decomposition was inhibited, and the lignified material, which was also relatively resistant to anaerobic decay, was preserved. Under true wet swamp conditions, decay of plant material was carried out to a much higher degree, resulting in only the most resistant fragments being left, thus forming durain.

Clarain is believed to have been formed under conditions between the relatively dry environment of vitrain and fusain and the true wet swamp environment of durain. It is characterized by a relatively high vitrinite content. The woody tissue was evidently macerated before burial because clarains contain spores and other materials embedded in a matrix of vitrinite.[30]

Cannel coal was probably formed in lakes and pools where spores accumulated by wind and water were deposited in the muck on the bottom.

Boghead coal is very similar to cannel coal, but it originated from different material. It consists mainly of algae that died and settled into the mud.[27]

The U.S. Bureau of Mines system of classification of coal types is slightly different. Comparison of the international and the U.S. Bureau of Mines terminology used for basic coal types is shown in Table 1.[29]

Because of the inclusion of this sediment and waterborne dissolved elements, coal beds will contain all the elements found in the eroded rocks from which the sediments are derived. A coal bed could therefore contain all the naturally occurring elements. Additional elements can be introduced into the coal after burial by infiltrating groundwater, either precipitating as minerals or through ion exchange in the clays or organic matter.

According to Zubovic,[31,32] the ability of any particular element to form from chemical emplacement processes in the coal-forming swamp is regulated by several geological processes in the source area. Rapid and extensive uplift of the source areas would tend to reduce the length of time necessary for chemical weathering and would result in a greater influx of clastic materials and a smaller amount of soluble inorganic matter into the depositional basin. Under these conditions, a larger amount of the trace elements would be associated with the nonauthigenic mineral matter of coal and a lesser amount with the organic fraction.

Tectonic and climatic activity in the depositional basin must maintain conditions for the accumulation of organic matter. If subsidence is too rapid, little or no vegetation will be able to grow and clastic materials will dominate the depositional sequence. If little or no subsidence takes place, the biochemical cycle will be essentially completed and most of the vegetative organic matter will be oxidized. For organic matter to accumulate in significant quantities, an optimum balance of tectonic and climatic conditions must exist.

The rate of accumulation of organic matter is important in determining the concen-

Table 1
COMPARISON OF INTERNATIONAL AND U.S. BUREAU OF MINES TERMINOLOGY USED FOR BASIC COAL TYPES[9]

International	U.S. Bureau of Mines
Humic coals	Banded coals
Fusain (charcoal-like)	Bright (<20% opaque matter)
Vitrain (black, vitreous)	Semisplint (20 to 30% opaque matter)
Clarain (striated, glossy)	Splint (>30% opaque matter)
Durain (nonstriated, matte)	
Liptobiolithic coals	Nonbanded coals
Cannel	Cannel (abundant spore remains)
Boghead	Boghead (abundant algal remains)

After Spackman, W., Paper presented at Short Course on Coal Characteristics and Coal Conversion Processes, Pennsylvania State University, University Park, October 29 to November 2, 1973.

tration of trace elements in the resulting coal. If the availability of trace elements from a source area remains constant, then the concentrations that will result in the organic matter will be inversely proportional to the rate of accumulation of the organic matter.

In addition, if the areal extent of organic deposition is large, then low concentrations of trace elements will result. This is why generally thick widespread coal beds have low trace-element concentrations and, conversely, why thin-bedded coals have high concentrations.

During the initial phase of organic accumulation in a basin, the trace elements may be obtained by the plants from the underlying soil profile. When the peat accumulates to a point where the root systems are within the peat, then the dominant source becomes the elemental flux from the surrounding borderland. The mineral matter that makes up the soil most probably also had its origin in the same area.[31,32]

The age of a coal seam is important because coal constitutents vary according to the different materials found in plant matter, and these change during the evolution of the plants. Spackman[29] cites as an example the difference in exinite content of coals formed before and after the Tertiary period. During the Jurassic and Cretaceous periods, the dominant plant group was the gymnosperms, which are today represented by pines, fir, spruce, etc. In their reproductive cycle, gymnosperms produced and released large numbers of spores into the air. These spores were in such abundance during the dominant period of the gymnosperms that they constituted a large proportion of a hydroaromatic plant component of coal called exinite. Angiosperms are plants that appeared during the Cretaceous period and are today the dominant plant group. These plants evolved in embryo sac retained in an ovule on the sporophyte generation of the plant, and spores were no longer produced in such great abundance. Thus, one of the coal types common in the older coals is relatively rare in tertiary coals.

Geological and geochemical differences also influenced the production and composition of coal seams. For example, Illinois seams in the U.S. developed in close association with marine water, producing large concentrations of pyritic and organic sulfur in the coal, whereas coals that developed in parts of the Dakotas (U.S.) experienced minimal marine water influence and contain small concentrations of sulfur.[33]

The important role of microorganisms during coal formation is discussed in detail by Parks[28] and Given.[34,35] Microorganisms were essential to the early stages of the coalition process because they performed the function of converting dead plant material to the partially decayed organic matter called peat. Reasons for partial rather than

Table 2
REMAINING COAL RESERVES IN THE WORLD

Continent and country	Productible coal 10^9 metric tons	Percent of continental total	Percent of world total
Asia			
U.S.S.R.	600	52.3	25.8
China	506	44.1	21. 8
India	32	2.8	1.4
Japan	5	0.4	0.2
Others	4	0.4	0.2
Total	1147	100.0	49.4
North America			
U.S.	753	94.4	32.5
Canada	43	5.4	1.8
Mexico	2	0.2	0.1
Total	798	100.0	34.4
Europe			
Germany	143	47.5	6.2
U.K.	85	28.2	3.7
Poland	40	13.3	1.7
Czechoslovakia	10	3.3	0.4
France	6	2.0	0.3
Belgium	3	1.0	0.1
Netherlands	2	0.7	0.1
Others	12	4.0	0.5
Total	301	100.0	13.0
Africa	35	—	1.5
Australia	29	—	1.3
South and Central America	10	—	0.4
World Total	2320		100.0

After Meyers, R. A., Coal Desulfurization, Marcel Dekker, New York, 1977.

total decay of the plant material are still somewhat open to debate; however, studies of present peat deposits in Florida (U.S.) indicate two important factors, the inherent resistance of certain plant materials such as lignins to decay and the chemical condition of the swamp water. For example, the pH, oxygen content, and mineral matter would have determined whether the dominant bacterial population of the water was aerobic or anaerobic and thus would have controlled the rate of decomposition.

II. COAL DEPOSITS AND RESERVES

On a global scale coal deposits are much more uniformly distributed than oil and gas reserves. Remaining coal reserves of the world by region and principal coal producing countries are shown in Table 2. Table 2 indicates the big reserves in the U.S.S.R., China, and the U.S., which together comprise 80% of the world total. Europe, another big energy consumer, has 13% of the world total. Obviously, these figures should be taken with some caution, and they might be subject to some changes. For example, the U.S. coal resources were estimated in 1974 at 3.6×10^9 metric tons.[37] The 1974 estimate of the amount of coal that can be mined economically was 0.396×10^9 metric tons. These reserves will be sufficient to satisfy projected energy requirements in the U.S. (see Figure 1).

Coal-bearing rocks underlie about 14% of the land area of the contiguous U.S.

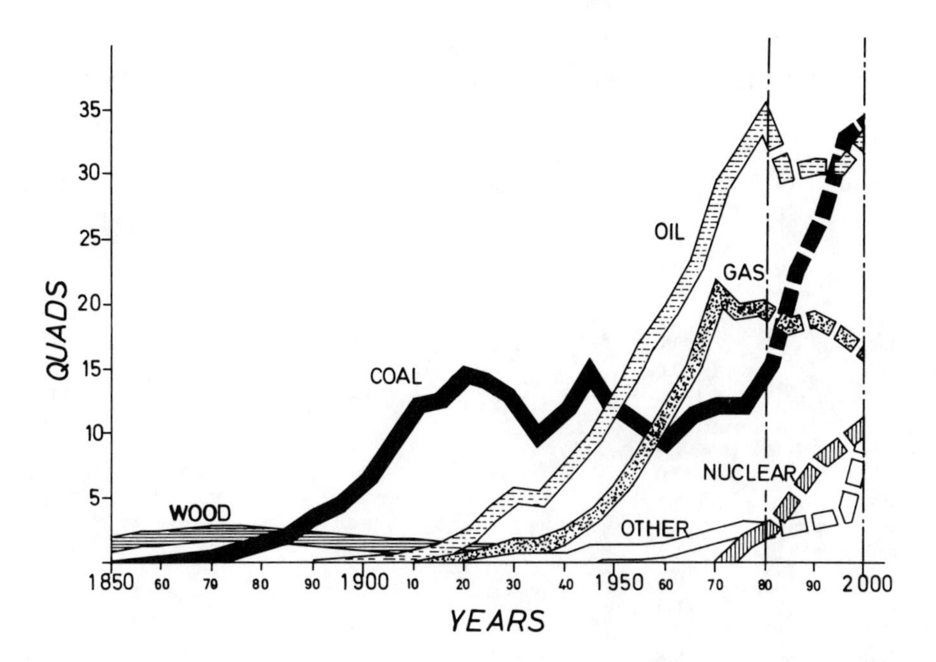

FIGURE 1. Different types of energy supply in U.S. projections from Energy Information Administration.

According to Averitt[37] bituminous, subbituminous, lignitic, and anthracitic coals represent 43.1, 28.1, 27.7, and 1.1%, respectively, of the identified U.S. coal resources in 1974. Geographical distribution of these reserves is shown in Table 3.

Reserves of bituminous coal, as estimated by the U.S. Geological Survey, generally are categorized according to coal thickness. Coals having a thickness of 14 to 28 in. are classified as thin, of 28 to 42 in. as intermediate, and of more than 42 in. as thick.

In this text coal of different U.S. regions will be mentioned many times. Table 4 shows a geological classification of the U.S. coals by rank and their distribution by regions.[30] Coals of the western and central states were formed about 150 million years later than those of the eastern and southeastern states; therefore, the plants that provide the organic material for these coals were different. Because the locations were different, the coals were also exposed to greatly differing geological conditions. Thus, it is not surprising that the coals of Appalachia differ greatly from those of the Dakotas. Basically, there are three differences between the coals of the eastern and western U.S. (1) The coals of the western regions were formed much more recently than the eastern coals and are thus generally of much lower rank; (2) the difference in geological location resulted in different mineral distribution between coals of the two regions, e.g., eastern coals generally have a higher sulfur content; and (3) both the difference in geologic time and location resulted in the appearance of different plants, so that different plant materials were incorporated into each of the coal.[33]

III. CHARACTERIZATION OF COAL

Many authors have tried to define coal; the definition of Schopf[38] seems acceptable. Coal is defined as a readily combustible rock containing more than 50% by weight and more than 70% by volume of carbonaceous material. This carbonaceous material resulted from the accumulation and slow decay of plant remains under water. Coal is formed through an apparently continuous series of alterations: living material > peat > lignite > subbituminous coal > anthracite. This succession of changes in the proper-

Table 3
IDENTIFIED COAL RESERVES OF THE U.S. (10^9 METRIC TONS)

State	Bituminous coal	Subbituminous coal	Lignite	Anthracite and semianthracite	Total
Alabama	12.1	—	1.8	—	13.9
Alaska	17.6	100.4	—	—	118.0
Arizona	19.2	—	—	—	19.2
Arkansas	1.5	—	0.4	0.4	2.3
Colorado	99.0	17.9	0.02	0.08	117.0
Georgia	0.02	—	—	—	0.02
Illinois	132	—	—	—	132
Indiana	29.8	—	—	—	29.8
Iowa	5.9	—	—	—	5.9
Kansas	17.0	—	—	—	17.0
Kentucky (eastern)	25.6	—	—	—	25.6
Kentucky (western)	32.7	—	—	—	32.7
Maryland	1.1	—	—	—	1.1
Michigan	0.2	—	—	—	0.2
Missouri	28.3	—	—	—	28.3
Montana	2.1	160.4	102.0	—	264.5
New Mexico	9.8	45.9	—	0.004	55.7
North Carolina	0.10	—	—	—	0.10
North Dakota	0	—	318.0	—	318
Ohio	37.4	—	—	—	37.4
Oklahoma	6.4	—	—	—	6.4
Oregon	0.05	0.25	—	—	0.30
Pennsylvania	58.0	—	—	17.1	75.1
South Dakota	0	—	2.0	—	2.0
Tennessee	2.3	—	—	—	2.3
Texas	5.5	—	9.3	—	14.8
Utah	21.0	0.2	—	—	21.2
Virginia	8.3	—	—	0.3	8.6
Washington	1.7	3.8	0.1	0.005	5.6
West Virginia	91.0	—	—	—	91.0
Wyoming	11.5	111.7	—	—	123.2
Other states	0.5	0.03	0.05	—	0.5
Total	677.7	440.4	433.7	17.9	1569.7

Note: Data as of January 1, 1974. Some figures shown in this table were rounded off to the nearest first or second decimal place and may therefore be slightly different from the original source. Dashes indicate that tonnage is too small to make a noticeable contribution and/or is included under other ranks.

Data from Averitt, P., Coal Resources of the United States, Bull. 1412, U.S. Geological Survey, Reston, Va., January 1, 1974.

ties and structure of coal is called metamorphism. The degree of metamorphism is called "rank".[39]

Transformation of the vegetable matter into peat and then into coals of different rank was the result first of decomposition due to bacterial action, then of dehydration, devolatization, and densification due to chemical and geodynamic processes. Heat, pressure, and long periods of time were required for the slow progressive change in rank from peat to lignite to subbituminous coal to bituminous coal. Strong folding apparently is necessary to produce anthracite, a higher rank of coal.[40]

According to Teichmuller and Teichmuller[41] time is an important factor in bringing about advances in the rank of coals. Pressure effects upon coal formation are related

Table 4
DISTRIBUTION OF COALS BY REGION IN THE U.S.

	Appalachian	Interior	North Great Plains	Rocky Mountain	Pacific	Gulf	Alaska
States	Pennsylvania, Ohio, Virginia, West Virginia, E. Kentucky, Alabama, Tennessee	East coast region, Indiana, Illinois, W. Kentucky, West coal region, Oklahoma, Missouri, Kansas, Arkansas	North and South Dakota, Montana, NE Wyoming	Many distinct basins of different geological history in SW Wyoming, Colorado, Utah, New Mexico, Arizona	Washington, Oregon, California	Parts of Arkansas, Texas, Louisiana, Mississippi, W. Alabama	Alaska
Age[a]	Carboniferous (300)	Carboniferous	Some Cretaceous, mostly early Tertiary (100—50)	Mostly Cretaceous, early Tertiary (130—60)	Tertiary (60—15)	Tertiary (70—30)	Cretaceous and early Tertiary (100—50)
ASTM rank	Principal deposits of high-volatile A and B, medium-volatile, and low-volatile bituminous anthracite	Principal deposits of subbituminous, high-volatile A, B, and C; minor deposits of low-volatile bituminous and anthracite	Principal deposits of lignite and subbituminous	Principal deposits of subbituminous and high-volatile A, B, and C bituminous; minor deposits of medium-volatile bituminous and anthracite	Principal deposits of lignite, subbituminous and high-volatile A, B, and C bituminous; minor deposits of anthracite	Principal deposits of lignite	Principal deposits of subbituminous and high-volatile A, B, and C bituminous

[a] Numbers in parentheses represent millions of years.

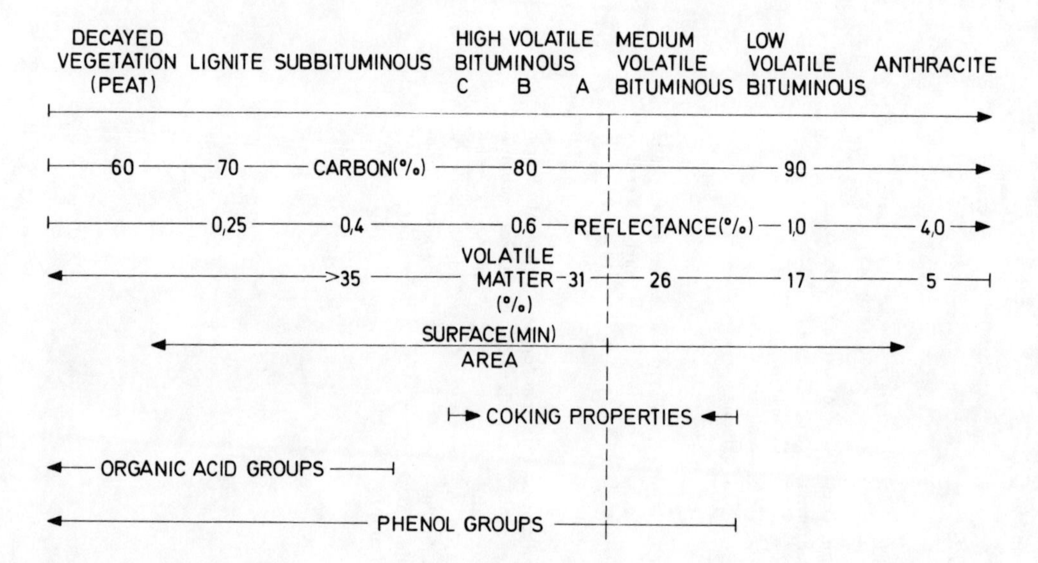

FIGURE 2. Progression of coal through the stages of development and some of the associated properties. (After Given, P. H.[89])

to its physical properties. Both porosity and moisture content of coal are decreased by pressure. The most important factor in the metamorphism of peat to coal is temperature.

The temperature gradient of the crust of the Earth is not constant; however, as an average, the temperature of the Earth increases by 2 to 5°C per 100 m of depth; temperatures of about 200°C would have been sufficient for anthracite formation.[27] In some cases frictional heat produced by tectonic movements such as shearing have brought about increases in coalification, but such instances cannot account for increases in coal rank with depth.[41]

Figure 2 shows progression of coal through the stages of development and some of the associated properties. The most important increases in the amount of fixed carbon and decreases in the amount of moisture and volatile matter that occur during the advance in rank from lignite to anthracite reflect also increases with an increase in rank and carbon content. Volatile matter, which decreases with increasing rank, consists of gaseous substances which are driven off by heat during the coalification process.

The surface area of coal is related to the plasticity and agglomerating characteristics of the coal on heating, an important factor in both coking and conversion processes. When exposed to rising temperatures, coals give off gases and condensable vapors, leaving a residue consisting mostly of carbon. Concurrently, the coal then softens and becomes plastic, and the particles cake to form a compact mass, which swells and resolidifies to form the porous coherent substance called coke. Minimum surface area and maximum plasticity are usually attained by high-volatile or medium-volatile bituminous coals, and only bituminous coals are suitable for coking. The dashed vertical line drawn through the region of minimum surface area indicates that coals consisting of about 85% carbon and 28% volatile matter should be very plastic an highly expanding. Van Krevelen[27] indicates that coals containing about 87% carbon and 29% volatile matter are most plastic and subject to expansion. Figure 2 also indicates that organic acid groups are essentially absent in bituminous coals and that phenol groups are rare in anthracite.

Actually, the relationship of a coal property to the degree of metamorphism cannot be realistically illustrated by a line; a band would be more appropriate. No coal bed is uniform in properties throughout; vertical samples taken from a single coal deposit will vary somewhat in analysis from all the others because of the increase in temperature with depth. Coals of the same rank from different basins would produce even wider bands.[33]

Volatile matter and fixed carbon are the more important contributors to the energy produced when coal is burned. The caloric or heating value of coal is expressed in British thermal units per pound (Btu/lb) BTU being the amount of heat required to raise the temperature of 1 avdp. lb of water 1°F at or near 39.2°F, its temperature at maximum density. The average heating value of bituminous coals in the U.S. is 13,100 Btu/lb, and of bituminous coals in the midwestern states is as follows: Illinois, 11,250 Btu/lb; Iowa, 9940 Btu/lb; Kansas, 12, 340 Btu/lb; Missouri, 11,320 Btu/lb; and Oklahoma, 12,910 Btu/lb. (1 Btu = 0.252 kCal).

The American Society for Testing and Materials (ASTM) classification of coals by rank is shown in Table 5, which details the parameters used to determine the rank of coals. According to this system, there are two ranks of lignite, three ranks of subbituminous coals, and five ranks of bituminous coals. The bituminous ranks include three levels of high-volatile-matter content indicated by A, B, and C, with A being the highest rank and containing the least volatile matter. These ranks are followed by medium- and low-volatile-matter bituminous coals. Anthracite coals are of higher rank than bituminous and are also separated into three levels. Only the bituminous coals are indicated to be agglomerating. The ASTM system, as well as the International System of Europe, uses volatile matter content to classify high-rank coals and calorific values for lower-rank cells. Coals having 69% or more fixed carbon on the dry, mineral-matter-free basis are classified according to fixed carbon, regardless of calorific value. This classification does not include a few coals principally nonbanded varieties, which have unusual physical and chemical properties and which come within the limits of fixed carbon or calorific value of the high-volatile bituminous and subbituminous ranks. All these coals either contain less than 48% dry, mineral-matter-free fixed carbon, or have more than 15,500 Btu/lb, calculated on the moist, mineral-matter-free basis. Distinctive features of coal of various ranks are shown in Table 6.[43] It should be noted that this classification system was designed primarily as a guide for the coking industry.

The International system assigns coal a three-digit number. The first digit indicates the rank or class of the coal, the second indicates the behavior of the coal when it is heated rapidly, and the third indicates the behavior of the coal when it is heated slowly, as in a coke oven. Table 7 shows a comparison of the International and ASTM systems of ranking coal.

Given[44,45] has summarized some important correlations of the various ranks of coal with aromatic carbon and the number of benzene rings. He notes that the aromaticity of coal increases with increasing rank and that much of the nonaromatic carbon is in hydroaromatic rings. His findings are shown in Table 8. According to Given,[44,45] the rank of coal ideally should be based upon the vitrinite it contains if the rank is to be used for scientific purposes.

The question of coal characterization is the one still being discussed. Maybe this point is best illustrated in the report of Hower et al.[46] The authors have conducted work to determine the extent to which the U.S. coals were adequately characterized. Questionnaires were sent to 81 agencies in the coal community selected to form a representative cross-section of the organizations concerned with coal characterization. Fifty-nine completed questionnaires were received. Respondents included representatives of the agencies with the longest experience in characterization and those in the best position to know the status of the knowledge of the composition and properties

Table 5
CLASSIFICATION OF COALS BY RANK (ASTM D 388)

Class	Group	Fixed-carbon limits[a]	Volatile-matter limits[a]	Calorific value[b] Btu/pound	Calorific value[b] kcal/g	Agglomerating character
Anthracitic	Meta-anthracite	≥98	<2			
	Anthracite	92—98	2—8			
	Semianthracite	86—92	8—14		7.2—7.8	Nonagglomerating
Bituminous	Low-volatile bituminous coal	78—86	14—22			
	Medium-volatile bituminous coal	69—78	22—31			
	High-volatile A bituminous coal	<69	≥31	≥14000	>7.8	Commonly, agglomerating
	High-volatile B bituminous coal			13000-14000	7.2—7.8	
				11500—13000	6.4—7.2	
	High-volatile C bituminous coal			10500—11500	5.9—6.4	Agglomerating
Subbituminous	Subbituminous A coal			10500—11500	5.9—6.4	Nonagglomerating
	Subbituminous B coal			9500—10500	5.3—5.9	
	Subbituminous C coal			8300—9500	4.6—5.3	
Lignite	Lignite A (lignite)			6300—8300	3.5—4.6	
	Lignite B (brown coal)			<6300	<3.5	

[a] Dry, mineral-matter-free basis.
[b] Moist, mineral-matter-free basis.

Table 6
COAL PHYSICAL APPEARANCE AND CHARACTERISTICS

Rank	Features	
	Physical appearance	Characteristics
Lignite	Brown to brownish black	Poorly to moderately consolidated; weathers rapidly; plant residues apparent
Subbituminous	Black; dull or waxy luster	Weathers easily; plant residues faintly shown
Bituminous	Black, dense, brittle	Is moderately resistant to weathering; plant structures visible with microscope; burns with short blue flame
Anthracite	Black, hard, usually with glassy luster	Very hard and brittle; burns with almost no smoke

Table 7
PARAMETERS OF INTERNATIONAL COAL CLASSIFICATION COMPARED WITH ASTM RANK

International class	ASTM rank	Volatile matter
		Dry, ash-free (%)
1A	Anthracite	3—6.5
1B	Anthracite	6.5—10
2	Semianthracite	10—14
3	Low-volatile bituminous	14—20
4	Medium-volatile bituminous	20—28
5	Medium-volatile bituminous	28—33
		Moist, ash-free (Btu)
6	High-volatile A bituminous	> 13,950
7	High-volatile B bituminous	12,960—13,950
8	High-volatile C bituminous or subbituminous A	10,980—12,960
9	Subbituminous B	10,260—10,980

After Yancey, H. F. and Geer, M. R., in *Coal Preparation,* American Institute of Mining, Metallurgical and Petroleum Engineers, New York, 1968, chap. 1.

of the coals comprising the U.S. coal fields. Analysis of the responses to the questionnaire resulted in the following conclusions:

1. The U.S. coals are inadequately characterized for their efficient and effective use, this being particularly true in relation to coal conversion technology and maintaining environmental quality.
2. The number of agencies conducting coal characterization programs is too small to meet the U.S. needs within the time frame required.
3. The scope of coal characterization programs should be expanded to develop a broader spectrum.

This is also true of coals at the other locations on this planet. Let us mention some of the work aimed to cure this situation.

Table 8
APPROXIMATE VALUES OF SOME COAL PROPERTIES IN DIFFERENT RANK RANGES

Component	Lignite	Subbituminous	Bituminous					Anthracite
			High-volatile			Medium-volatile	Low-volatile	
			C	B	A			
Carbon (mineral-matter-free) (%)	65—72	72—76	76—78	78—80	80—87	89	90	93
Oxygen (%)	3C	18	13	10	10—4	3—4	3	2
Oxygen, as COOH (%)	13—10	5—7	0	0	0	0	0	0
Oxygen, as OH (%)	15—10	12—10	9	?	7—3	1—2	0—1	0
Aromatic carbon atoms, % of total carbon	50	65	?	?	75	80—85	85—90	90—95
Average number of benzene rings per layer	1—2	?	2—3	2—3	2—3	2—3	5?	25?
Volatile matter (%)	40—50	35—50	35—45	?	31—40	31—20	20—10	10
Reflectance (%) (vitrinite)	0.2—0.3	0.3—0.4	0.5	0.6	0.6—1.0	1.4	1.8	4

After Given, P. H., Paper presented at Short Course on Coal Characteristics and Coal Conversion Processes, Pennsylvania State University, October 29 to November 2, 1973.

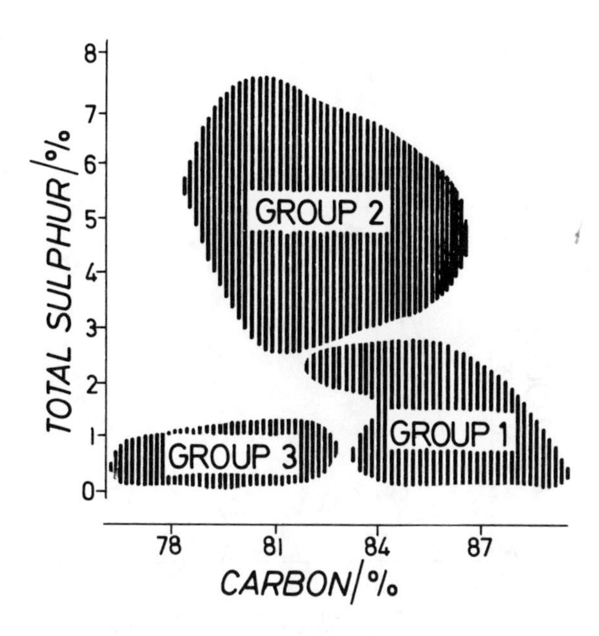

FIGURE 3. Clustering of coals by sulfur and carbon contents only. (After Yarzab, et al.[47]).

In their recent paper Yarzab et al.[47] have established a predictive base by means of which the behavior of coals in the liquefaction process could be computed. This has been achieved by the study of the conversion of 104 high-volatile bituminous coals at 400°C with tetralin in the back reactor. For the whole set of coals, volatile matter and vitrinite reflectance have the highest correlation coefficients with conversion (0.85 and −0.84, respectively). However, tests showed that the sample set contained more than one population. Cluster analysis partitioned the set into three reasonably homogeneous populations. A factor representing sulfur content was the major contributor from a set of variables in separating the coals into groups, with smaller contributions from factors related to rank and petrographic composition. Each of the groups contained samples mostly from one geological province, but 11 coals of relatively high sulfur content from the Eastern region were clustered in a group that contained also 25 coals from the Interior region. The ranges of conversion are distinctly different for the three groups, and the three regression equations developed for correlating conversion each require a different set of coal properties. The carbon and total sulfur contents of the coal samples are plotted against one another in Figure 3. The rings around the groups of points enclose points that belong to each of the three clusters (except that two points are wrongly clustered when wt% C and S_t are the only discriminants). It can be seen that the coals from the Eastern region that are found in Group 2 instead of Group 1 are those of high sulfur content relative to other coals in that region.

The classification discussed by Yarzab et al.[47] does segregate coals in a manner significant to their behavior in liquefaction under differing sets of conditions. Rank is a factor in this classification, but it is not dominant. Thus, the order of decreasing conversion level for the groups is $2 > 3 \gg 1$, whereas the order of increasing rank level is $3 < 2 < 1$ (Figure 3). In these circumstances, it is clearly desirable to examine closely the nature and scientific basis of the classification, to the extent possible at this time.

The major interest of the results obtained by Yarzab et al.[47] is in the area of coal classification. The various national and international systems for the classification of coals by rank were devised chiefly to assist those who wish to manufacture metallurg-

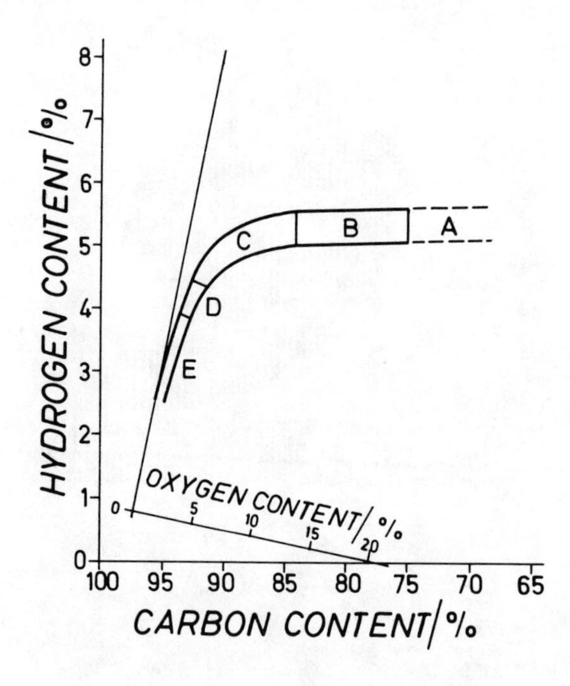

FIGURE 4. Seyler's coal chart: hydrogen, carbon, and oxygen relationship for different coals. (A) brown coal and lignite, (B) lignitous coal, (C) bituminous coal, (D) carbonaceous coal, (E) anthracite. (From Battaerd, H. A. J. and Evans[53].)

ical coke or to burn coal on a grate. It is hardly to be expected that such systems can effectively classify coals according to their behavior in processes of conversion to liquid or gaseous fuels. These systems rely on the volatile-matter yield, calorific value, and, in some cases, aspects of the coking behavior, for the division of coals into rank classes. The classifications are based on the concept of the degree of metamorphism. It is assumed that all coals follow a single, though broad, band of development. Although this assumption is still built into industrial classification systems, it has long been recognized by some to be inadequate because it takes no account of differing petrographic compositions. Owing to differences in structure of the various macerals in coals, the effects of metamorphic processes are different for different macerals. Therefore, because of differences in maceral contents, all coals can hardly fit a single band of development as defined by commonly determined coal properties, and deviations are particularly to be expected when the contents of macerals other than vitrinite are relatively high. Even then, use of the usual coal properties for purposes of classification may be misleading.[47]

There are three methods for coal classification which have been proved useful in the study of coalification.[48]

1. The Seyler or other similar diagrams such as those of both Mott[49] and Francis[4]
2. Statistical constitutional analysis according to van Krevelen[27]
3. The construction of model coal structures

These three methods may be combined with advantage. Seyler plotted the mass % of carbon against the mass % of hydrogen on a dry ash-free or mineral-matter-free basis (see Figure 4) for a wide range of coals, particularly from Great Britain. The

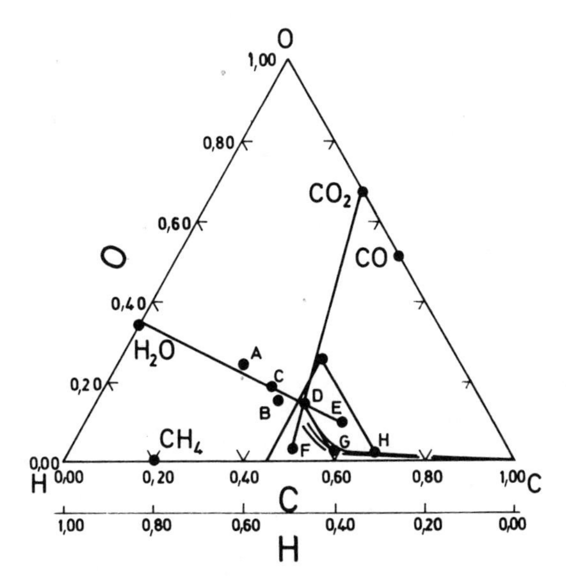

FIGURE 5. Ternary diagram with Seyler coalification band illustrating the chemistry of the formation and maturation of coals. (A) cellulose, (B) representative lignin, (C) representative peat, (D) brown coal/lignite, (E) inertinite rich coal, (F) exinite rich coal, (G) vitrinite rich coal, (H) anthracite. Dehydration and decarboxylation reactions are indicated by the lines CDE and DF, respectively. (After Stephens.[48])

plot defines a "coalification band" along which a great proportion of the coals lies in accordance with their degrees of metamorphism. Not all coals lay within the band. Coals of the same carbon content were found which incorporated both more and less hydrogen than indicated by the "coalification band". Microscopic examination suggested that those coals rich in hydrogen (perhydrous) were rich in the maceral exinite, while those poor in hydrogen (subhydrous) contained above-average amounts of the inertinite macerals.

Skew axes on the Seyler diagram indicate the calorific value of the coals and their volatile-matter yield under standard conditions. The swelling indices of the coals on carbonization are also indicated by contours. In short, the Seyler diagram is a comprehensive summary of the technological value of coals in terms of their chemical and petrographic compositions.

Van Krevelen and co-workers have developed the technique of structural constitutional analysis which combines elementary and functional-group analysis with data on several physical and chemical characteristics of coals to generate "structural parameters" by which coals can be compared in relative terms with one another. The characteristics considered included density, velocity of transmission of sound, heat of combustion, refractive index, and electrical conductivity. The principal "structural parameters" derived were indices of aromaticity, ring condensation, and molecular size. Subsequently, molecular models of coals which matched coals in their structural parameters were prepared and their technical characteristics and behavior compared with those of coals.

As pointed out previously, coal is the product of the degradation and metamorphism of ancient plant materials which consisted principally of cellulose and lignin. The first step in coalification is the formation of peat. Point A in Figure 5 lies at the position for the composition of cellulose, $(C_6H_{10}O_5)_m$, B at the position for a representative lignin, and C at the position for a representative peat. C lies between A and B indicating a contribution from both to the constitution of this peat; however, the contribution from the cellulose is significantly less than that of the lignin (Point B is obviously closer to C than to A) because cellulose undergoes biological degradation more readily than lignin.[48]

A representative lignite or brown coal has a composition which reports at Point D. A line from D through C virtually passes through the point for water. This indicates that the sum of all the reactions involved in the conversion of the various microscopic components of peat to the various microscopic components of a lignite or brown coal is one of dehydration.

If dehydration continues beyond the lignite/brown coal stage then the composition of the solid residue is driven towards Point E in the subhydrous region of the Seyler diagram classification, the region of composition in which coals relatively rich in the inertinite-type macerals report.

On the other hand, if decarboxylation replaces dehydration on about Point D then, as demonstrated by extending a line from the point for carbon dioxide through D, the composition of the residual solid is driven towards Point F in the perhydrous region of the Seyler diagram classification where coals relatively rich in the maceral exinite report.

A vectorial combination of dehydration and decarboxylation lines applied to lignite/brown coals near Point D drives the composition of the residue towards G in the Seyler coalification band. This is the reaction path followed during the formation of both bituminous coal and vitrinite, the principal maceral of coal.[48]

The subsequent maturation of bituminous coal from G onwards to anthracite requires mainly the loss of hydrogen. Bituminous coal is commonly found with adsorbed methane, indicating that this maturation step is achieved principally by the separation of this gas chemically from the bulk of the coal. However, for the composition of anthracite to conform at this stage with the rather closely defined coalification band, there must also be some loss of oxygen, very probably as water or carbon dioxide or carbon monoxide. These processes move the composition of the coal on the ternary diagram from G to H. The ultimate product of metamorphism, carbon as graphite (or diamond in the rarest of circumstances) requires chemically a continuation of the same kind of elimination of closely bound residual nitrogen and sulphur — presumably as ammonia and hydrogen sulfide — is of particular significance at this stage.[48]

Molecular models for coal have been published by Hill and Lyon,[50] Given,[34] and Wolk et al.[51] These models are representatives of possible or "average" structures, each an amalgam of its author's individual opinion with general opinion at the time of publication. Oka et al.[52] have more recently used a computer to generate model structures for coal-derived materials of relatively low molecular weight. Their technique requires specifically ^1H and ^{13}C NMR as well as molecular weight data.

Stephens[48] has described the advantage gained by combining these three methods into a single system based on a ternary diagram. This includes:

1. The chemical and petrographic evolution of coals to be illustrated in chemical terms (it also offers the opportunity of clarifying the relations between the evolution of hydrocarbons (oil and gas) and coals)
2. The chemistry of the technology of coal to be summarized clearly
3. The chemical structure of coals to be compared readily and systematically with all the structural types of organic compounds

Battaerd and Evans[53] have also advocated the use of ternary diagrams. Their work

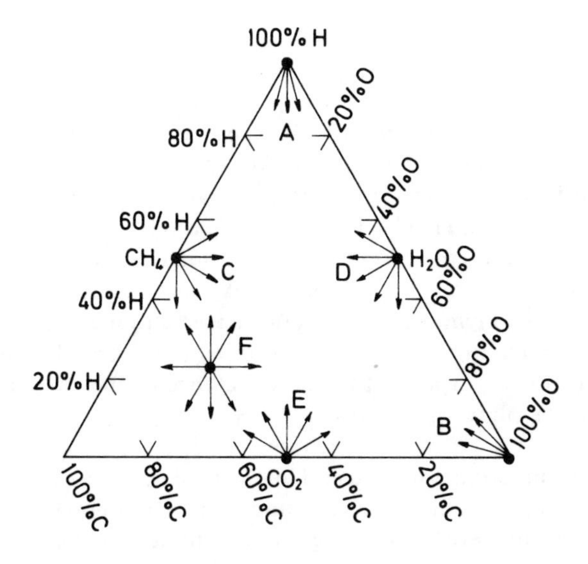

FIGURE 6. The bond-equivalent diagram. All percentages are bond-equivalent percentages. (A) dehydrogenation trajectories, (B) deoxidation trajectories, (C) demethanation trajectories, (D) dehydration trajectories, (E) decarboxylation trajectories, (F) decellulosation trajectories. (After Battaerd, H. A. J. and Evans.[53])

differs from the Stephens[48] in advocating plotting by bond equivalent percent rather than atomic percent and by applying the technique to coal technology rather than to coal geochemistry. Their work is presented with bond-equivalent diagram shown in Figure 6. The point midway between 100% C and 100% H represents methane, CH_4 as at this point C and H have combined in such a way as to satisfy their normal valences. Demethanation is represented by lines radiating out in all directions from this point. Likewise the point midway between 100% H and 100% O represents water, H_2O, and lines radiating out in all directions from this point represent dehydration. Carbon dioxide is represented by a point midway between 100% C and 100% O, and decarboxylation is represented by lines radiating out from this point. Similarly deoxidation is represented by lines radiating out from 100% O and dehydrogenation by lines radiating out from 100% H. Reactions involving all three elements are represented by lines radiating out from the appropriate point, e.g., decellulosation (removal of $C_6H_{10}O_5$) is represented by lines radiating out from the point 54.6% C, 22.7% H, 22.7% O.

According to Stephens,[48] plotting may be carried out efficiently by computer with a relatively simple program. The program first calculates a nominal chemical formula for each coal or coal-like material considered. These formulae have the configuration of $C_{1000}H_xO_y$. The triangular coordinates corresponding to these formulae along with those needed for any chemical are then calculated. The program prints a list of the coals and chemicals with their formulae, coordinates, and identity numbers. A plotter draws and labels a triangle of the requisite scale and graphs in points for each set of coordinates. For additional literature on coal characterization see References 54 to 57.

IV. PHYSICAL PROPERTIES

A. Optical Properties of Coal

The molecular structure of coal is directly related to molar refraction and absorption of light; therefore, optical properties of coals can be important indicators of their constitution. Of the optical properties, reflectance is the most conspicuous and easiest to determine.[27] The reflectance of coal has proved to be a rapid and accurate index of coal rank, and is largely uninfluenced by the petrographic variation which can affect the results of many of the chemical tests used for the same purpose. One of its principal advantages is that it can be made on extremely small amounts of material. Davis[58] has described in detail means and methods of reflectance determination.

Reflectance is defined as the proportion of normally incident light that is reflected by a plane, polished surface of the substance under consideration. The term "reflectivity" has been used synonymously with reflectance, although some authors have restricted use of the latter to the specific conditions that normally apply for measurements on coal. For most purposes it is the reflectance of the vitrinite component that is determined, for the following reasons:[58]

1. Vitrinite is the preponderant maceral in most coals.
2. Vitrinite often appears homogeneous under the microscope.
3. Particles of vitrinite are usually large enough to permit measurements to be made easily.
4. In the application of petrographic techniques to the industrial uses of coal, particularly in carbonization, interest is focused on the behavior of vitrinite which is the marceral principally responsible for the plastic and agglutinating properties of coal.

The reflectance in oil of an absorbing medium such as vitrinite is considered to be given by Beer's equation

$$r_o = \frac{(\mu - \mu_o)^2 + \mu^2 k^2}{(\mu + \mu_o)^2 + \mu^2 k^2} \quad (1)$$

where μ and k are the refractive and absorption indices of the vitrinite, respectively, and μ_o is the refractive index of the immersion oil (normally taken as 1.518 at the standard wavelength of 546 nm).

Ideally, μ and k should have been calculated previously, requiring that reflectances be determined separately in two different media (air and oil are convenient), using the following equations:

$$\mu = \frac{1/2(\mu_o^2 - 1)}{\mu_o(1 + r_o)/(1 - r_o) - (1 + r_a)/(1 - r_a)} \quad (2)$$

$$k^2 = \frac{r_a(\mu + 1)^2 - (\mu - 1)^2}{\mu^2(1 - r_a)} \quad (3)$$

where r_a and r_o are the measured reflectances of the vitrinite in air and oil, respectively, and μ_o is the refractive index of the oil at the temperature at which the determination is made. In these equations reflectances are expressed as fractions, not percentages.

In practice, it is acceptable to determine the reflectance in oil of vitrinite by direct comparison with a reflectance standard and without recourse to these equations, by assuming the refractive index of the immersion oil.[58]

Vitrinite usually behaves as a uniaxial negative substance, with the optical axis of the indicatrix normal to the bedding plane. Consequently, a surface that is perpendicular to the bedding displays a maximum refractive index, and thus reflectance, when the incident light is polarized in a direction parallel to the bedding, and a minimum reflectance when the light is polarized perpendicular to the bedding (the difference between the maximum and minimum reflectances is termed bireflectance). A surface parallel to the bedding displays maximum reflectance in all directions. On an oblique surface of coal, maximum reflectance is displayed parallel to the bedding and intermediate values between maximum and minimum are at right angles to that direction. Therefore a maximum reflectance value can always be obtained with a reflectometer by rotating any coal particle under the microscope until the bedding direction is parallel to the plane of vibration of the polarized incident light. Chiefly for this reason, many petrologists have preferred to use maximum reflectance as an indicator of rank. In the U.S., this procedure is followed almost exclusively, and is required in the method of the American Society for Testing and Materials.[58]

Very few reports have been published on the reflectance of macerals other than vitrinite. The existing data on reflectances for vitrinite, exinite, and micrinite show a decrease of carbon content in vitrinite. Fusinite is usually found to be the highest reflecting maceral.

Most of the reported work on the optical properties of coal constituents is done at either 530 or 546 nm (mercury line in the green region). Very little work has been published on the variation of optical properties of coal macerals with wavelength, although the small amount of data acquired has led to some interesting conclusions concerning the structure of coal.[58] The bulk of the reflectance data which have been assembled indicates that there is no stepwise or discontinuous change of reflectance with coal.

Davis[58] has shown that there is a good general relationship between reflectance and aromaticity. Many authors consider reflectance as the best single parameter of coal rank.

B. Mechanical Properties of Coal

The mechanical properties of coal have an important impact on nearly every stage of the coal utilization process from mining, through the various stages of preparation, and in the beginning stages of the final conversion or combustion process. Although various tests exist for characterizing specific mechanical properties such as grindability, or the swelling and plastic properties of coal that are important in the coking process, these tests usually measure a property that is important for one of the traditional preparation steps or end uses of coal. These tests do not in general provide mechanical properties data that is suitable for a general characterization of mechanical properties or is stated in terms that can be directly applied by the engineer who is designing new or innovative systems. Little fundamental understanding of the mechanical behavior of coal and its relationship to chemical and physical structure appears to exist.[59]

In discussing physical properties of coal one should recognize that coal is not a homogeneous, reproducible material. Therefore, the key to the understanding of physical properties of coal is through the relationship between the structure and properties. Past research into properties of coal, including strength, hardness, abrasiveness, particle shape, friction factor, and grindability is reviewed in the paper by Austin.[60] Although these properties vary with coal rank, petrographic and mineral matter variations within the rank have a substantial effect as well. Abrasiveness of coal, which adversely affects the wear rate of crushing and grinding equipment, was found to depend equally upon both coal rank and mineral content. An empirical formula is presented[60] which relates the Hardgrove Grindability Index to the maceral and mineral content of coal. Hardness values are tabulated for 26 minerals found in coal.

1. Pore Structure

Coal has an extensive pore structure as the result of the manner in which the coal molecules are linked together. Fresh coal usually contains a large amount of absorbed water. The moisture content ranges from 1 to 5% in bituminous coal, 20% in subbituminous, to nearly 45% by weight in lignite. High organic sulfur content coals tend to have a lower moisture content than low organic sulfur coals of the same rank. This was explained by the fact that since hydrogen bonding (by water) to sulfur groups is very weak or nonexistent, as evidenced by the lower water solubility of simple sulfur compounds compared to their oxygen analogs, high organic sulfur coals could be relatively more hydrophobic than low organic sulfur coals. This fact becomes important when considering the potential for chemical removal of organic sulfur from coal which requires penetration of the coal matrix with liquids in order to cause chemical reaction with the organic sulfur.[36]

There have been several excellent reviews discussing the problems associated with determining surface area and pore volume distributions in coals.[61-63] Coal is a microporous solid, and this has caused serious problems in establishing a suitable system to study the surface properties of coal. The classical low temperature adsorption of nitrogen at 77 K is unsuitable for coal because of activated diffusion. Of the adsorbates used, carbon dioxide, adsorbed at 195 K, seems to overcome most of the problems associated with other adsorbates, i.e., low adsorption temperatures or large molecular size preventing access to the fine pore structure.

In the paper by Debelak and Schrodt[64] a comparison was made between pore volume distribution evaluated from CO_2 adsorption isotherms calculated from the Cranston-Inkley method and the Medek method for four Kentucky coals. These distributions, shown in Figure 7, agree in the micropore range from approximately 0.6 to 2.0 nm. Most of the pore structure is in the micropore range. For the mercury penetration studies a rectilinear relation is found between penetrated volume and applied pressure from approximately 0.1 MPa (1 atm) up to 345 MPa (3400 atm). Surface areas were calculated from adsorption data by the BET, Langmuir, and Dubinin-Polanyi methods.

Several authors have characterized coal as a solid colloid having a certain porosity, to which it owes some of its properties — the ability to absorb gases and vapors, to swell in vapors and liquids, and to develop heat on wetting.[27] The porosity, or volume percentage occupied by pores, may be determined from density measurements by using helium and mercury as displacement liquids. Zwietering and van Krevelen[65] conclude from such experiments under pressure that coal contains two pore systems, a macropore system, which will allow entry of mercury under pressure, and a micropore system, which cannot be permeated by mercury even at high pressures. At normal pressure, mercury will not penetrate into the pore system if the pores are no larger than 10 μm in diameter.

Gan et al.[66] worked with 27 coals including samples representing all ranks and all major coal-fields of the U.S. Tests were performed on particles of 40 × 70 mesh size; therefore, they do not necessarily represent the whole coal. The pore structure of the coals was studied by determining (1) the total volume from helium and mercury densities, (2) the pore surface areas from absorption of nitrogen at 77 K and carbon dioxide at 298 K, (3) the macropore size distribution from mercury porosimetry, and (4) size distribution of pores below 30 nm from nitrogen isotherms measured at 77 K.

Because helium, as a small atom, can penetrate almost all the pores of the coal, tests were performed at normal pressure at 32.5°C. Application of some pressure, however, is required for penetration by mercury. Gan et al.[66] determined that negligible penetration of coal by mercury occurs above 60 psia; therefore, mercury density values were calculated from measurements of mercury displaced by the samples at 60 psia. Helium

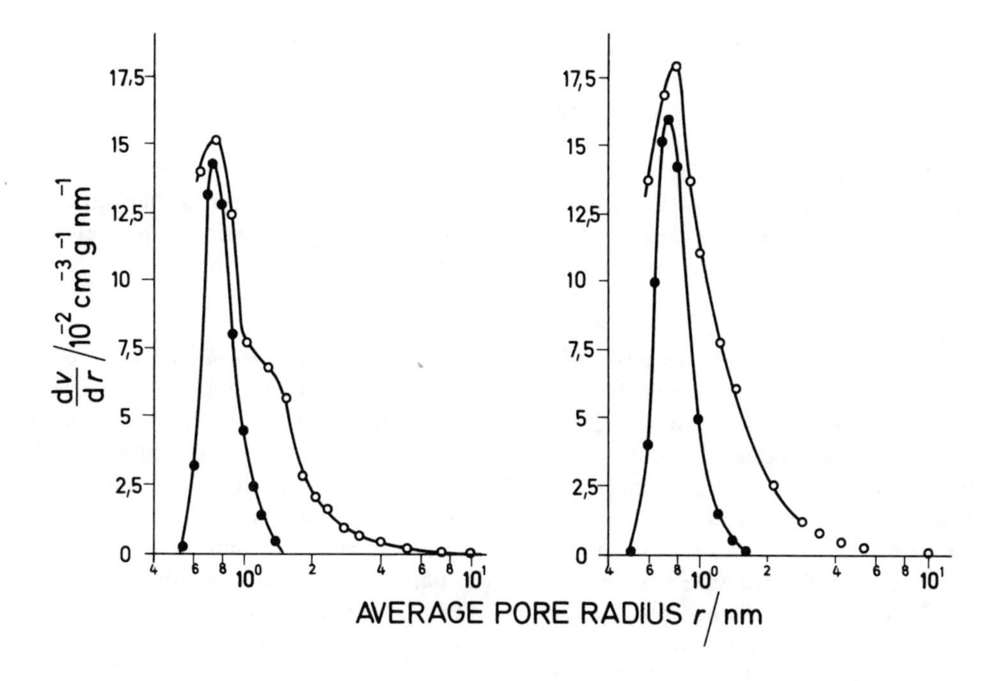

FIGURE 7. Pore volume distributions obtained from the Cranston-Inkley and Medek methods. (From Debelak and Schrodt[64]).

Table 9
HELIUM AND MERCURY DENSITIES AND TOTAL OPEN PORE VOLUME OF COALS

Sample no.	Carbon (%)	LTA mineral matter (%)	Helium density[a] (g/cm³)	Helium density[b] (g/cm³)	Mercury density[a] (g/cm³)	Mercury density[a] (g/cm³)	Total open pore volume[b] (g/cm³)	Open porosity (%)
1	90.8	19.2	1.67	1.53	1.51	1.37	0.076	10.4
2	89.5	6.6	1.38	1.33	1.29	1.25	0.052	6.5
3	88.3	5.4	1.36	1.32	1.29	1.25	0.042	5.3
4	83.8	2.5	1.30	1.28	1.25	1.23	0.033	4.1
5	81.3	10.3	1.35	1.27	1.15	1.07	0.144	15.5
6	79.9	14.0	1.36	1.25	1.24	1.14	0.083	9.5
7	77.2	18.2	1.41	1.27	1.19	1.06	0.158	16.7
8	76.5	25.3	1.49	1.29	1.33	1.13	0.105	11.9
9	75.5	10.9	1.38	1.30	1.07	1.00	0.232	23.2
10	71.7	9.9	1.42	1.35	1.24	1.17	0.114	13.3
11	71.2	9.0	1.46	1.40	1.28	1.22	0.105	12.8
12	63.3	13.7	1.55	1.45	1.41	1.31	0.073	9.6

[a] Mineral-matter-containing basis.
[b] Mineral-matter-free basis.

After Gan, H., Nandi, S. P., and Walker, P. L., Jr., *Fuel,* 51(4), 272, 1972.

and mercury densities, along with the total open pore volumes of 12 coals, are given in Table 9. Both helium and mercury densities decreased gradually, reaching a minimum at about 80% carbon, then rose rapidly as the carbon content increased to 90%. Total open pore volumes were calculated from the differences of the reciprocals of mercury and helium densities.

Low- and high-rank coals were observed to be very porous, whereas the intermediate-rank coals (78% carbon) were observed to exhibit relatively little porosity. Table 10 gives the total porosity of several American coals and the distribution of pore sizes.[67]

There seems to be no apparent correlation between total pore volume and rank; however, low-rank coals usually have a greater number of large pores than do high-rank coals. The pore and capillary structure of coal is important because the materials involved in chemical reaction with the coal must pass through these spaces.

2. Coal Fracture

Fracture processes are an important part of all coal handling operations. Because of the complexity of the structure of coal, its resistance to fracture depends to a large extent on the size of the unit of coal under consideration. In many proposed *in situ* gasification and liquefaction processes, the unit of importance is the coal bed itself. In such cases, we may be concerned with the fracture resistance of the coal on its weakest planes, the bedding planes and joints of the coal bed. At the opposite end of the scale, the processes of abrasion and erosion of structural materials by coal are influenced by the fracture of micron-size coal particles. Mining and comminution of coal involve fracture processes on all intervening size levels.[59]

Coal fracture processes depend on its hardness, strength, friability, grindability, and abrasiveness. These properties are defined according to Ensminger[33] in the following way.

The hardness of coals is related to rank. Hardness varies with rank by decreasing from a maximum in coals containing about 40% volatile matter to a minimum in coals containing about 15% volatile matter, then decreasing again in coals containing about 5% volatile matter. The minimum hardness value is attained at about 85 to 90% carbon content.

The strength of coal in crushing and grinding is a highly variable characteristic. Laboratory studies have shown that coals from the same horizon or from different horizons in the same bed give widespread results, probably due to different manipulative techniques and the presence of random cracks. Generally, however, the compressive strength is reached at 20 to 25% volatile-matter content. The compressibility of the lower-rank coals increases as volatile matter is lost during metamorphism, then decreases as the higher-rank coals become more hardened and compact.

Friability, which refers to the amount of breakage that results from the handling of coal, is affected by several coal characteristics such as toughness, elasticity, fracture characteristics, and strength. Highly friable coals result in increased surface area upon handling, allowing more rapid oxidation, making conditions more favorable for spontaneous combustion, and lowering coking quality. Laboratory tests have shown that friability and rank of coals are related: lignites are the least friable, and friability increases with rank up to the the low-volatile bituminous coals, then decreases in anthracite. Although a relationship does exist between rank and friability, it is only a general one. Friability of coals of the same rank may vary widely.

A general relationship also exists between grindability and rank: medium- and low-volatile groups are much easier to grind than are either high-volatile bituminous and subbituminous coals and anthracites. The ASTM uses the Hardgrove test of grindability, in which a 50-g portion of closely sized coal is ground for 60 revolutions in a ring-and-ball-type mill. A grindability index is calculated by a formula that compares the percent of <200-mesh material from the test with a base coal chosen as 100 grindability.

Abrasiveness can be an important coal characteristic for the industrial consumer as well as for miners because of the wear it produces on coal-handling equipment. Studies

Table 10
PORE DISTRIBUTION IN COALS

Sample No.	Rank	Pore size (cm^3/g)			Pore distribution (%)			
		V_t (4—3000 nm)	V_1 (30—3000nm)	V_2 (1.2—30 nm)	V_3 (0.4—1.2 nm)	V_3 (0.4—1.2 nm)	V_2 (1.2—30 nm)	V_1 (30—3000 nm)
1	Anthracite	0.076	0.009	0.010	0.057	75.0	13.1	11.9
2	Low-volatile bituminous	0.052	0.014	0.000	0.036	73.0	Nil	27.0
3	Medium-volatile bituminous	0.042	0.016	0.000	0.026	61.9	Nil	38.1
4	High-volatile bituminous	0.033	0.017	0.000	0.016	48.5	Nil	51.5
5	High-volatile B bituminous	0.144	0.036	0.065	0.043	29.9	45.1	25.0
6	High-volatile C bituminous	0.083	0.017	0.027	0.039	47.0	32.5	20.5
7	High-volatile C bituminous	0.158	0.031	0.061	0.066	41.8	38.6	19.6
8	High-volatile B bituminous	0.105	0.022	0.013	0.070	66.7	12.4	20.9
9	High-volatile C bituminous	0.232	0.043	0.132	0.070	24.6	56.9	18.5
10	Lignite	0.114	0.088	0.004	0.022	19.3	3.5	77.2
11	Lignite	0.105	0.062	0.000	0.043	40.9	Nil	59.1
12	Lignite	0.073	0.064	0.000	0.009	12.3	Nil	87.7

After Walker, P. L., Jr., Paper presented at Short Course on Coal Characteristics and Coal Conversion Processes, Pennsylvania State University, University Park, October 29 to November 2, 1973.

of coal abrasiveness have demonstrated that it is caused more by the impurities in coal than by the coal material itself. Therefore, a well-cleaned and -prepared coal will be much less abrasive than unwashed coal.[33]

The coal properties also define its elastic and nonelastic ("plastic") deformations which have serious impact on coal utilization processes. A description of the elastic behavior of coal is a fundamental requirement for modeling almost every physical process in coal from the behavior of explosion generated waves in a coal bed to the distribution of stresses around a crack tip. The complexity of the structure of coal means that descriptions of elastic behavior must be in terms of the average behavior of a sample of size appropriate to the process in question. Characterization of elastic behavior on a wide variety of size scales is required in order to suitably characterize the behavior of coal for all processes of interest. It should be kept in mind that the elastic behavior of coal can be expected to be quite complex. It is generally nonlinear and anisotropic. In fact it is not truly elastic, but visco-elastic even at low temperatures, so that time-dependent elastic moduli such as those used to characterize engineering polymers are a more appropriate means of behavior than time-independent elastic moduli.[59]

The irreversible inelastic or "plastic" deformation behavior of coal also has a role in determining the fracture properties as well as in the processes of erosion and abrasion of structural materials by coal.

The elevated-temperature mechanical behavior of coal is important to the operation of the front end stages of coal conversion systems. Hot extrusion coal feeders for high-pressure gassifiers are one example of this. The elevated temperature properties are also essential for modeling the behavior of certain transition zones around an *in situ* gasifier cell. Description of the mechanical behavior of coal above 300°C is complicated by the fact that polymeric structure of the coal is both thermally softening and reacting chemically when it is heated into this temperature range. Therefore the behavior depends in a complex way on heating rate, time, and environment. Nevertheless, a precise description of coal behavior as it is heated is an important factor for process and equipment design.

C. Electrical Conductivity

Electrical conductivity of coal has attracted some interest only recently. The electrical conductivity of coal and its pyrolysis products is of interest as a means to locate and identify regions of differing physical properties during *in situ* coal gasification. We shall describe here some of the measurements reported in the literature.

In the report by Diebold,[68] data on electrical conductivity of coal slag in general and Rosebud coal ash in particular are presented. In these measurements, in order to eliminate effects due to thermal cycling, two samples of Rosebud bottom ash with identical thermal histories were measured twice, giving each repetition of each sample the identical thermal cycle. Results are shown in Figure 8. Several characteristics are to be noted. The most important result is that the conductivity is reproducible between samples and between runs. The result is quite encouraging because it indicates that conductivity measurements on coal slag are meaningful. Two regions of instability are noticeable in Figure 8. The first occurs at a reduced temperature of 6.1 K^{-1} and the second at 7.15 K^{-1}. These instabilities regularly appear in the Rosebud ash and can be described generally as occurring at 6.2± 0.2 K^{-1} (1610 K) and 7.2 ± 0.2 K^{-1} (1390 K). They may be due to the mode of devitrification taking place within the sample.

The report by Duba,[69] presents the measurements of electrical conductivity of coal taken from sites in Wyoming, U.S. where the Lawrence Livermore Laboratory was conducting experiments related to the *in situ* gasification of coal. The char samples used in this study were the products of a series of laboratory experiments in which

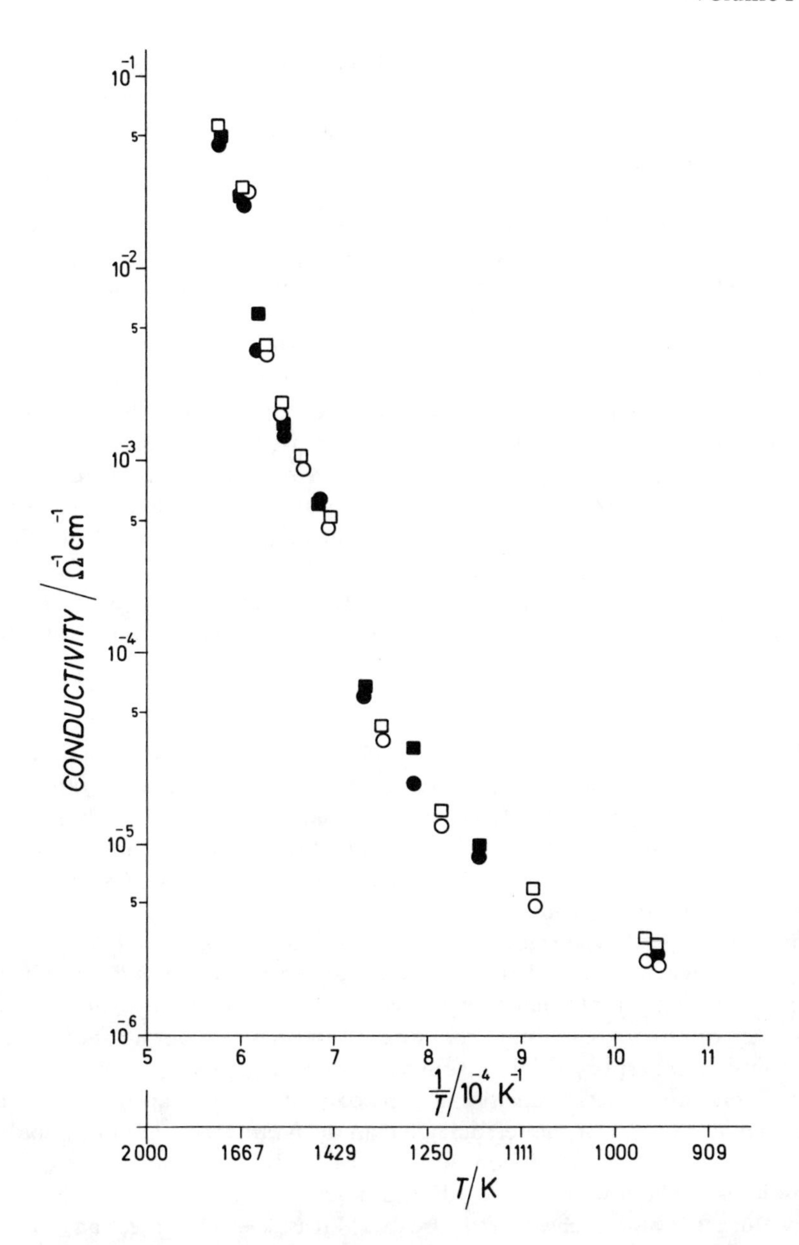

FIGURE 8. Comparison of the electrical conductivity of two similar samples of Rosebud bottom ash under repeated identical thermal cycling in air. (From Diebold, F. E.[68])

coal from the Roland Seam of the Wyodak Mine (Wyoming) was pyrolyzed to 1000°C in an argon environment. The heating rate during pyrolysis was 3.33°C/min. The samples were irregular flakes about 3 mm on a side and up to 1 mm thick. For most samples, electrical conductance was measured at 1 kHz on a capacitance-measuring assembly. Samples that had been heated above 700°C had to be measured on a volt-ohm meter. Samples were measured at 24 ± 2°C after opposite surfaces were coated with conductive silver paint.

Because of their very fragile nature, the samples were measured in the as-received condition. Thickness was calculated using from 4- to 8-μm measurements weighted by visual estimation of the amount of material having a given thickness. Areas were estimated by placing the sample on millimeter-ruled graph paper. The average sample was approximately 3 mm on a side and 0.5 mm thick. Coupling with the capacitance bridge was accomplished by placing one silver-painted surface of the sample in contact with platinum foil. The foil was then clipped to one arm of the bridge, and the circuit was completed using a manual probe contact at several points on the surface of the sample. Differences in conductivity measured from point to point on a given sample were always less than 10%; capacitance was more variable.

The obtained results showed that the conductivity of coal is a strong function of its thermal history. After a large initial decrease, owing to water loss on drying to 110°C, conductivity appears to decrease further on heating from 110 to 300°C. It then begins to increase slowly with heating to 515°C, after which it increases rapidly. The large increase in the conductivity of the char recovered after pyrolysis at temperatures greater than 600°C is probably related to the increased carbon content of the char. Chemical analysis indicates that this char has 80% total carbon, whereas the initial material had only 67% total carbon after drying at 110°C. This dependence of the conductivity of coal char on carbon content is consistent with the results of Pope and Gregg,[70] which showed a dependence of σ on the carbon content of the vitrain component of coal. This enormous change in electrical conductivity of coal during pyrolysis may be exploited during *in situ* coal gasification for locating and monitoring the underground reaction zone.

D. Magnetic Properties of Coal

The first study of the magnetic properties of coal was made by Wooster and Wooster.[71] They found that coal was basically diamagnetic at room temperature but the diamagnetism varied significantly from sample to sample. In some specimens they also found paramagnetic inclusions. Later Honda and Ouchi[224] made a more detailed study of the magnetic susceptibility of coal and found that the specimens stripped of their major impurities by strong acid washing, such as iron sulfides and clays, were diamagnetic; however, when they corrected the observed magnetic susceptibility for the diamagnetic component (calculated from chemical analyses) they found a residual paramagnetic susceptibility that could not be removed by washing in hydrochloric acid. The possibility of ferromagnetism in coal was also recognized, but none of these investigators examined coal specifically for sources of ferromagnetism.

Iron is the major magnetic transition ion in coal; thus, iron-bearing minerals are the probable sources of most of the ferromagnetism or paramagnetism in raw coal. Coals often contain several percent of iron, and most of this iron is in the form of pyrite. Fe^{2+} in pyrite is in the low spin state $3d^6$ and has no magnetic moment. However, as shown by Burgardt and Seehra[72] and also by Marusak et al.[73] the Van Vleck paramagnetism of FeS_2 is significantly larger than its diamagnetism, and thus pyrite has a small residual paramagnetism ($\approx 1.75 \times 10^{-7}$ EMU/g). Pyrite can oxidize to pyrrhotite, a strongly ferromagnetic mineral. Although this mineral is not generally found in fresh coal, it can be present in old specimens which have been exposed to air.

Mössbauer spectral measurements have shed some light on the problem of iron in coal. In 1967 Lefelhorz et al.[74] found that the Mössbauer spectra showed, in addition to nonmagnetic pyrite, a two-line spectrum that they attributed to high-spin Fe^{2+} in sixfold coordination. They further suggested that these lines may be due to organically bound iron. Even at liquid nitrogen temperature, they were unable to detect the six-line hyperfine structure that is characteristic of magnetically ordered compounds and thus they ruled out the presence of significant ferromagnetism in the coals they studied. More recently, Levinson and Jacobs[75] studied the Mössbauer spectra of several West Virginia coals as a technique to measure pyrite. Their measurements indicated that essentially all the iron in their coal specimens was in the form of iron sulfide. Montano[76] has also studied the Mössbauer spectra of West Virginia coals; he observed FeS_2, $FeSO_4 \cdot H_2O$, and $Fe_2(SO_4)_3 \cdot 9H_2O$ formed by oxidation subsequent to mining, but found no evidence of organically bound iron.

Huffman and Higgins[77] have studied by Mössbauer spectroscopy an impressive number of iron-bearing mineral phases that are present in coal. They did not consider or observe the presence of any ferromagnetic minerals, and they were unable to detect any organically bound iron in their coal specimens. Smith et al.[78] were the first to report detection of the presence in coal of small amounts of Fe^{3+} by Mössbauer spectroscopy, but they did not discuss the mineral species. They ascribed the observed Fe^{2+} to mineral species rather than to organically bound iron.

Many measurements indicate that coal is magnetically very heterogeneous. It is however desirable to have a good knowledge of the sources of magnetism in coal. Let us just mention that several magnetic separation methods[79-81] have recently been tried with varying degrees of success to reduce the sulfur content of coal.

In the work by Alexander et al.[82] magnetic susceptibility and other static magnetic parameters have been measured on a number of bituminous coals from various locations in the U.S. It was found that the paramagnetic Curie constant correlates negatively with carbon concentration on a moisture-free basis. The major contribution to the total paramagnetism comes from the mineral matter rather than from free radicals or broken bonds. Analysis of the data indicates that the specific paramagnetism is generally lower in the mineral matter found in high-ash compared to low-ash coal. A substantial number of the coal specimens tested also had a ferromagnetic susceptibility which appeared to be associated with magnetite. Magnetite and α-iron spherules, possibly of meteoritic or volcanic origin, were found in several specimens.

According to Alexander et al.[82] measured values of magnetic susceptibility, χ, of coal in low applied magnetic fields ($H_a < 6$ kOe) can be represented by equation:

$$\chi = \chi_p + \chi_d + \sigma/H_a = \chi_o + \sigma/H_a \quad (4)$$

where χ_p is the paramagnetism in the coal, χ_d is the diamagnetism, σ is the ferromagnetic saturation magnetization, and H_a is the applied magnetic field. The resultant magnetic susceptibility, χ_0, is the intercept on the susceptibility axis of a χ vs. $1/H_a$ plot and can be obtained even if ferromagnetic material is present in the sample. However, this equation assumes that all the ferromagnetic minerals particles are saturated; at low magnetic field (<6kOe, $6/4\pi \times 10^9$ A/m), this condition is strictly true only for needle-like crystal inclusions. A spherical ferromagnetic particle is completely unsaturated and acts like a paramagnetic particle. It can be shown that the internal demagnetizing field in a spherical ferromagnetic particle is equal and opposite to the applied magnetizing field for small external fields of < 6 kOe. The cancellation of the internal and applied fields produces an apparent temperature-independent paramagnetic susceptibility which is a positive intercept on a χ vs. 1/T plot. In between needle-like and spherical particles there may be irregularly shaped particles that are only partially saturated. Therefore, the Owen-Honda equation that applied to all shapes of ferromagnetic inclusions in coal should be rewritten as follows:[82]

$$\chi = \chi_p + \chi_d + \sigma/H_a + [f(H_a)/H_a + \chi_s] \quad (5)$$

where χ_s is the apparent paramagnetic susceptibility of those mineral particles that have a large demagnetization factor and therefore have not started to saturate at low magnetic fields, $f(H_a)/H_a$ is a term describing the susceptibility of partially saturated irregularly shaped ferromagnetic particles, and σ/H_a is the susceptibility of the completely saturated ferromagnetic particles.

The sum of the temperature-independent paramagnetic susceptibility terms shown in this equation (χ_d plus the terms in square brackets) is designated as χ_I. Thus, as χ_p is inversely proportional to the temperature we can write, for a particular field and after correcting for σ/H_a:

$$\chi = \chi_I + C/T \tag{6}$$

where C is the Curie constant. By plotting χ vs. $1/T$ one can determine χ_I and C.

E. Thermal Properties of Coal

The most important thermal property of coal is undoubtedly its heating value or caloric value. The measurement of this property is a well-known practice in so many laboratories that we do not need to discuss it here. Other thermal properties of coal which are of interest include thermal diffusivity, thermal conductivity, and heat capacity. Values for these properties are required for the design of equipment involving the thermal treatment of coal, in processes such as carbonization, gasification, and coal liquefaction. The thermal diffusivity α is related to the thermal conductivity κ the heat capacity C, and the density ϱ, by the definition

$$\alpha \equiv \kappa/C\rho \tag{7}$$

The thermal diffusivity α is the proportionality constant in the right-hand term of the unsteady state conduction equation (Fourier's second law):

$$\partial T/\partial t = \alpha \nabla^2 T \tag{8}$$

where T is the temperature and t the time. Thus it can be measured experimentally in an unsteady state conduction apparatus.[83]

The thermal conductivity of any material is the proportionality constant κ in Fourier's first law, which relates the steady state heat flux q in a material to the temperature gradient, ΔT:

$$Q = -\kappa \nabla T \tag{9}$$

Since nearly all materials, including coal and air, obey this law, κ can be determined for solid coals or beds of coal particles by measuring q and ΔT in a suitable experimental apparatus. The task is simplified if the temperature gradient vector is made one dimensional.

The heat capacity of coal can be measured by standard calorimetric methods. Since the heat capacity is measured per unit mass rather than per unit volume, samples of any size or shape may be used. This method specifies that the test be carried out on dried samples. The heat capacity of coal of any desired moisture content may be calculated from the heat capacity of the dry coal and the water, assuming the heat capacities are additive. Additional literature on physical properties of coal can be found in References 84 to 95.

V. CHEMICAL PROPERTIES

The analysis of coal consists of two categories of determination. These are the proximate analysis and the ultimate analysis (See Volborth et al.[96])

Proximate analysis consists of the determinations of (1) moisture in a specially designed oven at 105°C, (2) ash by heating the coal powder to 750° C in an aerated muffle furnace, (3) volatile matter by rapid heating in a closed crucible to 950°C with a minimum access of air and measuring the weight loss by weighing the coke button so produced, and (4) fixed carbon which is calculated by subtracting the sum of the 3 other results minus moisture from 100. In the proximate analysis family of determinations only the moisture can be chemically defined if one assumes that it is all water. This however, is not necessarily true, since methane, ethane, nitrogen, ammonia, carbon dioxide, and other volatile components can be assumed to be present in varying amounts in gases that evolve at temperatures below 100°C. Thus, while moisture as determined in coal may contain other volatile components, the moisture content and type have an effect on the capacity of some coals to adsorb and hold methane.

Volatile matter is an important parameter for assessing coal quality. Its correct estimation is also essential as it can help in correlating many other properties of coal, i.e., rank, calorific value, total carbon and hydrogen, coke yield, petrographic composition, etc. Many formulae have already been developed based on these correlations.

While calculating volatile matter on unit-coal basis, with mineral matter and moisture considered as diluents it is presumed that they do not affect the pyrolitic behavior of a coal to the extent of altering the volatile matter expressed on unit-coal basis. Thus, the present concept is that the pyrolysis of coal and that of the associated mineral matter proceed independently, and the products of pyrolysis from these two do not interact.

Unless there is an unusual variation in petrographic composition, coal samples drawn from a seam at any particular place show more or less the same volatile matter when expressed on unit-coal basis. In a very thick seam it may vary up to 2%. But it is observed that wherever there is high carbonate content in a coal, the unit-coal volatile matter is found to be much higher than that of other coal samples having little or no carbonate.[97]

The majority of standard methods for the determination of volatile matter in coal and coke allow for either single or multiple determinations. (British Standards Institution, BS 1016, Part 3, 1973; British Standards Institution, BS 1016, Part 4, 1973; Standards Association of Australia, AS K152, Part 3, 1968; Standards Association of Australia, AS K152, Part 4, 1968; International Organization for Standardization, ISO 562-1974 (E); German Institute for Standardization, DIN 51720, June 1978). The primary experimental requirements are heating to 885 or 900°C within 3 or 4 min of crucible insertion, and a uniform heat zone (900°C ± 5°C or ± 10°C) within the test area of the furnace prior to crucible insertion.[98]

According to the ISO standard 562 the volatile matter is the loss in mass when coal is heated to 900°C in the absence of air and the gaseous combustion products are allowed to escape. This loss in mass results from the decomposition of the organic coal substance with the minerals, and the decomposition products may occur; the inherent moisture is also driven off.

Since moisture associated with the coal substance is determined by a separate analysis, an appropriate correction to the weight loss can be made, but the percentage of inorganic decomposition products cannot be determined as a separate entity. Generally it is assumed that the effect of the decomposed mineral matter can be disregarded if the mineral matter content is low. In the formulae for calculation to other reference bases, only a few reactions are taken into account without knowing their real magnitude or significance.[99]

It is important to recognize the difference between mineral-matter content and ash yield. The ash yield is defined as the amount of residue of a fuel after incineration in air at 815°C ISO standard 1171. Under these conditions the mineral matter is decomposed, and the reactions which occur result in a decrease or an increase in weight. A decrease in weight predominates for most hard coals so that their ash yield is less than their mineral-matter content. If the reactions resulting in an increase in weight predominate, the ash yield is higher than the mineral-matter content.[99]

The conditions prevailing during the determination of volatile matter differ from those during the determination of ash yield. Owing to the exclusion of air for volatile matter determination, reducing conditions exist; also the duration of heating under these conditions is much shorter than that for ashing, so that partially different reactions will take place.

The analysis moisture and the water of hydration are driven off under both conditions. The decomposition of carbonates and sulfates, however, will be incomplete during the determination of volatile matter because of the short time of reaction. The greatest difference between both conditions affects the transformation of pyrite. Under ashing conditions it is transformed into ferric oxide. Under reducing conditions it is probably incompletely transformed into ferrous sulfide, because of the very short duration of the test. Therefore, iron disulfide as well as ferrous sulfide will be present after the determination.[99]

The quantitative relation between the volatile-matter yield, as determined analytically, and the mineral-matter and ash yield can be derived according to the following considerations as described by Scholz.[99]

Assume a coal of V_o wt % volatile matter (dry mineral-matter-free) and mineral matter of V_m wt % volatile decomposition products. With these components a blend should be made of M wt % mineral matter and (100 − M) wt % coal. Accordingly the following equations for the experimentally determined V_{total} of the blend on a dry basis:

$$V_{total} = \frac{100 - M}{100} V_o + \frac{M}{100} V_m \tag{10}$$

$$V_{total} = V_o + (V_m - V_o)\, 0.01\, M \tag{11}$$

Equations 10 and 11 can be applied accordingly to all characteristics which are inherent in the coal substance and the mineral matter, e.g., carbon, hydrogen, and calorific value.

If M substituted by $f_A A$, Equation 11 becomes

$$V_{total} = V_o + [(V_m - V_o)\; 0.01\, f_A]\, A \tag{12}$$

where f_A is the mineral-matter factor and A the ash yield.

Thus, the correlation between the experimentally determined V_{total} and the ash yield, A, can be expressed by a rectilinear equation.

VI. CHEMICAL COMPOSITION

Coal is a very heterogeneous solid originating from plant material. It contains, in varying amounts, essentially all elements of the periodic table combined into nearly all of the minerals normally encountered in the crust of the Earth. The organic matrix comprises most of the coal weight and consists mainly of carbon, with smaller amounts of hydrogen, oxygen, nitrogen, and sulfur. Coal can be considered a rock structure containing both macroscopic and microscopic petrographic features. It has also been viewed as an organic chemical substance containing the classical organic functional groups, e.g., mainly carbonyl and hydroxid, aromic and heterocyclic ring units and aliphatic bridges. From another standpoint, coal is a solid colloid which has a large volume porosity and can adsorb gases and vapors as well as liquids. Furthermore, the

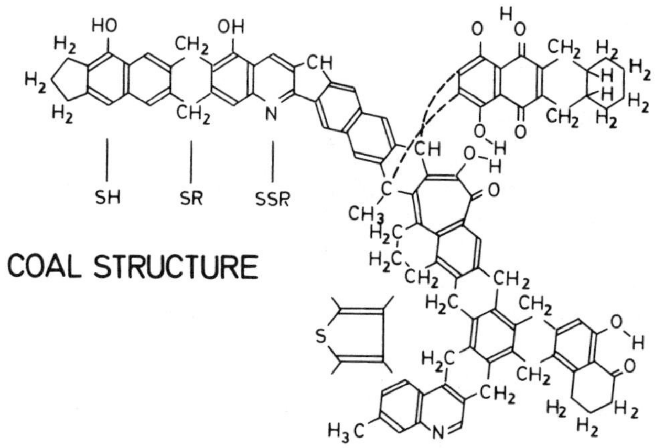

FIGURE 9. Model of organic coal matrix (From Meyers, R. A.[36])

organic coal matrix may be characterized as a cross-linked polymer (formed from the cellulosic polymer present in plant material) which, in the absence of degradation, is essentially insoluble and nonvolatile.[36]

Coal is composed basically of two types of material: inorganic cyrstalline minerals and phytogenic noncrystalline macerals. Both occur in coals as grains, particles, and fragments ranging in size from 2 μm^3 to several centimeters and larger.[33] We shall discuss to some extent both types of coal material.

A. Coal Structure

Strictly speaking, there is no such thing as a coal molecule; however many molecular subgroups have been identified. These groups are present in significant quantities in all types of coals and lignites.

The most prominent of these molecular groupings is the benzene ring; condensed ring varieties, such as naphthalene, anthracene, and larger-ring compounds also are abundant. Studies made in the past 15 years indicate that there also may be a significant number of straight-chain hydrocarbon subgroups. The subgroups are interlinked or bonded in an almost infinite variety of ways.

In any typical representation of a coal molecule, atoms of elements besides carbon and hydrogen are scattered throughout the structure. One of the most common of these elements is oxygen. This element can be present in heterocyclic rings, or it can be a part of a functional group, such as an acid or an aldehyde. The weight percentage of oxygen diminishes from about 25% in lignites to nearly 0% in anthracite.

Other elements commonly found in coals are nitrogen and sulfur. These elements are incorporated into the molecular structure in a manner similar to that of oxygen. But there seems not to be any correlation between element quantity and coal rank. In addition, chlorine is present occasionally as a substituent atom.[100]

Given[34] has proposed an admittedly arbitrary model (see Figure 9) for the organic coal matrix involving methylene bridges of the 9,10-dihydroanthracene type, aromatic structures including benzopyridine, benzoquinone, and benzotropolone as major constituents, and nonaromatic units such as cyclohexanone, cyclohexane, and the like. The major feature of this model involves the bonding of aromatic nuclei by two methylene linkages.

This structure was suggested on the basis of spectroscopic and carbon/hydrogen ratio data, as well as a knowledge of the chemical reactions of coal with bromine and oxygen. A complete determination of the actual structure formula for coal is an elusive goal, since coal structure must be highly variable among various coals and may include all known organic structures, as well as some yet to be defined.[36]

There are some interesting papers providing insight into the chemical nature of coals and its genesis, i.e., Murchison and Westoll,[101] Maxwell et al.[102] and Hayatsu et al.[103]

The lignite, bituminous, and anthracite coals represent three stages in the continuous process of coalification which took place over periods of millions of years. As the original plant materials, consisting largely of well-ordered polymers (e.g., cellulose and lignin) were degraded, lighter, hydrogen-rich compounds were formed and trapped leaving a macromolecular residue deplated in hydrogen and rearranged as a completely disordered macromolecular material. As coalification progressed, the aromatic character of the coal increased with rings fusing and becoming cross-linked. Thus the degree of condensation increases in the order: lignite, bituminous coal, and anthracite. A large number of aromatic units indigenous to the original coals were identified in the work by Hayatsu et al.[103] Of particular interest is the demonstrated chemical nature of organic nitrogen, oxygen, and sulfur in the coal. There has been a lack of reliable information about these important elements which are troublesome in processes for conversion of coals to liquids or gases. The fact that no organic sulfur compounds were isolated from lignite suggests that sulfur has not yet been incorporated into stable aromatic systems and thus might be easier to remove from lignite than from a higher rank coal. Table 11 shows some of the compounds identified in coal using GC-MS and high-resolution MS by Hayatsu et al.[103]

As another interesting study of coal structure let us mention work by Camier and Siemon,[104] who studied brown-coal structure by X-ray crystallography. In their view the micellar theory of brown-coal structure must be regarded as questionable. It is evident that at one time all materials on the coalification series from lignin to bituminous coal were considered to contain crystallites and/or micelles. Surprisingly, although the micelle theory for bituminous coals has been rejected, and lignin and humic substances have been found to be amorphous, the concept of brown coal containing micelles and crystallites has persisted. While it is possible that brown coal may contain some form of structure unit of micelle size, it is clear that the structure of brown coal needs to be studied in the light of modern techniques.

Coal chemistry is discussed in detail also by Ensminger.[33] In order to understand the chemistry of coal it is helpful to consider first the chemicals found in plants: carbohydrates, lignins, proteins, waxes, and resins.

Cellulose, the major constituent of the cell wall, is the most abundant member of the carbohydrate group. The middle lamella and the primary cell wall in woody plants become lignified, but the secondary cell wall consists of unmodified cellulose, which is made up of a long chain of saturated hexagonal rings linked by oxygen bridges. Related compounds are starch, pectin, alginic acid, and chitin.

Lignin is a polymer having the same skeleton as phenylpropane. Lignin in conifers is synthesized from coniferyl alcohol, whereas the starting material in deciduous trees is sinapin alcohol, which contains one more methoxyl group than does coniferyl alcohol.

Lignanes are a group of compounds in woody plants structurally similar to lignin. Lignins and lignanes are phenolic.

Proteins are polymers of amino acids joined by peptide (amide) bonds. Most amino acids are related to aliphatic compounds; a few contain aromatic or heterocyclic rings. Plants proteins are among the compounds least resistant to decomposition processes, but amino acids are involved to some extent in coalification.

Fats and waves are esters derived from fatty acids and glycerine and from fatty acids and wax alcohols, respectively. Fats and sterols are especially involved in the formation of boghead coal. The products of wax alcohols and waxy acids are found in cuticles, spore exines, and coke along with lignin and tannins.

Table 11
COMPOUNDS IDENTIFIED BY GC-MS AND HIGH-RESOLUTION MS

1. Straight-chain hexane
2. 2-hexane
3. Dimethylbutane, methyl-cyclopentane
4. Cyclohexane
5. C_7-alkene (B)
6. Benzene
7. Thiophene
8. C_7-alkane (B), C_7-alkene (B)
9. C_7-alkadiene (B) or C_7-alkyne (B)
10. Cyclohexane
11. C_7-alkane (B)
12. Dimethylcyclopentane
13. 2- and 3-Methylhexanes
14. Heptene
15. 2,3-Dimethyl-2-pentene
16. Methylcyclohexane
17. Dimethylhexane
18. Heptyne
19. Trimethylpentane
20. Methylheptane
21. Methylheptene
22. Trimethylcyclopentane
23. 1-Methylcyclohexene
24. Toluene
25. Dimethylcyclohexane
26. Methylthiophene
27. C_9-alkene (B)
28. Ethylcyclohexane
29. Trimethylcyclohexane
30. *n*-propyl and/or isopropylcyclohexane
31. C_4-alkylcyclopentane (?)
32. C_9-alkane (B), C_9-alkene
33. C_9-alkyne (?) and/or C_9-alkadiene (?)
34. Ethylbenzene
35. Dimethylthiophene
36. *m*, and *p*-xylene
37. *o*-xylene
38. C_9-alkene (B)
39. Tetramethylcyclohexane
40. C_{10}-alkene (B)
41. C_4-alkylcyclohecane
42. Diethylcyclohexane (?)
43. C_{10}-alkane (B), C_{10}-alkene (B)
44. C_{10}-alkene
45. Ethyloctane (?)
46. Trimethylthiophene
47. Propylbenzene
48. Methyl ethylbenzene
49. Trimethylbenzene
50. C_{11}-alkene (B)
51. C_4-alkylbenzene
52. C_{11}-alkene (B) and C_4-alkylbenzene
53. Tetramethylbenzene
54. 1-methyl-4-isopropyl-3-cyclohexene (?)
55. Methylindan
56. C_{12}-alkene (B)
57. Dimethylindan
58. C_5-alkylbenzene
59. Tetralin
60. C_6-alkylbenzene
61. C_{13}-alkene, C_6-alkylbenzene
62. Naphthalene
63. C_{13}-alkane (B), C_{14}-alkene (B)
64. 2-Methylnaphthalene
65. 1-Methylnaphthalene
66. C_3-alkyldecalin
67. C_4-alkyldecalin
68. Trimethylindan
69. Tetramethylindan and/or trimethyltetralin
70. C_9-alkycyclohexane
71. Biphenylene
72. 2-Ethylnaphthalene
73. 1-Ethylnaphthalene
74. Dimethylnaphthalene
75. Cadinane (4,10-dimethyl-7-isopropyldecalin)
76. Dihydrocadinene (T), C_9-alkylcyclohexane
77. Selinane and eremophilane (hydronaphthalenes)
78. Dihydroselinene (T) and/or dihydroeremophilene (T)
79. Dihydrocadinene (T)
80. C_5-alkylindan
81. Methylcenaphthene, 2(?)-isopropylnaphthalene
82. Diphenylmethane
83. 3- or 4-Methylbiphenyl
84. C_{10}-Alkylbenzene
85. Tetramethylindan
86. $C_{15}H_{28}$-sesquiterpenoid hydrocarbon (?)
87. C_5-alkyltetralin
88. Methyl-ethylnaphthalene and/or trimethylnaphthalene
89. Trimethylnaphthalene
90. Fluorene
91. 1,6-Dimethyl-4-isopropyl-1, 2-dihydronaphthalene (T)
92. Iso-butylnaphthalene, trimethylnaphthalene
93. 1-Methyl-4-isopropyl-naphthalene
94. Eudalene (1-methyl-7-isopropyl-naphthalene)
95. C_5-alkyltetralin (?)
96. 1-Methyl-2-propylnaphthalene (T)
97. Cadalene (1,6-dimethyl-4-isopropyl-naphthalene)
98. Tetramethylnaphthalene
99. 1,4-Dimethyl-6(?)-isopropyl-naphthalene, C_6-alkyltetralin (T)
100. C_6-alkyltetralin (T)
101. 1,2,5,7-Tetramethyl-naphthalene
102. Pristane
103. Methylfluorene
104. Pentamethylnaphthalene
105. Dibenzothiophene
106. Trimethyloctahydrophenanthrene

Table 11 (continued)
COMPOUNDS IDENTIFIED BY GC-MS AND HIGH-RESOLUTION MS

107. Methyltetrahydrophenanthrene (T)
108. Phenanthrene
109. $C_{14}H_{23}$(m/e 191, base peak), $C_{20}H_{36}$(M^+) tricyclic terpenoid (?)
110. Dimethyltetrahydrophenanthrene (T)
111. Ethyltetrahydrophenanthrene (T)
112. Anthracene
113. Naphthofuran
114. Methyldibenzothiophene
115. C_9-alkyltetralin (?)
116. $C_{20}H_{32}$(abietadiene (?)
117. $C_{19}H_{30}$(tricyclicditerpenoid (?)
118. 2- and/or 3-methylphenanthrene
119. 1- and/or 9-methylpehnanthrene
120. 1,7-Dimethylphenanthrene
121. Dimethyldibenzothiophene
122. C_{10}-alkyltetralin (or C_{11}-alkylindan) (?)
123. Dehydroabiethene (4,20-dimethyl-13-isopropyl-8H-phenanthrene)
124. Dimethylphenanthrene and/or dimethylanthracene
125. Dehydroabiethane
126. $C_{20}H_{32}$(tricyclicditerpenoid) (?)
127. Fluoranthene
128. C_{11}-alkyltetralin (?)
129. Abietatetraene (T) (trimethylisopropyl-6H-phenanthrene)
130. 1,2,3,4-Tetrahydroretene (T)
131. Methyl-ethylphenanthrene and/or trimethylphenanthrene
132. Pyrene
133. Simonellite
134. Retene (1-methyl-7-isopropylphenanthrene)
135. 1,2-Benzofluorene
136. 2,3-Benzofluorene
137. 3,4-Benzofluorene
138. Methylbenzofluorene
140. Tetramethylphenanthrene and/or tetramethylanthracene
141. Chrysene and/or triphenylene

Note: B, branched; T, identification tentative; ?, identification uncertain.

After Hayatsu, R., Winans, R. E., Scott, R. G., Moore, L. P., and Studier, M. H., *Fuel*, 57, 541, 1978.

Typical resin compounds are diterpene resin acids as alicetic and pimaric acids which are "isoprene" derivates structurally related to steroids. A variety of other terpenes (both acyclic and cyclic) derived from two or more isoprene units are frequently associated with resinous plant exudates. Many of these materials are susceptible to a variety of acid-catalyzed transformations. Rearranged as well as polymerized products would be expected to form during coalification.[33]

Coal is a complete mixture of various macromolecules; thus, to establish a reasonable chemical characterization of coal, certain essentials must be known:[33,105]

1. The molecular weight distribution of the macromolecules — This parameter should correlate with reactivity; however, little work has been done in this field.
2. The nature of the constituent molecules — Data to date indicate that the constituent molecules are polycyclic aromatic units of various sizes containing a variety of functional groups; however, the manner of linkage and the number found in an average molecule are not known with certainty.
3. The nature of the cross-links — Much evidence indicating the presence of short methylene bridges between aromatic units has been collected. Gaps in the knowledge of cross-links and their relative reactivities make designing coal conversion systems very difficult. Many other variables such as rank, ash content, and reaction conditions affect the reactivity of coal and must also be considered in designing such systems.
4. The functional groups present — Some current research, involving direct determination of aromatic and aliphatic carbon concentrations in coal by solid-phase ^{13}C NMR spectroscopy, supports data previously obtained by indirect techniques concerning the groups; however, some more recent work suggests that functional groups in coal are not randomly distributed, as is assumed in all current attempts at forming an idea of coal structure. Also, in regard to sulfur, much is yet un-

known about the different forms of organic sulfur in coal and the manner in which these forms are distributed. Additionally, almost nothing is known about the organometallic compounds in coal.[105]

The true structure of coal is presently unknown; most workers agree that the structure is so compilated and variable that it is impossible at this time to form an accurate model of it. Despite, however, the present inadequate knowledge of coal constituents and their confirmation, many significant attempts have been made to gain insight into the structure of coal, and through these attempts the knowledge of its general character has been greatly improved.

According to Ensminger,[33] the organic chemistry of coal can be best understood by studying functional groups such as reactive oxygen, nitrogen, sulfur groups, and other organic structural elements.

Given[34] suggests that aromatic and hydroaromatic compounds, when joined together, form the basic coal structure. This structure cannot be accurately described at the present time; however, it has been observed that the rings are joined together in a variety of patterns — single rings, condensed double rings, condensed triple rings, etc. Clusters thus formed are believed to be held together by bridges: (1) short aliphatic groups, principally methylene, probably not many longer than four carbon atoms; (2) ether linkages; (3) sulfide and disulfide; and (4) biphenyl types.

On the basis of available experimental data, hydroxyl groups in coals are mainly phenolic or acidic; there is little evidence to indicate the presence of alcoholic or weakly acidic hydroxyl groups. Hydroxyl oxygen may be present in concentrations of 9% in brown coals, but is usually about 8% with 65% carbon and 27% oxygen. As rank increases, the hydroxyl content decreases slightly until, at 80% carbon and 12.5% oxygen, a more rapid decrease is initiated, resulting in a hydroxyl oxygen concentration of less than 1% at 90% carbon content.

Research has shown that carboxyl groups are absent in most coals above the lignite stage of development, but are present in brown coals and lignites. At 83% carbon content, all carboxyl groups disappear. Methoxyl groups are among the first lost during metamorphism of coal; as a result, they usually do not appear in significant amounts in true coals. Carbonyl groups are found at all levels of coalification; however, analytical methods and their results have been quite variable.

During the metamorphic process, in coals containing up to 70 to 80% carbon, the methoxyl groups are the first lost; then the carboxyl and the carbonyl groups decrease rapidly. At 81 to 89% carbon content, the hydroxyl groups decrease rapidly. At greater than 92% carbon content, almost all oxygen is in nonreactive, stable forms. Data on hydroxyl and carbonyl oxygen concentrations of various coals support this conclusion. It has been further suggested that hydroxyl and carbonyl groups may account for 70 to 90% of the oxygen in bituminous coals. Nitrogen in coal is almost completely in cyclic structures.

It is estimated that the nitrogen content of the vegetable matter that was converted to coal to be between 0.6 and 1.0%. Because of the coalification process, the nitrogen content of the material is believed to increase up to 45% volatile matter content, then decrease slowly with increase in rank. The researchers believe that the polar character of nitrogen and oxygen influence the swelling and softening capability of the coal by affecting the balance of cohesive forces.

Hayatsu et al.[103] report that their results obtained by aqueous $Na_2Cr_2O_7$ oxidation of bituminous coal support the idea that "coal is predominantly an aromatic material". Wiser[106] suggests that, in bituminous coal, 70 to 75% of the carbon present is in the aromatic form. Six-membered rings are in the majority, but there are some five-membered ones. Fifteen to twenty-five percent of the carbon is involved with hydroaromatic structures.[106]

The distribution of carbon in aromatic, hydroaromatic, and methyl forms in coal was determined from pyrolysis by Mazumdar et al.[107] The three forms are shown to comprise 97.5 to 98.0% of the total carbon in low-rank (<86% carbon) bituminous coals and 99.4% in high-rank (90.4% carbon) bituminous coals. The addition of carboxyl groups to the amounts shown for low-rank bituminous brings the total up to the maximum of 98%.

Almost all the work up until the present has been performed under the widely accepted theory that coal is a largely aromatic structure. However, this view is not shared by everyone. Chakrabartty and Berkowitz[108] state that the skeletal carbon arrangements of coal appear to be made up largely of nonaromatic structures. They conclude from their research on oxidation of coal with sodium hypohalite that these structures are modified bridged tricycloalkane systems or polyamantanes. Chakrabartty and Berkowitz[108] hold that the characterization of coal as a nonaromatic structure is not inconsistent with existing experimental evidence.

B. Macerals

Macerals are the fragmentary organic remains of plants that died, were altered to peat by partial decay, and, because of exposure to heat in the crust of the earth through time, were converted to their present state in coal. The carbonaceous, combustible fraction of coal is made up of macerals; thus, by definition, macerals comprise more than half of the coal mass.[29,33]

Probably the most widely used method of determining the maceral content of coal is by measurement of reflected light, a technique developed to a point of usefulness by worker of the Prussian Geological Survey.[28] Each maceral reflects characteristic amounts of light when direct light is applied to polished coal surfaces. It is generally agreed that the reflectance of the macerals increases with the increasing rank of coal, but there has been some controversy as to whether the increase occurs in a series of sudden jumps at certain ranks, or is a continuous change process. In any case, reflectance measurements offer an objective method of classifying the petrographic constituents of coals.[28]

Another method of determining the petrographic composition of coals is the transmitted-light technique. This method is frequently referred to as the American system of coal petrography, in contrast to the German system mentioned above. The transmitted-light technique has not been as widely adopted as the reflected-light technique, probably because of the difficulties encountered in performing the required thin-section preparation and analysis.

We shall describe the common coal macerals and their distinguishing characteristics according to Spackman.[29,33]

Vitrinite macerals are from the humic fraction of coal and are produced by relatively slow alteration of point cell wall substances. A woody texture and brown color characterize vitrinites in coals of early metamorphic stages. In bituminous and anthracite coals, the vitrinites appear textureless. Low-rank vitrinites show colors from buff and cream to yellow, tan, and pale orange red; in vertically incident light, they appear grey and have less than 5% reflectance. Vitrinites of intermediate rank are tan, orange red, reddish brown, and deep red, and they reflect 0.5 to 2.5% of vertically incident light. High-rank vitrinites in anthracites are opaque and have a reflectance of 2.5 to 6.0%.

Semivitrinites generally have the same characteristics as vitrinites, but show greater reflectance. They may have a remnant cell structure, serrate margins, and greater relief than do vitrinites. Unlike vitrinites, they may be inert in coking.

Fusinite macerals are from the charcoal-like fraction of coal and are produced by rapid charring and alteration of cell wall material. They are opaque in thin section and appear white in vertically incident light. They frequently include residual cell wall structures.

Semifusinite macerals are characterized by optical properties intermediate between those of vitrinite and fusinite.

Micrinite macerals consist of attritus and are formed by granulation and metamorphism of material from cell walls. They appear translucent and yellowish brown to brown in lignites when observed in thin section and dark gray in incident light. In high-rank coals they are opaque and white. They range from 1 to 6 μm in diameter.

Micrinite macerals appear white in incident light, but are not as reflective as fusinite. They never contain remnant cell walls and range from ten to hundreds of micrometers in their greatest dimension.

Exinite macerals are derived from waxy secretions such as plant cuticles and spore and pollen exines. In low- to intermediate-rank coals, they are yellow in thin section and dark gray to black in the medium range. In very high-rank coals, either they have the same optical properties as vitrinite, or they disappear.

Resinite macerals are hydrogen-rich and are formed from resinous secretions and excretions. They usually appear yellow in thin section and are oval, irregular, or rod-shaped. Like exinite, they either take on the optical properties of vitrinite or disappear in high-rank coals.

Sporinite, cutinite, and alginite are macerals easily identified by their shape. They are composed of the material that their name indicates: spores, cutin, or algae. The term exinite includes both sporinite and cutinite.

Sclerotinite macerals consist of fossilized remains of fungal sclerotia. Sclerotinites are opaque and highly reflective.

Van Krevelen[27] groups the macerals into three major categories: (1) macerals whose origin is definitely due to woody and cortical tissues (vitrinite, fusinite, semifusinite); (2) macerals whose origin is definitely due to plant material other than woody tissues (exinite, sporinite, cutinite, resinite, alginite); and (3) a maceral whose origin has not yet been traced to a specific vegetable tissue (micrinite).

Fusinite, semifusinite, sclerotinite, and micrinite behave as inert substances on heating and are collectively referred to as "inertinite". Sporinite, cutinite, resinite, and alginite are rich in hydrogen and become very plastic on heating. They may be included in the exinites.[27]

Vitrinite is the best coke-forming maceral. In anthracite it is relatively inert, and in low-rank, high-volatile coals it will neither cake nor soften. However, if the volatile matter content of the coal is from about 19 to 33%, vitritinite is responsible for the actual coking properties. Coking coals usually have characteristic contents of vitrinite, inertinite, and exinite.[27,33]

Given[39] suggests that macerals must be physically separated to study their properties. Because vitrinites sometimes occur in pure bands, they are the easiest to study. Lithotypes containing 60 to 70% sporinite can be found in European coals; however, North American coals containing sporinite concentrations above 40% are rare. Higher concentrations of the macerals can be obtained by gravitational separation of appropriate lithotypes.

Given[39] gives the order of maceral density as fusinite > micrinite > vitrinite > sporinite. Relatively few European coals and even fewer American coals have been separated into their component macerals. It appears that only one American coal[109] has been found to contain all four macerals. Thus, little is known of the properties of macerals and their behavior in processes other than coking.

Given, Peover, and Wyss[110] separated vitrinites, exinites, and fusinites from the same coals and analyzed them for their content of carbon, hydrogen, oxygen, ash, and petrographic purity. They note that exinites closely resemble vitrinites in many of their chemical characteristics; however, in this study three significant differences appeared: (1) Exinite is relatively insoluble in pyridine, as compared with vitrinite, whereas nitrobenzene and trichlorobenzene appear to be better solvents for exinite than for vitrinite; (2) exinite did not yield humic-acid-like products on oxidation as do vitrinites; and (3) the extent of reduction of exinite with lithium, as well as the variation of the extent with rank, was less in all cases than the reduction of vitrinites.

It is suggested that the solubility and oxidation differences between exinite and vitrinite are due to different intermolecular forces and molecular weights. Exinites are thought to have a more open hydroaromatic character. (Interlayer spacing for exinites has been shown to be 0.47 nm, whereas, the spacing in vitrinites is 0.35 nm.) Because of the closer molecular arrangements of vitrinites, different intermolecular forces from those in exinite are believed to be in effect. No explanation is given for the difference in reducibility.[110]

Reggel et al.[109] studied vitrinite, exinite, micrinite, and fusinite from Hernshaw bed high-volatile A bituminous coal, Bonne County, West Virginia. By comparing the analytical and dehydrogenation data, the researchers formed the following conclusions. All four macerals are similar in sulfur content, but fusinite has the lowest nitrogen content. Exinite and vitrinite contain similar amounts of carbon, but the hydrogen-carbon ratio is much higher in exinite; in fact, exinite generally shows a higher content of hydrogen and volatile matter and a higher hydrogen-carbon ratio than any other maceral. Vitrinite and micrinite are similar in volatile matter, fixed carbon (incorporated into organic compounds), carbon, and hydrogen contents, but the hydrogen-carbon ratio for vitrinites is normally a little higher. The hydrogen-carbon ratio of fusinite is very low, and fusinite has a much lower volatile-matter and hydrogen content than does vitrinite. It is concluded from the low hydrogen-carbon ratio of fusinite that this maceral must be more aromatic than vitrinite. Vitrinite and exinite are hydroaromatic, whereas micrinite, having a hydrogen-carbon ratio close to that of vitrinite, contains some structure that is not aromatic but also does not give off hydrogen during dehydrogenation.[109]

C. Mineral Matter

According to Gluskoter[111,112] the term "mineral matter in coal" is widely used, but its meaning varies appreciably. The term usually includes all inorganic noncoal material found in coal as mineral phases and also all elements in coal that are considered inorganic. Therefore, all elements in coal except carbon, hydrogen, oxygen, nitrogen, and sulfur are included in this broad definition. Four of these five organic elements also are found in coals in inorganic combination and therefore are part of the mineral matter. Carbon is present in carbonates $Ca(Fe,Mg)CO_3$; hydrogen in free water and water of hydration; oxygen in water, oxides, carbonates, sulfates, and silicates; and sulfur in sulfides (primarily pyrite and marcasite) and sulfates.[111]

The changes in the mineral matter content in coal during ashing have long been recognized. A number of workers have suggested schemes for calculating the true mineral matter content from determinations made during the chemical analyses of coal. Parr[113] reported one of the earliest of such schemes in which only the total sulfur and ash contents were considered in developing the conversion formulae. This is still the most widely used procedure. According to Parr the mineral matter in coal is given by the formula:

$$\%\ \text{Mineral matter} = 1.08 \times \%\ \text{ash in coal} + 0.55\ \%\ \text{sulfur in coal} \qquad (13)$$

where the factor 1.08 is the empirically derived value for the water of hydration of the minerals usually found in ash. Ash in coal is high temperature ash (HTA).

Based on similar considerations, King et al. in 1936 derived a more accurate mineral-matter formula:

$$\% \text{ MM} = 1.09 \times \% \text{ ash} + 0.5 \times \% \text{ pyritic S in coal} + 0.8 \text{ CO}_2 \text{ in coal}$$
$$- 1.1 \times \% \text{ SO}_3 \text{ in coal} + 0.5 \times \% \text{ Cl in coal} \quad (14)$$

These formulae have been derived from accurate stoichiometric relations considering decomposition of hydrated minerals, oxidation, and formulation of Fe_2O_3 from FeS_2, etc.

There are obvious objections to the use of these formulae, particularly the Parr, but the most serious, which apply equally to all such formulae, are that they have to assume that the clay minerals in coals have a constant average water of decomposition, and that they ignore the possible presence of quartz, which yields no water of decomposition on heating. The King-Maries-Crossley (KMC) formula has been most commonly used in Britain, where all commercially significant coals are of Carboniferous age and of a restricted range of rank.

It follows from the above that what is needed in coal analysis is a directly determined mineral-matter content. The one method of obtaining this, that has received general support in the literature, is based on demineralization with hydrochloric and hydrofluoric acids.[115-117]

Miller et al.[118] have described a new method for determining the mineral-matter content of coals, based on low-temperature ashing in an oxygen plasma. Unwanted side-reactions can occur during ashing, particularly oxidation of pyrite to hematite and fixation of organic sulfur as sulfate. These effects cannot be totally eliminated, but procedures are offered that minimize them, and are presented as accurate and rapid means of determining the mineral-matter content of bituminous coals. Data are presented to illustrate the magnitude of the effects under the recommended conditions of ashing. Duplicate determinations of the mineral matter contents of 13 coals by acid demineralization and low-temperature ashing (LTA) are reported in support of the new procedures. Statistical analysis of the data shows that the precision (± 2 SD) of the LTA procedure under the defined conditions is 0.20. Acceptance of the procedure is suggested for routine use with most coals of bituminous or anthracite rank. The procedure as proposed is unsuitable for use with HVC coals from Western regions of the U.S. or with lignites and subbituminous coals.

Pollack[119] has used LTA and normative analysis (NA), which calculates minerals from elemental analysis, and in this case is based on the major inorganic elements in coal.

Using the average absolute percent differences and their standard deviations as criteria, the NA and Parr formula values vary from the LTA by about the same amount and to the same degree, when samples showing differences larger than 50% are excluded. Considerable differences in some samples could result if the elemental analyses were run on freshly ground samples and the LTA on samples that had been stored for a period of time. A significant amount of pyrite could have been oxidized to iron sulfates and have thus increased the LTA.

The stoichiometry and accuracy of mineral-matter evaluation can be further improved by reconstituting the main minerals based on the analysis of ash.

Kiss and King[120] have discussed the estimate of mineral matter in brown coals, i.e., lignite in American terminology. The dominant mineral groupings in these coals are quartz (SiO_2); a variety of weathered clays, mostly kaolinite (Al_2O_3). · $2SiO_2$ · $2H_2O$) and aluminum hydroxide (AlOOH) together with some accessory minerals containing potassium and titanium; and the iron disulfide minerals pyrite and/or marcasite (FeS_2).

Owing to the acid nature of the coal (pH 4.5 to 5.5) there are no carbonate minerals present although siderite, $FeCO_3$ does occur in interseam sediments. The mineral group can therefore be represented by expressing the results of chemical analysis on a coal dry basis as the sum of

$$SiO_2 + Al_2O_3 + TiO_2 + K_2O + FeS_2 \equiv \text{Minerals} \quad (15)$$

The error in this expression is in ignoring the water of constitution of clays which is approximately 15% the weight of clay or aluminum hydroxide.

The soluble mineral NaCl is best regarded as part of the inorganics group. Thus, the correct way of representing this group is

$$Na + Ca + Mg + Fe + NaCl \equiv \text{Inorganics} \quad (16)$$

There are no significant errors associated with expressing the chemical analysis results this way.

Although the amount of mineral matter in coals varies considerably, it is normally large enough to be significant in any way the coal is used. In a study of 65 Illinois coals, Rao and Gluskoter[121] found the mineral-matter content to range from 9.4 to 22.3%, corresponding to an ash content of 7.3 and 15.8%, respectively.. O'Gorman and Walker[122,123] found an even larger range (9.05 to 32.26%) in mineral matter content in 16 whole coal samples from a wide distribution of locations in North America.

Interest in mineral-matter content of coal intensified as electric power plants become larger and the boilers began to operate at higher temperatures. Problems of fireside boiler-tube fouling and corrosion, which became increasingly severe at the higher temperatures, were related to the sulfur, chlorine, alkali, and ash content of the coals. Within the past several years, the general public has become more interested in both air and water pollution. Therefore, both the coal consumer and the producer need a more thorough knowledge of the mineral matter in coal and of the products and by-products of the mineral matter produced when coal is combusted.

More recently, there has been much concern about the possible effects of the mineral matter in coal on processes used to convert coal to other fuels such as gasification, liquefaction, and production of clean solid fuels. Not only is removing and disposing of the mineral matter a problem, but also the possible chemical effects such as catalyst poisoning, which might be expected in the methanation of gas from coal, should be considered.[111]

Minerals found in coal are classified by Nelson[124] as syngenetic or epigenetic. Syngenetic minerals were included in the coal during the biochemical changes associated with coalification, whereas epigenetic minerals were deposited in the cleats and cracks of the coal after the coalification process was complete. The amounts and kinds of minerals found in coal vary widely and undoubtedly depend on the conditions of formation. Table 12 lists some of the common coal minerals.

Clay minerals are the most commonly occurring inorganic constituents of coals and of the strata associated with the coals. Much of the work on clays reported in the literature is concerned with these strata, not the coals themselves. Many different clay minerals have been reported in coals, but the most common are illite $[(OH)_4K_2(Si_6 \cdot Al_2)Al_4O_{20}]$, kaolinite $[(OH)_8Si_4Al_4O_{10}]$, and mixed-layer illite-montmorillonite. Rao and Gluskoter,[121] in an investigation of 65 coals from the Illinois Basin, reported a mean value of 52% for clay in the mineral matter.

Analysis of mineral matter from coals of the Illinois Basin revealed six common minerals, identified trace minerals, and allowed development of a model of the depositional environment for the Herrin (No. 6) coal. Coal samples were low-temperature-

Table 12
MINERAL COMPOSITION OF SOME KENTUCKY COALS AS DETERMINED BY INFRARED SPECTROMETRY

Classification	Mineral constituents	Chemical formula	Sample no. 1	Sample no. 2	Sample no. 3
Silicates	Kaolinite	$Al_2Si_2O_5(OH)_4$	3—40	1—10	1—10
	Illitte	$(OH)_4K_y(Si_{8-y}\cdot Al_4O_{20}$	Trace	1—10	1—10
	Muscovite	$KAl_2(AlSi_3(O_{10}(OH)_2$	—	—	—
	Chlorite	$(Mg,Fe)_6(AlSi)_4O_{10}(OH)_8$	Trace	1—10	1—10
	Montmorillonite	$Na_y(Al_{2-y}Mg_y)Si_4O_{10}(OH)_2$	—	Trace	—
	Mixed-layer illite-montmorillonite		Trace	1—10	—
	Plagioclase	$(Na,Ca)Al(SiAl)Si_2O_8$	—	—	—
Calbonates	Calcite	$CaCO_3$	—	—	—
	Aragonite	$CaCO_3$	—	—	—
	Dolomite	$CaMg(CO_3)_2$	—	—	—
	Ankerite	$CaCO_3\cdot(Mg,Fe,Mn)CO_3$	—	—	—
	Siderite	$FeCO_3$	—	1—10	30—40
Oxides	Quartz	SiO_2	40—50	1—10	1—10
	Hematite	Fe_2O_3	—	10—20	—
	Rutile	TiO_2	1—10	—	—
Sulfates	Gypsum	$CaSO_4\cdot 2H_2O$	1—10	10—20	1—10
	Jarosite	$KFe_3(OH)_6(SO_4)_2$	—	—	—
	Thenardite	Na_2SO_4	—	1—10	—
Sulfides	Pyrite	FeS_2	1—10	10—20	1—10
	Marcasite	FeS_2	—	—	—

After Ensminger, J. T., in *Environmental Health and Control Aspects of Coal Conversion*, Braunstein, H. M., Copenhover, E. D., and Pfuderer, H. A., Eds., Rep. ORNL/EIS-94, Oak Ridge National Laboratories, Tenn. 1977.

ashed, followed by quantitative X-ray diffraction analysis for mineral determination, and computation of mean concentration and standard deviation of each mineral. Most common were the clays (illite, kaolinite, and mixed-layer clays), pyrite, quartz, and calcite. Trace minerals include siderite, dolomite, feldspar, gypsum, marcasite, and sphalerite. Iron sulfate minerals resulting from pyrite oxidation were common in all but extremely fresh samples. The distribution of minerals in the Herrin (No. 6) coal may best be interpreted as a response to both a general northeast to southwest paleoslope and the presence of a distributary channel system contemporaneous with the coal swamp. Calcite, illite, and "expandable" clay minerals respond to the paleoslope with illite preferentially deposited landward, and expandable clays and calcite deposited basinward. Concentrations of kaolinite, quartz, and pyrite are influenced by the channel system with high kaolinite and quartz, and low pyrite bordering the channel. O'Gorman and Walker[123] also found that the clay minerals make up the greater part of the mineral matter in most of the coals that they studied.

Pyrite is the dominant sulfide mineral in coal. Marcasite has also been reported from many different coals. Pyrite and marcasite are dimorphs, minerals that are identical in chemical composition (FeS_2) but differ in crystalline form; pyrite is cubic while marcasite is orthorhombic. Other sulfide minerals that have been found in coals, and sometimes in significant amounts, are sphalerite (ZnS) and galena (PbS). Sulfates are not common and often are not present at all in coals that are fresh and unweathered. Pyrite is very susceptible to oxidation and decomposes to various phases of iron sulfate minerals at room temperature.

Sulfides and sulfate minerals make up 25% of the mineral matter content of Illinois

Table 13
MINERALS ASSOCIATED WITH U.S. BITUMINOUS COAL

Group	Species	Formula
Shale	Muscovite	$(K,Na,H_3OCa)_2(Al,Mg,Fe,Ti)_4$
	Hydromuscovite	$(Al,Si)_8O_{20}(OH,F)_4$
	Illite	
	Bravaisite	
	Montmorillonite	
Kalin	Kaolinite	$Al_2(Si_2O_5)(OH)_4$
	Livesite	
	Metahalloysite	
Sulfide	Pyrite	FeS_2
	Marcasite	
Carbonate	Ankerite	$(Ca,Mg,Fe,Mn)CO_3$
	Ankeritic calcite	
	Ankeritic dolomite	
	Ankeritic chalybite	
Chloride	Sylvine	KCl
	Halite	NaCl

After Meyers, R. A., *Coal Desulfurization*, Marcel Dekker, New York, 1977.

coals.[111] Mineral matter in Illinois coal is also discussed by Ball[125-128] and Ball and Cady.[129]

The carbonate minerals, in general, vary widely in composition because of the extensive solid solution of calcium, magnesium, iron, manganese, etc. that is possible within them. There is also a wide range of mineral compositions for the carbonate minerals in coals.

The list of iron sulfate minerals found in weathered Illinois coals include: Szomoluokite ($FeSO_4 \cdot H_2O$); Rozenite ($FeSO \cdot 4H_2O$); Melanterite ($FeSO_4 \cdot 7H_2O$); Coquimbite ($Fe(SO_4) \cdot 9H_2O$); Rosmerite ($Fe_2(SO_4) \cdot Fe_2(SO_4)_3 \cdot 12H_2O$); Jarosite ($(Na,K)Fe_3(SO_4)_2(OH)_6$[111] Carbonates form readily in nonacid seas. Of the carbonates, dolomite and ankerite are probably encountered most often, and quartz, which is found in almost all coals, sometimes occurs in concentrations as high as 20%.

The major minerals present in U.S. bituminous coals are listed in Table 13.[36] It is very difficult to totally separate the minerals from the coal for analysis since the mineral component of coal is intimately associated with the organic coal matrix. The listing in Table 13 is based on optical microscopy, spectroscopy, and analysis of the ash content obtained after combustion of the organic coal matrix.

Ash is the residue remaining after complete incineration of coal. It is related to mineral matter, but is different in chemical composition from and found in less quantity than, the original mineral matter. The heat of the combustion process induces changes in the mineral portion of coal such as loss of water of constitution by silicate minerals, loss of carbon dioxide from carbonate minerals, oxidation of iron pyrites to iron oxide, and fixation of oxides of sulfur by bases such as calcium and magnesium. The conditions of incineration determine the extent to which these changes take place; therefore, standard procedures have been prescribed for determining ash content. These vary somewhat between countries, but generally consist of burning a 1- to 2-g sample of coal in a well-ventilated muffle furnace at temperatures of 700 to 850°C.[130]

Ash consists mainly of compounds of silicon, aluminum, iron, and calcium with measurable contributions from the compounds of magnesium, titanium, sodium, and

Table 14
OXIDES OF MINERALS ASSOCIATED WITH BITUMINOUS COAL ASH

Constituent oxide	Percent
Silica (SiO_2)	20—60
Aluminium oxide (Al_2O_3)	10—35
Ferric oxide (Fe_2O_3)	5—35
Calcium oxide (CaO)	1—20
Magnesium oxide (MgO)	0.3—4
Titanium oxide (TiO_2)	0.5—2.5
Alkali (Na_2O + K_2O)	1—4
Sulfur trioxide (SO_3)	0.1—12

potassium. Analyses of ash are commonly reported as oxides; however, the constituents occur in ash predominantly as a mixture of aluminum oxides are derived primarily from the silicates, whereas iron oxide comes from pyrite, which forms ferric oxide and sulfur oxides when burned. Calcium and magnesium oxides are formed from the decomposition of carbonate minerals, and sulfates are produced by reactions among carbonates, pyrite, and oxygen.[130] Table 14 shows coal ash composition; oxides and minerals associated with U.S. bituminous coal are shown (see Meyers[36]). Coal associated minerals of the U.S. are discussed in detail in the publication by Bucklen et al.[131]

A study of coal mineral matter by O'Gorman and Walker[122] presented the minerals detected in 57 coal samples from across the U.S. From the semiquantitative data, kaolinite was the most common clay appearing in all the samples with a median value of about 30%. Other constituents in order of abundance were pyrite, quartz, illinite, mixed layer illite-montmorillonite, and calcite. In Table 15 a summary of some of the properties of these minerals is given which include the chemical formula, the density, surface area, and the weight loss on heating to 1000°C in an inert atmosphere.

Chang-Ching[132] has made a systematic determination of the free and combined silica, alumina, titania and acid-insoluble iron in the Miocene coals of Taiwan. From the analytical estimates of the combined silica and alumina, the mole ratio of silica to alumina can be roughly calculated, which may form a useful indicator of the type of clay minerals that compose the mineral matter in the coal samples investigated. A comparision of the analytical results for total silica, alumina, titania, and iron formerly published and newly determined was also made. Some of his results are shown in Table 16. Additional literature on different aspects of coal chemistry can be found at the end of this chapter (References 133 to 216).

VII. RADIOACTIVITY

Radioactive nuclides in the nature can be divided into two groups:

1. Radioactive nuclides with very long half-lives as ^{40}K, ^{50}V, ^{87}Rb, ^{115}In, ^{138}La, ^{142}Ce, ^{152}Gd, ^{174}Hf, $^{190,192}Pt$, ^{204}Pb, $^{235,238}U$. These radionuclides have persisted since the formation of Earth.
2. Radioactive nuclides with the short half-lives on the geological time scale which are being continuously produced by cosmic-ray radiation. The list includes: ^{10}Be, ^{16}Al, ^{36}Cl, ^{14}C, ^{32}Si, ^{39}Ar, ^{3}H, ^{22}Na, ^{35}S, ^{7}Be, ^{37}Ar, ^{32}P.

Most of the activity of rocks and soils is due to uranium, thorium (with their decay products), ^{40}K and ^{87}Rb.

Table 15
MAJOR MINERALS IN COAL

Mineral	Formula	Density (g/cm^3)	N_2 Surface area (m^2/g)	Wt loss on heating to 1000°C
Kaolinite	$Al_2Si_2O_5(OH)_4$	2.6	13.6	14
Illite	$K_{1-1.5}Al_4(Si_{7-6.5}Al_{1-1.5}O_{20})(OH)_4$	2.6—2.9	54.7	8.5
Montmorillonite	$(Na,Ca)_{0.33}(Al,Mg)_2Si_4O_{10}(CH)_2 \cdot NH_2O$	2—3	22.2	5
Pyrite	FeS_2	5.0	1.0	26.7
Calcite	$CaCO_3$	2.7—2.9	1.0	44
Quartz	SiO_2	2.65	1.0	0

After O'Gorman, J. V. and Walker, P. L., Jr., *Fuel*, 52, 71, 1973.

Table 16
MAIN INORGANIC CONSTITUENTS IN THE MIOCENE COALS OF TAIWAN (WT %, DRY COAL BASIS)

	Sample no.				
Inorganic constituents	1	2	3	4	5
SiO_2					
Free	1.3	2.1	3.2	4.7	2.6
Combined	0.95	1.9	5.2	0.49	2.8
Total SiO_2	2.3	4.0	8.4	5.2	5.4
Total Si					
Present study	1.1	1.9	3.9	2.5	2.5
Direct ashing	1.1	1.3	3.1	0.6	12
Al_2O_3					
Total Al_2O_3	0.75	1.9	3.8	0.17	2.0
Total Al					
Present study	0.4	1.0	2.0	0.09	1.1
Direct ashing	0.75	1.0	1.3	0.4	0.55
Fe_2O_3					
Acid-insoluble Fe_2O_3	0.45	0.3	0.5	0.18	0.25
Acid-insoluble Fe	0.3	0.2	0.35	0.13	0.15
Total Fe (previously studied)	2.5	4.4	2.05	0.7	3.1
TiO_2					
Total TiO_2	0.05	0.08	0.17	0.14	0.09
Total Ti					
Present study	0.03	0.05	0.10	0.08	0.05
Direct ashing	0.03	0.05	0.08 5	0.04	0.25
Mole ratio					
Combined SiO_2 total Al_2O_3	1.7	1.7	2.3	5.0	2.4

After Chang-Ching, Y., *Fuel*, 57, 731, 1978.

There are three naturally occuring radioactive series: thorium (4n), uranium-radium (4n + 2) and uranium-actinium (4n + 3) series. The expressions (4n), (4n + 2), and (4n + 3) describe the mass number of any number in respective series, with n being an integer. For example, ^{232}Th isotope is a member of 4n series because of its mass number, A, can be expressed as A = 232 = 458. The three radioactive series are given in Tables 17 to 19.

For the three series listed the mode of decay, the main energies of the α, β and γ emission, and the half-lives of the various isotopes are given. The mode of radioactive

Table 17
THORIUM SERIES (4n)

Nuclide	Half-life	Type of decay and particle energies (MeV) (%)	Gamma energies — internal conversion (IC)
^{232}Th	$1.4 \cdot 10^{10}$ years	α: 4.007 (76), 3.948 (24) 3.948 (24)	γ: 0.059
^{228}Ra	6.7 years	β^- ∼0.04 (100)	
^{228}Ac	6.13 hr	β^-: 2.10 (≅12), 1.76 (∼12), 1.18 (∼35), others of low energy (41)	γ: 0.057 to 1.46 (many lines)
^{228}Th	1.91 years	α: 5.421 (71), 5.338 (28) ∼ 5.2 (0.4)	γ: 0.084 (2%), others (very weak)
^{224}Ra	3.64 days	α: 5.681 (95) 5.445 (4.9)	γ: 0.241 (3.2% + 0.5% (IC)
^{220}Rn	51.5 sec	α: 6.28 (∼100)	
^{216}Po	0.16 sec	α: 6.775 (∼100) β: 0.04	
^{216}At	3.10^{-4} sec	α: 7.79 (0.04)	
^{212}Pb	10.6 hr	β^-: 0.57 (∼12), 0.33 (∼80), others of low energy (8)	γ: 0.30 (∼3%) + 1% IC), 0.24 (∼36% + 36% IC), 0.12 (∼0% + 3% IC)
^{212}Bi	60.5 min	α: 6.08 (10), 6.04 (25) others (1)	γ: 1.62, 1.08, 0.79, 0.73 (together ∼ 10%) 0.04 (0% +∼ 25% IC) others weak)
		β^-: 2.25 (54), 1.52 (5), others of low energy (5)	
^{212}Po	$3 \cdot 10^{-7}$ sec	α: 8.78 (∼64), others of higher energy (very weak)	
^{208}Tl	3.1 min	β^-: 1.79 (19), 1.52 (7), 1.25 (9), 1.03 (1)	γ: 2.62 (36%), 0.86 (4%) 0.58 (31%), 0.51 (∼8% +∼ 1% IC), 0.28 (∼3% +∼ 1% IC)
^{208}Pb	Stable		

decay is mainly by the emission of α- and β-particles with de-excitation by the emission of γ-rays and the mass of the decay product is either the same as that of the precursor (β-decay) of four mass-units less (α-decay). Hence the usual reference is 4n, 4n + 2 and 4n + 3 series. The uranium isotopes all decay by the emission of high-energy α-particles (about 4 MeV) and low-energy γ-ray and X-rays (about 0.1 MeV).

Of the uranium isotopes, ^{238}U has a half-life similar to that of the age of the Earth, i.e. 4, 5 + 10^9 years. The half-life of ^{235}U is slightly less, being 8, 8 × 10^8 years. Therefore the ratio $^{238}U/^{235}U$ has not remained constant throughout the life of the Earth, but at any point in time the ratio has a fixed value regardless of the source of uranium.

Variations in this natural $^{238}U/^{235}U$ ratio have only been observed in the Oklo uranium deposits of Gabon, where, owing to the formation of a natural reactor, there has been a significant depletion of ^{235}U. In general this phenomenon is regarded as being an isolated occurrence and for our purposes it is accepted that the $^{238}U/^{235}U$ ratio is constant. Uranium has a third important isotope, ^{234}U, which is a decay product of ^{238}U.

Table 18
URANIUM — RADIUM (4n + 2) SERIES

Nuclide	Half-life	Type of decay and particle energies (MeV) (%)	Gamma energies — internal conversion (IC)
^{238}U	$4.51 \cdot 10^9$ years	α: 4.2 (100)	γ: 0.048 (0% + 23% IC)
^{234}Th	24.1 days	β⁻: 0.19 (65), 0.10 (35)	γ: 0.091 (α_{IC} = 2.5), 0.063 (α_{IC} = 0.2), 0.029(α_{IC} = 10)
Protactinium-234 m	1.18 min	β⁻: 2.31 (~90), 1.50 (~9), 0.58 (~1) IT: 1	1.00, 0.75, others
Protactinium-234	6.66 hr	β⁻: 1.13 (13), 0.53 (27), 0.32 (32), 0.16 (28)	γ: 1.68, 1.43, 1.24, 0.924, 0.877, 0.803, 0.732, 0.603, 0.566, 0.368, 0.333, 0.293, 0.225, 0.153, 0.099, 0.043
^{234}U	$2.5 \cdot 10^5$ years	α: 4.768 (72), 4.717 (28)	γ: 0.051 0% + 28% IC)
^{230}Th	$8 \cdot 10^4$ years	α: 4.682 (76)	γ: 0.067 (0% + 24% IC)
^{226}Ra	1620 years	α: 4.777 (94.3)	γ: 0.188 (~4% + ~2% IC)
^{222}Rn	3.825 days	α: 5.48 (~100)	
^{218}Po	3.05 min	α: 6.00 (~100) β⁻: (0.02)	
^{218}At	1.3 sec	α: 6.70 (~0.02), 6.65 (~0.001) β⁻: (very weak)	
^{218}Rn	$1.9 \cdot 10^{-2}$ sec	α: 7.13 (very weak)	γ: 0.61 (very weak)
^{214}Pb	26.8 min	β⁻: 0.65 (~44), 0.59 (~56)	γ: 0.35, 0.30, 0.24
^{214}Bi	19.9 min	α: ~5.5 (0.04), β⁻: 3.26 (19), 1.88 (9), 1.51 (40), 1.0 (23), 0.4 (9)	: 1.76[a], 1.12[a], 0.61[a] others up to 2.43
^{214}Po	$1.6 \cdot 10^{-4}$ sec	α: 7.68 (~100)	
^{210}Ti	1.3 min	β⁻: 0.061 (20), 0.015 (80)	γ: very weak
^{210}Pb	22y	β⁻: 0.061 (20), 0.015 (80)	γ: 0.0465 (~5% + ~75% IC)
^{210m}Bi	5.0 d	α: 5.06 ($1.7 \cdot 10^{-4}$) β⁻: 1.17 (~100)	
^{210}Po	138.4 days	α: 5.305 (~100)	γ: 0.8 (very weak)
^{206}Ti	4.2 min	β⁻: 1.51 ($1.7 \cdot 10^{-4}$)	
^{206}Pb	Stable		

[a] Most abundant.

Data on radionuclide concentrations in coal will be discussed in detail in Chapter 2. Here we may just mention the work by Styron et al.[217,218] They have measured the activities of different radionuclides in U.S. coals. According to their report data on radionuclide concentrations in coal samples indicated excellent precision for the analysis of ^{234}U and ^{238}U. Duplicate analyses of each coal sample for ^{234}U and ^{238}U showed precisions of 5.1 and 7.0%, respectively.

Measurements of ^{234}U and ^{238}U concentrations in coal samples ranged from 0.04 to

Table 19
URANIUM — ACTINIUM SERIES (4n + 3)

Nuclide	Half-life	Type of decay and particle energies (MeV) (%)	Gamma energies — internal conversion (IC)
^{235}U	$7.1 \cdot 10^{8}$ years	α: 4.56 (7), 4.52 (4), 4.35 (84), 4.18 (6)	γ: ∼0.2000 (>4%), 0.185 (55%), ∼0.165, (>4%), 0.143 (12%), 0.110 (5%), 0.095 (9%)
^{231}Th	25.6 hr	β^-: 0.30, others	γ: 0.084, others
Protactinium-231	$3.4 \cdot 10^{4}$ years	α :5.046 (10), 5.017 (23), 5.001 (24), 4.938 (22), 4.722 (11), others	γ: 0.29, 0.027, many others
^{227}Ac	22 years	α: 4.94 (1.2) β^-: 0.046 (99)	
^{227}Th	18.2 days	α: 6.036 (23), 5.976 (24), 5.958 (3.5), 5.865 (3), 5.755 (21), 5.712 (5), 5.708 (8.7), 5.699 (4), others (∼7.8)	γ: 0.24 (∼10%), 0.05 (∼7.5%), many others of low energy (weak), all highly converted
^{223}Fr	22 min	α: 5.34 (very weak) β^-: 1.15 (∼1%)	γ: 0.31, 0.21, 0.08
^{223}Ra	11.7 days	α: 5.742 (10.5), 5.712 (50), 5.602 (24), 5.429 (2.4), others (13.1)	γ: 0.34 (2.8%), 0.32 (2.3%), 0.27 (10%), 0.15 (5.5%), 0.14 (4%), 0.12 (2%), all highly converted
^{219}Rn	3.9 sec	α: 6.818 (82), 6.547 (13), 6.419 (5)	γ: 0.40 (∼5%), 0.27 (∼9% + ∼4% IC)
^{215}Po	$1.8 \cdot 10^{-3}$ sec	α: 7.360 (∼100) β^-: (0.0005)	
^{215}At	$\sim 1 \cdot 10^{-4}$ sec	α: 8.00 (very weak)	
^{211}Pb	36 min	β^-: 1.39 (∼80), 0.5 (∼20)	γ: 0.83 (13%), 0.43 (6%), 0.40 (6%)
^{211}Bi	2.16 min	α: 6.617 (∼83) 6.273 (17) β^-: 0.3%	γ: 0.35 (13% + 4% IC)
^{211}Po	0.52 sec	α: 7.44 (∼0.3) 6.90 (weak), 6.57 (weak)	γ: 0.89 (weak) 0.57 (weak)
^{207}Tl	4.79 min	β^-: 1.44 (100)	
^{207}Pb	Stable		

1.20 pCi/g; ^{210}Po, from 0.03 to 0.58 pCi/g. Combining values for all mines studied gives a mean concentration for ^{234}U of 0.31 ± 0.32 pCi/g, for ^{238}U of 0.30 ± 0.32 pCi/g, for ^{210}Po of 0.23 ± 0.14 pCi/g. These radionuclide concentrations in Western coal from operating mines are roughly comparable to values for Eastern coal as reported by Bedrosian et al.,[219] Martin et al.,[220] and an EPA survey, and are slightly below the national average (0.6 pCi/g) reported by Swanson et al.[222] A few Western coal reserves, however, are reported to have significantly higher concentrations, on the order of 10

Table 20
MASSES OF THE VARIOUS DAUGHTERS IN SECULAR EQUILIBRIUM WITH 1 g OF ^{238}U

Isotope	Mass (g)
^{238}U	1.0
^{234}Th	1.4×10^{-11}
^{234m}Pa	4.8×10^{-16}
^{234}U	5.4×10^{-6}
^{230}Th	1.8×10^{-5}
^{226}Ra	3.3×10^{-7}
^{222}Rn	2.2×10^{-10}
^{218}Po	1.2×10^{-15}
^{214}Pb	1.0×10^{-14}
^{214}Bi	7.4×10^{-15}
^{210}Pb	4.1×10^{-9}
^{210}Bi	2.7×10^{-12}
^{210}Po	7.4×10^{-11}

to 100 times.[223] These coals with high concentrations of uranium represent a valuable source of uranium and if used in combustion at all they would likely be used with dual resource recovery in mind.

Values for ^{210}Po suggest that a secular equilibrium exists for isotopes of the uranium decay chain in these samples of coal, i.e., the concentrations of ^{238}U and ^{210}Po are essentially equal. Masses of the various daughters in secular equilibrium with 1 g of ^{238}U are shown in Table 20.

REFERENCES

1. **Cameron, C. C. and Wright, N. A.** Some Peat Bogs in Washington County, Maine: Their Formation and Trace-Element Content, Interdisciplinary Studies of Peat and Coal Origins, Microform Publ. 7, The Geological Society of America, Boulder, 1977, 50.
2. **Casagrande, D. J. and Erchull, L. D.,** Organic Geochemistry of Okefenokea Peats: Metal Constituents, Interdisciplinary Studies of Peat and Coal Origins, Microform Publ. 7, The Geological Society of America, Boulder, 1977, 72.
3. **Duel, M.,** Biochemical and geochemical origins of ash-forming ingredients in coal, in Meet. Am. Chem. Soc., Div. Gas and Fuel Chemistry, Chicago, Ill., September 7 to 12, 1958.
4. **Francis, W.,** *Coal, Its Formation and Composition,* Edward Arnold, London, 1954.
5. **Francis, W.,** *Coal,* Edward Arnold, London, 1961.
6. **Manskaya, S. M., Kodina, L. A., Generalova, N. V., and Kravtsova, R. P.,** Interaction of germanium with lignin structure during the early stages of coal formation, (In Russian) *Geokhimiya,* 5, 600, 1972.
7. **Medvedev, K. P.,** Role of macro and trace elements in coal formation processes, *Khim. Tverd. Topl.,* 4, 3, 1971.
8. **Azizov, T. M. and Vlasov, V. I.,** Geochemical features of the accumulation of coal measures in the Chu region, *Izv. Akad. Nauk. Kaz. S.S.R. Ser. Geol.,* 30, 83, 1973.
9. **Bloxam, T. W.,** *Organic Geochemistry of Some Carboniferous Shales from the South Wales Coalfield,* Rep. 71, Great Britain Institute Geological Science, London, 1971, 1.
10. **Breger, I. A.,** Geochemistry of coal, *Econ. Geol.,* 53, 828, 1958.
11. **Clarke, F. W. and Washington, H. S.,** The Composition of the Earth's Crust, Prof. Paper 127, U.S. Geological Survey, Reston, Va., 117, 1924.
12. **Gluskoter, H. J.,** Inorganic geochemistry of Illinois agglomerating coals, in Proc. Coal Agglomeration and Conversion Symp., Morgantown, W.Va., 1975, 9.

13. **Goldschmidt, V. M.**, in *Geochemistry,* Muir, A., Ed., Clarendon, Oxford, 1954, 730.
14. **Hak, J. and Babcan, J.**, The geochemistry of germanium and berillyium in coals of the Sokolov basin, Geochem. in Czech., in Trans. 1st Conf. Geochem., Ostrava, September 20 to 24, 1967, 163.
15. **Hidalgo, R. V.**, Inorganic Geochemistry of Coal, Pittsburgh Seam, Ph.D. dissertation, West Virginia University, Morgantown, 1974, 125.
16. **Nicholls, G. D.**, The geochemistry of coal-bearing strata, in *Coal and Coal Bearing Strata,* Murchison, D. and Westoll, T. S., Eds., New York, 1968, 269.
17. **Renton, J. J.**, Inorganic geochemistry/mineralogy of West Virgina agglomerating coals, in Proc. Coal Agglomeration and Conversion Symp., Morgantown, W. Va., 1975, 21.
18. **Renton, J. J. and Hidalgo, R. V.**, Some Geochemical Considerations of Coal, Coal-Geology, Bull. No. 4, West Virginia Geological and Economic Survey, 1975, 38.
19. **Saprykin, F. Ya., Kler, V. R., and Kulachkova, A. F.**, Geochemical characteristics of rare element concentration in various types of coal and bituminous shale-bearing formations, *Geol., F.,* 1971; Abstr. 2K57, *Uglenosn. Formatsii Ikh. Genezis,* 88, 1970.
20. **Saprykin, F. Ya., Kier, V. P., and Kulachkova, A. F.**, Geochemical characteristics of the concentration of rare elements in the diverse types of coal-bearing formations, *Uglenos. Form. Ikh. Genezis, Dokl. Vses. Geol. Ugol. Soveshch.,* 4, 126, 1973.
21. **Shakhov, F. N. and Efendi, M. E.**, Geochemistry of the coals of the Kuznetsk Basin, *C. R. Acad. Sci. U.R.S.S.,* 51, 139, 1946.
22. **Swaine, D. J.**, Trace elements in Coal, in *Recent Contributions to Geochemistry and Analytical Chemistry,* Turgarinov, A. E., Ed., John Wiley & Sons, New York, 1976.
23. **Yadrenkin, V. M. and Shugurov, V. F.**, Comparative lithological-petrographic and geochemical characteristics of coal-bearing formations of the Jureika, Lower Tunguska, and Bakhty sections of the Tunguska syncline, *Tr. Sib. Nauch. Issled. Inst. Geol., Geofiz. Miner. Syr'ya,* 188, 98, 1974.
24. **Yudovich, Ya. E.**, Geochemistry problems of coal inclusions (for example, the Lena coal-bearing basin hear Yakutsk, *Lithol. Polez. Iskop.,* 5, 52, 1968.
25. **Yudovich, Ya. E., Ed.**, *Geokhimiya Iskopaemykh Uglei (Neorganisheskie komponenty),* Izdatel'stvo Nauka, Leningrad, 1978.
26. Panel on the trace element geochemistry of coal resources development related to health, in Trace Element Geochemistry of Coal Resources Development Related to Environmental Quality and Health, National Academy Press, Washington, D.C., 1980.
27. **Van Krevelen, D. W.**, *Coal,* Elsevier, Amsterdam, 1961.
28. **Parks, B. C.**, Origin, petrography, and classification of coal, in *Chemistry of Coal Utilization,* Suppl. Vol., Lowry, H. H., Ed., John Wiley & Sons, New York 1.
29. **Spackman, W.**, What is coal? Thc range of U.S. coals available, paper presented at Short Course on Coal Characteristics and Coal Conversion Processes, Pennsylvania State University, University Park, October 29 to November 2, 1973.
30. **Given, P. H.**, Organic chemistry of coals, paper presented at Short Course on Coal Characteristics and Coal Conversion Processes, Pennsylvania State University, University Park, October 29 to November 2, 1973.
31. **Zubovic, P.**, Physiochemical properties of certain minor elements as controlling factors of their distribution in coal, *Adv. Chem. Ser.,* 55, 221, 1966.
32. **Zubovic, P.**, Geochemistry of trace elements in coal, in Symp. Proc. Environmental Aspects of Fuel Conversion Technology. II, EPA-600/2-76-149, U.S. Environmental Protection Agency, Washington, D.C., 1976, 47.
33. **Ensminger, J. T.**, Coal: Origin, classification, and physical and chemical properties, in *Environmental, Health and Control Aspects of Coal Conversion,* Rep. ORNL/EIS-94, Braunstein, H. M., Copenhover, E. D., and Pfuderer, H. A., Eds., Oak Ridge National Laboratories, Oak Ridge, Tenn., 1977.
34. **Given, P. H.**, *Fuel,* 39, 147, 1960.
35. **Given, P. H. and Miller, R. N.**, Determination of forms of sulphur in coals, *Fuel,* 57, 380, 1978.
36. **Meyers, R. A.**, *Coal Desulfurization,* Marcel Dekker, New York, 1977.
37. **Averitt, P.**, Coal Resources of the United States, Bull. 1412, U.S. Geological Survey, Reston, Va., January 1, 1974, 131.
38. **Schopf, J. M.**, A definition of coal, *Econ. Geol.,* 51 521, 1956.
39. **Given, P. H.**, How may coals be characterized for practical use? Paper presented at Short Course on Coal Characteristics and Coal Conversion Processes, Pennsylvania State University, University Park, October 29 to November 2, 1973.
40. **Burchett, R. R.**, Coal Resources of Nebraska, Rep. NP 23879, University of Nebraska, Lincoln, 1977.
41. **Teichmuller, M. and Teichmuller, R.**, Geological causes of coalification, in *Coal Science,* Gould, R. F., Ed., American Chemical Society, Washington, D.C., 1966.

42. Yancey, H. F. and Geer, M. R., Impurity properties, in *Coal Preparation,* American Institute of Mining, Metallurgical and Petroleum Engineers, New York, 1968, chap. 1, 1.
43. Gilluly, J., Waters, A. C., and Woodford, A. O. *Principles of Geology,* W. H. Freeman, San Francisco, 1968.
44. Given, P. H., Inorganic species in coal and their potential catalytic effects on liquefaction processes, Appendix, in A Research and Development Program for Catalysis in Coal Conversion, NTIS No. PB 242-412, A-122-A-140, National Technical Information Service, Springfield, Va., 1974.
45. Given, P. H., Problems in the chemistry and structure of coals as related to pollutants from conversion processes, in Symp. Proc. Environmental Aspects of Fuel Conversion Technology, EPA-650/2-74-118, St. Louis, Mo., 1974, 27.
46. Hower, J. M., Davis, A., Dolsen, C. P., and Spackman, W., Survey of Selected Agencies Conducted to Determine the Extent to Which the Nation's Coals are Adequately Characterized, Ref FE-20308-TR5, National Technical Information Service, Springfield, Va., 1977.
47. Yarzab, R. F., Given, P. H., Spackman, W., and Davis, A., Dependence of coal liquefaction behavior on coal characteristics, IV. Cluster analyses for characteristics of 105 coals, *Fuel,* 59, 81, 1980.
48. Stephens, J. F., Methods for coal classification, *Fuel,* 58, 489, 1979.
49. Mott, R. A. The origin and composition of coals, *Fuel (London),* 21(6), 129, 1942.
50. Hill, G. R. and Lyon, L. B., *Ind. Eng. Chem.,* 54, 36, 1962.
51. Wolk, R., H., Stewart, N. C., and Silver, H. F., *Am. Chem. Soc. Div. Fuel Chem. Prepr.,* 20, 116, 1975.
52. Oka, M., Chang, H. C., and Gavalas, G. R., *Fuel,* 56, 3, 1977.
53. Battaerd, H. A. J. and Evans, D. G., An alternative representation of coal composition data, *Fuel,* 58, 105, 1979.
54. Kuhn, J. K., Fiene, F., and Harvey, R., Geochemical Evaluation and Characterization of a Pittsburgh, No. 8 and a Rosebud Seam Coal, Report METC/CR-7B/8, National Technical Information Service, Springfield, Va., 1978.
55. Parr, S. W., The classification of coal, *Univ. Ill. Eng. Sta. Bull.,* 180, 62, 1928.
56. Augenstein, D. A. and Sun, S. C., Methodology for the Characterization of Anthracite Refuse, The Pennsylvania State University Special Research Report SR-86, Pennsylvania State University, University Park, 1971, 77.
57. Sun, C. C. and Campbell, J. A. L., Anthracite lithology and electrokinetic behavior, in *Advances in Chemistry,* Series 55, Gould, R. F., Ed., American Chemical Society, Washington, D. C., 1966, 363.
58. Davis, A., The reflectance in coal, in *Analytical Methods for Coal and Coal Products,* Vol. 1, Karr, C., Ed., Academic Press, New York, 1978.
59. Turner, A. P. L., Conrad, H., Dickinson, J. M., Keefer, D. W., Lytle, J. M., Ohr, S. M., Powell, W. R., Sanner, W. S., and West, C. A., Unpublished report, 1980.
60. Austin, L. G. Mechanical and Comminutice Properties of Coal, Materials Problems and Research Opportunities in Coal Conversion, Vol. 2, NTIS PE-248 081, Staehle, R. W., Ed., Ohio State University, Columbus, April 1974, 49.
61. Marsh, H., *Fuel,* 44, 253, 1965.
62. Spencer, D. H. T., *Porous Carbon Solids,* Bond, R. L., Ed., Academic Press, New York, 1967, 87.
63. Parfitt, G. D. and Sing, K. S. W., Eds., Characterization of Powdered Surfaces, *Academic Press,* London, 1976.
64. Debelak, K. A. and Schrodt, J. T., Comparison of pore structure in Kentucky coals by mercury penetration and carbon dioxide absorption, *Fuel,* 58, 732, 1979.
65. Zweitering, P. and van Krevelen, D. W., Chemical structure and properties of coal. IV, Pore structure, *Fuel,* 33, 331, 1954.
66. Gan, H., Nandi, S. P., and Walker, P. L., Jr., Nature of porosity in American coals, *Fuel,* 51(4), 272, 1972.
67. Walker, P. L., Jr., Physical structure of coals, paper presented at Short Course on Coal Characteristics and Coal Conversion Processes, Pennsylvania State University, University Park, October 29 to November 2, 1973.
68. Diebold, F. E., Characterization of coal for open cycle MHD generation system, in MHD Power Generation Research, Development and Engineering, Quarterly Report FE-1811-20, April-June 1976.
69. Duba, A. G., Electrical conductivity of coal and coal char, *Fuel,* 56, 441, 1977.
70. Pope, M. I. and Gregg, S. J., *Fuel,* 40, 123, 1961.
71. Wooster, W. A. and Wooster, N., The Magnetic properties of coals, in *Proc. Conf. Ultra-fine Structure of Coals and Cokes,* British Coal Utilities Research Association, London, 1944, 322.
72. Burgardt, P. and Seehra, M. S., *Solid State Commun.,* 22, 153, 1977.
73. Marusak, L. D., Cordero-Montaloo, C., and Mulay, L. N., *Mat. Res. Bull.,* 12, 1009, 1977.
74. Lefelhorz, J. F., Friedel, R. A., and Kohman, T. P., Mössbauer spectroscopy of iron in coal, *Geochim. Cosmochim. Acta,* 31, 2261, 1967.

75. Levinson, L. M. and Jacobs, I. S., Mössbauer spectroscopic measurement of pyrite in coal, *Fuel,* 56, 453, 1977.
76. Montano, P. A., Mössbauer spectroscopy of iron compounds found in West Virginia coals, *Fuel,* 56, 397, 1977.
77. Huffman, G. P. and Higgins, F. E., Mössbauer studies of coal and coke: quantitative phase identification and direct determination of pyritic and iron sulfide content, *Fuel,* 56, 592, 1978.
78. Smith, G. V., Liu, J. H., Saporoschenko, M., and Shiley, R., Mössbauer spectroscopic investigation of iron species in coal, *Fuel,* 57, 41, 1978.
79. Trindade, S. C., Howard, J. B., Kolm, H. H., and Power, G. J., Fuel, 53, 178, 1974.
80. Good, J. A. and Cohen, E., *Cryogenics,* 16, 579, 1976.
81. Sladek, T. A. and Cox, C. H., in Coal Benefication with Magnetic Fluids, 85th Natl. Meet. American Chemical Engineers, Philadelphia, Pa., June 4 to 8, 1978.
82. Alexander, C. C., Thorpe, A. N., and Senftle, F. E., Basic magnetic properties of bituminous coal, *Fuel,* 58, 857, 1979.
83. Evans, D. G. and Allardice, D. J., Physical property measurements on coal, especially brown coal, in *Analytical Methods for Coal and Coal Products,* Vol. 1, Karr, C., Ed., Academic Press, New York, 1978.
84. Bondarenko, S. T., Some electrical characteristics of Solid Fossil Fuels, in *Application of Electric Current for Direct Action on a Seam of Fuel in Shaftless Underground Gasification,* Meerovich, E. A., Ed., Academy of Science, U.S.S.R., 1959; UCRL-TRANS-11050, 1976.
85. Budde, K., Singer, W., Zobel, F., Dressel, S., Schaefer, D., and Pauli, W., Coke with Reduced Electrical Conductivity, German Patent 231178; Patent 2722476. 1978, 7.
86. Singer, W., Budde, K., Zobel, F., Schaefer, D., and Krug, H., Brown Coal High-Temperature Coke with Poor Electrical Conductivity, Germany (East) Patent 132977, 1978, 9.
87. Gontsov, A. A., Kurbatova, E. G., Molchanov, V. I., and Novogorodova, S. V., Effect of the grinding medium on the breakdown of coals, *Mater. Vses. Simp. Mekhanoemiss. Mekhanokhim. Tverd. Tel.,* 5th, 3, 122, 1977.
88. Ondov, J. M., Ragaini, R. C., and Biermann, A. H., Emissions and particle size distributions of minor and trace elements at two western coal-fired power plants equipped with cold size electrostatic precipitators, *Environ. Sci. Technol.,* 13, 946, 1979.
89. Hulett, L. D., Carter, J. A., Cook, K. D., Emery, J. F., Klein, D. H., Lyon, W. S., Nyssen, G. A., Fulkerson, W., and Bolton, N. E., Trace element measurements of the coal-fired Allen steam plant, particle characterization, in Pap. Coal. Util. Symp. Focus SO2 Emiss. Control, 1974, 207.
90. Cady, G. H. and Leighton, H. M., The physical constitution of Illinois coal and its significance in regard to utilization, in Proc. Ill. Min. Inst., 1933, 93.
91. Ergun, S., Donaldson, W. F., and Berger, I. A., Some physical and chemical properties of vitrains associated with uranium, *Fuel,* 39, 71, 1960.
92. Fischer, G. L. et al., Size-dependence of the physical and chemical properties of coal fly ash, *Am. Chem. Soc. Div. Fuel Chem. Prepr.,* 22, 149, 1977.
93. Dalmon, J. and Raask, E., Resistivity of particulate coal minerals, *J. Inst. Fuel,* 45, 201, 1972.
94. Vdovenko, M. I. and Zhukova, T. S., Determination of the surface tension of coal algas in the plastic state, *Probl. Teploenerg, Prikl. Teplofiz.,* 7, 20, 1971.
95. Pavlyuk, Yu. S., Effect of several mineral and organic additives on the bulk density of coal charges, *Met. Koksokhim.,* 37, 3, 1974.
96. Volborth, A., Miller, G. E., Garner, C. K., and Jarabek, P. A., Oxygen determination and stochiometry of some coals, *Am. Chem. Soc. Div. Fuel Chem. Prepr.,* 22, 1977; Symp. New Tech. Coal Anal., presented at 174th Am. Chem. Soc. Natl. Meet., Chicago, Ill. 1977, 9.
97. Choudhury, S. S. and Ganguly, P. C., Effect of carbonate minerals on volatile matter of coals, *Fuel,* 57, 175, 1978.
98. Doolan, K. J., Knott, A. C., Belcher, C. B., Multiple determination of volatile matter in solid fuels, *Fuel,* 59, 260, 1980.
99. Scholz, A., Correlation between the content of volatile matter and mineral matter in hard coals, *Fuel,* 59, 197, 1980.
100. Kasper, S. Coal conversion chemistry — see it in common terms, *Ind. Res. and Dev.,* 164, January 1981.
101. Murchison, D. and Westoll, T. S., Eds., *Coal and Coal Bearing Strata,* Oliver and Boyd, London, 1968.
102. Maxwell, J. R., Phillinger, C. T., and Eglinton, G., *Q. Rev.,* 25, 571, 1971.
103. Hayatsu, R., Winans, R. E., Scott, R. G., Moore, L. P., and Studier, M. H., Trapped organism compounds and aromatic units in coal, *Fuel,* 57, 541, 1978.
104. Camier, R. J. and Siemon, S. R., X-ray crystallography and brown-coal structure, *Fuel,* 57, 508, 1978.

105. Larsen, J. W., Summary — an evaluation of the current level of understanding of the fundamental organic chemistry of coal, in NSF Workshop on the Fundamental Organic Chemistry of Coal, University of Tennessee, Knoxville, July 17 to 19, 1975, 1.
106. Wiser, W. H., Some chemical aspects of coal liquefaction, paper presented at Short Course on Coal Characteristics and Coal Conversion Processes, Pennsylvania State University, University Park, October 29 to November 2, 1973.
107. Mazumdar, B. K., Ganguly, S., Sanyal, P. K., and Lahiri, A., Aliphatic structures in coal, in *Advances in Chemistry*, Ser. 55, Gould, R. F., Ed., American Chemical Society, Washington, D.C., 1966, 475.
108. Chakrabartty, S. K. and Berkowitz, N., Studies on the structure of coals. III. Some inferences about skeletal structures, *Fuel*, 53(4), 240, 1974.
109. Reggel, L., Wender, I., and Raymond, R., Catalytic dehydrogenation of coal. VII. A comparison of exinite, micrinite and fusinite with vitrinite, *Fuel (London)*, 49, 281, 1970.
110. Given, P. H., Poever, M. E., and Wyss, W. F., Chemical properties of coal macerals. I. Introductory survey and some properties of exinites, *Fuel*, 39, 323, 1960.
111. Gluskoter, H. J., Mineral matter and trace elements in coal, in *Advances in Chemistry*, Ser. 141, Babu, S. P., Ed., American Chemical Society, Washington, D.C., 1973, 1.
112. Gluskoter, H. J., Introduction to the occurrence of mineral matter in coal, in *Ash Deposits and Corrosion Due to Impurities in Combustion Gases*, Bryers, R. W., Ed., Hemisphere Publishing Washington, D.C., 1978, 3.
113. Parr, S. W., *The Analysis of Fuel, Gas, Water and Lubricants*, McGraw-Hill, New York, 49, 1932.
114. King, J., Maries, M. B., and Crossley, H. E., Formulae for the calculation of coal analyses to a basis of coal substance free of mineral matter, *J. Soc. Chem. Ind. (London)*, 57, 277, 1936.
115. Radmacher, W. and Mohrhauer, P., *Brennst. Chem.*, 36, 236, 1955.
116. Bishop, M. and Ward, D. L., *Fuel*, 37, 191, 1958.
117. Tarpley, E. C. and W. H. Ode, *U.S. Bur. Miners Rep. Invest.*, p. 5470, 1959.
118. Miller, R. N., Yarzeb, R. F., and Given, P. H., Determination of the mineral-matter contents of coals by low-temperature ashing, *Fuel*, 58, 4, 1979.
119. Pollack, S. S., Estimating mineral matter in coal from its major inorganic elements, *Fuel*, 58, 76, 1979.
120. Kiss, L. T. and King, T. N., The expression of results of coal analysis, *Fuel*, 56, 340, 1977.
121. Rao, C. P. and Gluskoter, H. J., Occurrence and Distribution of Minerals in Illinois Coals, Circular 476, Illinois State Geological Survey, Urbana, 1973, 56.
122. O'Gorman, J. V. and Walker, P. L., Jr., Mineral matter characteristics of some American coals, *Fuel*, 50, 135, 1971.
123. O'Gorman, J. V. and Walker, P. L., Jr., Thermal behavior of mineral fractions separated from selected American coals, *Fuel*, 52, 71, 1973.
124. Nelson, H. W., Mineral constituents in coal and their behavior during combustion, in *Corrosion and Deposits in Coal and Oil Fired Boilers and Gas Turbines*, American Society Mechanical Engineers, New York, 1959, 7.
125. Ball, C. G., Preliminary microscopic investigation of Illinois coal, *Ill. State Acad. Sci. Trans.*, 24, 327, 1931.
126. Ball, C. G., Koalinite in Illinois coal, *Econ. Geol.*, 29, 767, 1934.
127. Ball, C. G., Mineral matter of No. 6 Bed coal at West Frankfort, Franklin County, Illinois, Report of Investigations, 33, Illinois State Geological Survey, Urbana, 1935, 106.
128. Ball, C. G., Possible relations of mineral matter in coal to the time of coalification, *Illinois State Acad. Sci. Trans.*, 28, 181, 1935—1936.
129. Ball, C. G. and Caddy, G. H., Evaluation of ash correction formulae based on petrographic analysis of mineral matter in coal, *Econ. Geol.*, 30, 72, 1935.
130. Ode, W. H., Coal analysis and mineral matter, in *Chemical Coal Utilization*, Lowry, H. H., Ed., John Wiley & Sons, New York, 1962, 202.
131. Bucklen, O. D., Cockrell, C. F., Donahue, B. A., Leonard, J. W., McPadden, C. R., Meikle, P. G., Mih, L. C., and Shafer, H. E., Coal Associated Minerals of the U.S. VII. Uses, Specifications and Processes Related to Coal-Associated Minerals, PB-168, 116, RDR-8(7), Coal Research Bureau, National Technical Information Service, Springfield, Va., 1965.
132. Chang-Ching, Y., Determination of the main inorganic constituents in the Miocene coals of Taiwan, *Fuel*, 57, 731, 1978.
133. Korbuly, J. and Nagy, F., Experiments to increase the bulk density of coking coal admixtures at Duna iron works. *Publ. Hung. Min. Res. Inst.*, 15, 193, 1972.
134. McIntosh, M. J., Chemical heterogeneity of coal, *Fuel*, 55, 59, 1976.
135. Parr, S. W., The Chemical Composition of Illinois Coals, Bull. 16, Illinois State Geological Survey Yearbook for 1909, Urbana, 1909, 205.

136. **Parr, S. W. and Wheeler, W. F.,** Unit coal and the composition of coal ash, Bull. 37, University of Illinois Engineering Experiment Station, 1908, 68.
137. **Schlyer, D. J. and Wolf, A. P.,** A study of coal oxidation by charged particle activation analysis, in 4th Int. Conf. on Nuclear Methods in Environmental and Energy Research, BLN 27607, Columbia, Mo., 1980.
138. **Tippmer, K.,** Production of reducing gases by the partial oxidation of hydrocarbons and coal, *Erdoel Kohle, Erdgas, Petrochem.*, 29, 153, 1976.
139. **Tingey, G. and Morrey, J. R.,** Coal Structure and Reactivity, Battelle Energy Program Report, Battelle, Pacific Northwest Laboratory, Richland, Wash., 1973.
140. **Zinov'ev, Yu.Z., Klassen, V. I., and Litovko, V. I.,** Effect of some inorganic salts on the settling out of coal suspensions with pretreatment of the liquid physe in a magnetic field, *Nov. Methody Povysh. Eff. Obogashch. Polez. Iskop. Publ. Nauka, Moscow, S.S.S.R.*, p. 44, 1968.
141. **Borio, R. W. et al.,** Study of means for eliminating corrosiveness of coal to high temperature surfaces of steam generating units. II, *Combustion*, 39, 12, February 1968.
142. **Hill, V. L. and Howes, M. A. H.,** Metallic corrosion in coal gasification pilot plants, *Mater. Perform.*, 17, 22, 1978.
143. **Sedor, P., Diehl, E. K., and Bernhardt, D. H.,** External corrosion of superheaters in boilers firing high alkali coals, *J. Eng. Power*, 82, (Ser. A), 181, 1960.
144. **Segnit, E. R.,** Corrosion of firebrick by brown coal ash, *J. Aust. Ceram. Soc.*, 4, 25, 1968.
145. **Burkov, P. A., Aleksandrov, I. V., Borodin, B. A., Ivashkin, V. S., and Kamneva, A. I.,** Study of the flammability of brown coals of the Kharanorsk deposit, *Khim. Tverd. Topl.*, 1, 9, 1976.
146. **Mitsuo, O., Chang, H. C. and Gavalas, G. R.,** Computer assisted molecular structure construction for coal derived compounds, *Fuel*, 56, 3, 1977.
147. **Sukhov, V. A., Zamyslov, V. B., Shumeiko, V. P., Voitkovskii, Yu.B., and Lukovnikov, A. F.,** Behavior of iron compounds in thermally treated coal production and their role in the oxidation process, *Khim. Tverd. Topl. (Moscow)*, 6, 30, 1978.
148. **Menkovskii, M. A. and Kazantseva, K. I.,** State of oxidation of coals, *Zh. Khim.*, 1969; Abstr. 22041, Nauch. Konf. Vuzov S.S.S.R. Uchastiem Nauch.-Issled. Inst. Fiz. Gorn. Porod. Prots., Sekts, Khim. Biokhim. Gorn. Porod., 1969, 11.
149. **Dreesen, D. R., Wangen, L. E., Gladney, E. S., and Owens, J. W.,** Solubility of trace elements in coal fly ash, *DOE Symp. Ser.*, 45, 240, 1978.
150. **Sofiev, I. S., Gumarov, R. Kh., and Khalmukhamedova, R. V.,** Effect of some factors on sorption properties of coals, From Zh. *Geol. K.*, 1968; Abstr. 11K34, *Uglei Sredn. Azii Puti Ikh Ispol'z*, 102, 1968.
151. **Kuhn, J. K., Kidd, D., Thomas, J., Cahill, R., Dickerson, D., Shiley, R., Kruse, C., and Shimp, N. F.,** Volatility of coal and its by-products, in Environmental Aspects of Fuel Conversion Technology III, Environmental Protection Agency, Washington, D.C., 1977.
152. **Panin, V. I. and Glushnev, S. V.,** Nature of aluminium oxide distribution in Moscow basin coals, *Khim. Tverd. Topl.*, 2, 87, 1974.
153. **Elejalde, C. and Martin, A.,** Relation between amino acid content and trace elements in Spanish coals, *Bol. Real Soc. Espan. Hist. Nat. Secc. Geol.*, 66, 339, 1968.
154. **Foster, W. D. and Feicht, F. L.,** Mineralogy of concretions from Pittsburgh coal seam. With special reference to analcite, *Am. Mineral.*, 31, 357, 1946.
155. **Finn, C.P.,** An occurrence of Barytes in the Parkgate (South Yorkshire), *Seam. Trans. Inst. Min. Eng.*, 80, 25, 1930—1931.
156. **Arro, H., Mahlapuu, A., and Ratnik, V. E.,** Role of bound calcium oxide during formation of ash deposits, in *Mater. Konf. Protesessam Miner. Chasti Energ. Topl.*, Tallin, Politekh. Inst. Tallin, S.S.S.R., 1969, 95.
157. **Marier, P. and Dibbs, H. P.,** Catalytic conversion of sulfur dioxide to sulfur trioxide by fly ash and the capture of sulfur dioxide and sulfur trioxide by calcium oxide and magnesium oxide, *Thermochem. Acta*, 8, 155, 1974.
158. **Dixon, K., Skipsey, E., and Watts, J. T.,** The distribution and composition of inorganic matter in British coals, III. The composition of carbonate minerals in the coal seams of the East Midlands coalfields, *J. Inst. Fuel*, 43, 229, 1970.
159. **Kralik, J.,** Mineralogy of carbonates from the coal seams of the Ostava-Karvina district, *Cas. Mineral. Geol.*, 15, 313, 1970.
160. **Pringle, W. J. S. and Bradburn, E.,** The mineral matter in coal. II. The composition of the carbonate minerals, *Fuel*, 37, 166, 1958.
161. **Radd, F. J. and Wolfe, L. H.,** A Study of the Formation and Structure of Sulfide Growths, Carbonate Growth and Binder Layers in the Pittsburgh Seam Coal, Research Report No. 756-0-1-72, Research and Development Department, Central Research Division, Continental Oil Company, 1972, 52.

162. Gluskoter, J. H., Clay minerals in Illinois coals, *J. Sediment. Petrol.*, 37, 205, 1967.
163. Guin, J., Tarrer, A., Taylor, L., Prather, J., and Green, S., Jr., Mechanisms of coal particle dissolution, *Ind. Eng. Chem. Pred. Res. Dev.*, 15, 490, 1976.
164. Guin, J. A., Tarrer, A. R., Lee, J. M., VanBrankle, H. F., and Curtis, C. W., Further studies of catalytic activity of coal minerals in coal mineral residue, as catalysts and sulfur scavangers, *Ind. Eng. Chem. Proc. Res. Dev.*, 18, 631, 1979.
165. Akers, D. J., McMillan, B. G., and Leonard, J. W., Coal Minerals Bibliography, Final Report, Report FE-2692-5, National Technical Information Service, Springfield, Va., June 1977 to July 1978.
166. Coal Research Bureau, Coal Associated Minerals of the United States, I to VII, Coal Res. Dev. Rep. 8, Department of the Interior, Office of Coal Research, West Virginia University, Morgantown, 1965.
167. Lee, J. M., VanBrackle, H. F., Lo, Y. L., Tarrer, A. R., and Guin, J. A., Coal minerals catalysis in liquefaction, *Coal Proc. Technol.*, 4, 1, 1978.
168. Kolesnikov, V. V. and Komots'kii, M. K., Coal mineral content and physical-mechanical properties of county rocks in deep-seated horizons of the Krasnodorn area, *Geol. Zh. (Ukr. ed.)*, 30, 62, 1970.
169. Petrakis, L. and Grandy, D. W., Free radicals in coals and coal conversion. II. Effect of liquefaction processing conditions on the formation and quenching of coal free radicals, *Fuel*, 59, 227, 1980.
170. Ovcharenko, F. D. and Gordienko, S. A., Physicochemical investigation of the complex of divalent transition metal ions humic acids obtained from Ukrainian brown coal and peat, *Agrokem. Talajtan*, 18, 25, 1969.
171. Szalay, A., Accumulation of uranium and other trace metals in coal and organic shales and the role of humic acids in these geochemical enrichments, *Ark. Mineral. Geol.*, 5, 23, 1969.
172. Szalay, A. and Szilagyi, M., Accumulation of microelements in peat humic acids and coal, Adv. Org. Geochem., Proc. 4th Int. Meet., 1969, 567.
173. Brown, H. R. and Swaine, D. J., Inorganic constituents of Australian coals, *Coal Res. Commonw. Sci. Ind. Res. Org.*, 1964, 15
174. Crossley, H. E., The inorganic constituents of coal; occurrence and industrial significance, *Chem. Age (London)*, 55, 629, 1946; *Inst. Fuel Bull.*, 67, 57, 1946.
175. Finney, H. R. and Farnham, R. S., Mineralogy of the inorganic fraction of peat from two raised bogs in Northern Minnesota, in Proc. 3rd Int. Peat Congr., Quebec, 1968, 102 .
176. Roesler, H. J., Beuge, P., Schroen, W., Hahne, K., and Braeutigam, S., Inorganic components of lignites and their significance in lignite exploration, Freiberg, *Forschungs Hefte*, C331, 53, 1977.
177. Miller, R. N., Geochemical Study of the Inorganic Constituents in Some Low-Rank Coals, Ph.D. thesis, Pennsylvania State University, University Park, 1977.
178. Zubovic, P., Stadnichenko, T., and Sheffey, N. B., The Association of Minor Elements with Organic and Inorganic Phases of Coal, Prof. Paper 400-B, B84-B87, U.S. Geological Survey, Reston, Va., 1960.
179. Brown, R. L., Caldwell, R. L., and Fereday, F., Mineral constituents of coal, *Fuel*, 31, 261, 1952.
180. Gauger, A. W., Coal and its mineral matter with reference to carbonization. I, II, III, *Blast Furnace Steel Plant*, April (310—312), May (406—409), June (508—510), 1936.
181. Gauger, A. W., Barrett, E. P., and Williams, F. J., Mineral Matter in Coal — A Preliminary Report, paper presented at Trans. Am. Inst. Min. Metal. Eng., New York, 108, 1934, 226.
182. Granaf, B., Mineral Matter Effects in Coal Liquefaction, Rep. SAND-79-0306, First Quarterly Report, October 1 to December 31, 1978, 1979.
183. Hughes, R. E., Mineral Matter Associated with Illinois Coals, Ph.D. thesis, University of Illinois, Urbana-Champaign, 1971, 145.
184. Kuhl, J., Wilk, A., and Smolinska, U., Effect of some components of the mineral matter in coal on its mechanical cleaning, *Przegl. Gorn.*, 35, 51, 1979.
185. Millot, J. O., The mineral matter in coal. I. The water of constitution of silicate constituents, *Fuel*, 37, 71, 1958.
186. O'Gorman, J. V., Mineral matter and trace elements in North American coals, *Diss. Abstr. Int. B*, 1972, 193.
187. Parks, B. C., Mineral matter in coal, in Proc. Conf. on the Origin and Constitution of Coal, Nova Scotia Department of Mines, 1952, 272.
188. Sprunk, G. C. and O'Donnel, J. H., Mineral Matter in Coal, Tech. Paper 648, U.S. Bureau of Mines, Washington, D.C., 1942, 67.
189. Thiessen, G., Composition and origin of the mineral matter in coal, in *Chemistry of Coal Utilization*, John Wiley & Sons, New York, 1945, 485.
190. Thiessen, G., Ball, G. C., and Grotts, P. E., Coal ash and coal mineral matter, *Ind. Eng. Chem.*, 28, 355, 1936.
191. Binns, G.J. and Harrow, G., On the occurrence of certain minerals at Netherseal Colliery, Leicestershire, *Trans. Inst. Min. Eng.*, 13, 1896, 252.

192. **Chung, K. E.**, Products from Coal Minerals, Office of Coal Research, R&D Report No. 53, Development of a Process for Producing an Ashless, Low-Sulfur Fuel from Coal. VI. Part 3, Contract No. 14-01-0001-495, NTIS PB-237, 764, National Technical Information Service, Springfield, Va., 1974.
193. **Finkelman, R. B. and Stanton, R. W.**, Identification and significance of accessory minerals from a bituminous coal, *Fuel,* 57, 763, 1978.
194. **Finkelman, R. B., Stanton, R. W., Cecil, C. B., and Minken, J. A.**, Modes of occurrence of selected trace elements in several Appalachian coals, *Am. Chem. Soc. Div. Fuel Chem. Prepr.,* 24, 236, 1979.
195. **Kemežys, M. and Taylor, G.H.**, Occurrence and Distribution of minerals in some Australian coals, *J. Inst. Fuel,* 37, 389, 1964.
196. **Miller, W. G.**, Relationships between Minerals and Selected Trace Elements in Some Pennsylvania Age Coals of Northern Illinois, M. S. thesis, University of Illinois, Urbana-Champaign, 1974.
197. **Nuhfer, E. B.**, Fluorescent Minerals Associated with Coal, M. S. thesis, West Virginia University, Morgantown, 1967.
198. **Paulson, L. E., Beckering, W., and Fowkes, W. W.**, Separation and identification of minerals from Northern Great Plains province lignite, *Fuel,* 51, 224, 1972.
199. **Rekus, A. G., and Haberkorn, A. R.**, Identification of minerals in single particles of coal by x-ray diffraction powder method, *J. Inst. Fuel,* 39, 474, 1966.
200. **Selvig, W. A. and Seaman, H.**, Sulfur Forms and Ash-Forming Minerals in Pittsburgh Coal, Mining and Metallurgic Investigations, Coop. Bull. 43, U.S. Bureau Mines, Washington, D.C., 1929.
201. **Warne, S. St. J.**, Identification and evaluation of minerals in coal by differential thermal analysis, *J. Inst. Fuel,* 38, 207, 1965.
202. **Wright, C. H. and Severson, D. E.**, Experimental evidence for catalyst activity of coal minerals, *Am. Chem. Soc. Div. Fuel Chem. Prepr.* 16, 68, 1972.
203. **Karr, C., Jr., Estap, P. A., and Kovach, J. J.**, Spectroscopic evidence for the occurence of nitrates in lignites, *Am. Chem. Soc. Div. Fuel Chem. Prepr.,* 12, 1, 1968.
204. **Breger, I. A.**, Role of organic matter in the accumulation of uranium. Organic geochemistry of the coal-uranium association, Form. Uranium Ore Deposits, Proc. Symp., 1974, 99.
205. **Bonnett, R. and Czechowski, F.**, Gallium porphyrins in bituminous coal, *Nature (London),* 283, 465, 1980.
206. **Belly, R. T. and Brock, T. D.**, Ecology of iron-oxidizing bacteria in pyritic materials associated with coal, *J. Bacteriol.,* 117, 726, 1974.
207. **Dove, L. P.**, Sphalerite in coal pyrite, *J. Mineral. Soc. Am.,* 6, 1921, 61.
208. **Grady, W. C.**, A descriptive Approach to Microscopic Varieties of Pyrite in West Virginia Coals, Coal Research Bureau Technical Report No. 126, College of Mineral and Energy Resources, West Virginia Univesity, Morgantown, 1976.
209. **Grady, W. C.**, Microscopic variates of pyrite in West Virginia coals, *Trans. Soc. Min. Eng. AIME,* 262, 268, 1977.
210. **Greer, R. T.**, Nature and distribution of pyrite in Iowa coals, in *Proc. Electron Microsc. Soc. Am.* Bailey, G. W., Ed., Claitor's Press, Baton Rouge, 1976, 620.
211. **Greer, R. T.**, Colloidal pyrite growth in coal, *Colloid Interface Sci.,* 5, 411, 1976.
212. **Nelson, H. W., Snow, R. D., and Keyes, D. B.**, Oxidation of pyritic sulfur in bituminous coal, *Ind. Eng. Chem.,* 25, 1355, 1933.
213. **Tucker, W. M.**, Pyrite deposits in Ohio coal, *Econ. Geol.,* 14, 198, 1919.
214. **Cobb, J. C. and Russell, S. J.**, Sphalerite mineralization in coal seams of the Illinois Basin, *Geol. Soc. Am. Abstr. Progr.,* 8, 816, 1976.
215. **Hatch, J. R., Avcin, M. J., Wedge, W. K., and Brady, L. L.**, Sphalerite in Coals from Southeastern Iowa, Missouri, and Southeastern Kansas, Open-file Report 76-796, U.S. Geological Survey, Reston, Va., 1976, 26.
216. **Hatch, J. R., Gluskoter, H. J., and Lindhal, P. C.**, Sphalerite in coals from the Illinois basin, *Econ. Geol.,* 71, 613, 1976.
217. **Styron, C. E.**, Preliminary assessment of the impact of radionuclides in Western coal and health and environment, in Technology for Energy Conservation, NTIS, Rep. MLM-2497/OP, National Technical Information Service, Springfield, Va. 1978.
218. **Styron, C. E., Casella, V. R., Farmer, B. M., Hopkins, L. C., Jenkins, P. H., Phillips, C. A., and Robinson, B.**, Assessment of the Radiological Impact of Coal Utilization, Rep. MLM-2514, UC-90a, National Technical Information Service, U.S. Dept. Commerce, Springfield, Va., 1979.
219. **Bedrosian, P. H., Easterly, D. G., and S. L. Cummings**, Radiological Survey Around Power Plants Using Fossil Fuel, Report No. EERL71-3, U.S. Environmental Protection Agency, Washington, D.C., 1970.

220. **Martin, J. E., Harward, E. D., and Oakley, D. T.**, Comparison of radioactivity from fossil-fuel and nuclear power plants in Environmental Effects of Producing Electric Power. I. Appendix, 15 Committee Print, Joint Committee on Atomic Energy, 91st Congr. U.S., 1st Session, Washington, D.C., November 1969, 773, see also **Martin, J. E.**, et al., Radioactivity from Fossil Fuel and Nuclear Plants, in IAEA Symp. SM-146/19, New York, 1970.
221. **Strong, R. A. et al.**, Physical and Chemical Survey of Coals from Canadian Collieries — Nova Scotia; Inverness County, Mem. Ser. 74, Bur. Min. — Mines and Geology Branch, Department of Mines and Resources, Ottawa, Canada, 1939, 55.
222. **Swanson, V. E., Medlin, J. H., Hatch, J. R., Coleman, S. L., Wood, G. H., Jr., Woodruff, S. D., and Hildebrand, R. T.**, Collection, Chemical Analysis, and Evaluation of Coal Samples in 1975, Open file Report 76-468, U.S. Geological Survey, Reston Va., 1976, 503.
223. **Vine, J. D.**, Geology of uranium in coaly carbonaceous rocks, U.S. Geol. Surv. Professional Paper, 365-D, 113—167, 1962.
224. **Honda, H. and Ouchi, K.**, Magnetochemistry of Coal, I. Magnetic susceptibility of coal, *Fuel,* 36, 159, 1957.

Chapter 2

TRACE ELEMENTS IN COAL: OCCURRENCE AND DISTRIBUTION

I. GENERAL CONSIDERATIONS

Numbers of elements occur in coal in wide concentration ranges. Torrey[1] has compiled concentration values for different elements reported in U.S. coals. Table 1 shows his compilation. The estimated values for U.S. and worldwide average concentrations in coal are shown in Table 2.[2] As an additional illustration, Table 3 shows element concentrations in coal from Nebraska. Elemental concentrations of coals reflect the elemental composition of coal-forming material, properties of environment, and processes which were active during the coal formation period and later.

The modern investigations of trace elements in coals were pioneered by Goldschmidt,[4] who developed the technique of quantitative chemical analysis by optical emission spectroscopy and applied it to coal ash. In these earliest works, Goldschmidt was concerned with the chemical combinations of the trace elements in coals. In addition to identifying trace elements in inorganic combinations with the minerals in coal, he postulated the presence of metal organic complexes and attributed the observed concentrations of vanadium, molybdenum, and nickel to the presence of such complexes in coal.

Goldschmidt[5] also introduced the concept of a geochemical classification of elements, in which the elements are classified on the basis of their affinities and tendencies to occur in minerals of a single group. The chalcophile elements are those which commonly form sulfides. In addition to sulfur, they include Zn, Cd, Hg, Cu, Pb, As, Sb, Se, and others. When present in coals, these elements would be expected to occur, at least in part, in sulfide minerals. Sulfides other than pyrite and marcasite have been noted in coals, but, except in areas of local concentration, they occur in trace or minor amounts.

The lithophile elements are those that generally occur in silicate phases and include among others: Si, Al, Ti, K, Na, Zr, Be, and Y. These would be expected to occur in coals in combination with the silicate minerals: kaolinite, illite, other clay minerals, quartz, and stable heavy detrital minerals.

The carbonate minerals in coals occur primarily as epigenetic fracture fillings (cleat filling). Magnesium, iron, and manganese are often associated with the sedimentary carbonate minerals and would reasonably be expected to be associated with the cleat fillings in coal.[6]

The paper by Miller[7] reports trace element-mineral correlations with slightly different results. Correlations with correlation coefficient bigger than 0.5 are shown in Table 4.

A large number of silicate, sulfide, and carbonate minerals have been identified from coal seams, and the elements composing them necessarily occur in coals in inorganic combination. However, mineralogical investigations of coals have not generally been quantitative, and whether an element occurs only in inorganic combination or perhaps is also present in inorganic combination has not commonly been considered.

Nicholls[8] approached this problem by plotting the analytical data for the concentration of a single element in coal or in coal ash against the ash content of the coal. Diagrams depicting a number of such points for a single coal seam or for a group of coal seams in a single geographic area were interpreted for degree of inorganic or organic affinity of the element. Nicholls[8] concluded that boron is largely, almost entirely, associated with the organic fraction in coals; elements, as barium, chromium

Table 1
RANGE OF TRACE ELEMENTS IN U.S. COALS

Element	Concentration range (%)		
Na	0	—	0.20
Mg	0.1	—	0.25
Al	0.43	—	3.04
Si	0.58	—	6.09
Cl	0	—	0.56
K	0.02	—	0.43
Ca	0.05	—	2.67
Ti	0.002	—	0.32
Fe	0.32	—	4.32
Zn	0	—	0.56
	(ppm)		
Be	0	—	31
B	1.2	—	356
F	10	—	295
P	5	—	1430
Sc	10	—	100
V	0	—	1281
Cr	0	—	610
Mn	6	—	181
Co	0	—	43
Ni	0.4	—	104
Cu	1.8	—	185
Ga	0	—	61
Ge	0	—	819
As	0.5	—	106
Se	0.4	—	8
Br	4	—	52
Y	<0.1	—	59
Zr	8	—	133
Mo	0	—	73
Cd	0.1	—	65
Sn	0	—	51
Sb	0.2	—	9
La	0	—	98
Hg	0.01	—	1.6
U	<10	—	1000
Pb	4	—	218

After Torrey, S., Ed., *Trace Contaminants in Coal,* Noyes Data Corporation, Park Ridge, N.Y., 1978.

cobalt, lead, strontium, and vanadium, are, in the majority of cases, associated with the inorganic fraction; and a third group including nickel, gallium, germanium, molybdenum and copper, may be associated with either or both fractions.

A much more ambitious series of investigations of the organic-inorganic affinities of trace metals in coals was undertaken and reported on by Zubovic and co-workers[9,10] at the U.S. Geological Survey.

The Illinois State Geological Survey[6,11] has been extensively investigating trace elements in coal. As a part of this study four sets of float-sink samples were analyzed

Table 2
AVERAGE CONCENTRATIONS OF ELEMENTS IN COAL

Element	U.S. average	Worldwide average
	Concentration (%)	
Sulfur	2.0	2.0
Phosphorous	—	0.05
Silicon	2.6	2.8
Aluminium	1.4	1.0
Calcium	0.54	1.0
Magnesium	0.12	0.02
Sodium	0.06	0.02
Potassium	0.18	0.01
Iron	1.6	1.0
Manganese	0.01	0.005
Titanium	0.08	0.05
	Concentration (ppm)	
Antimony	1.1	3.0
Arsenic	15	5.0
Barium	150	500
Beryllium	2.0	3
Bismuth	0.7	5.5
Boron	50	75
Bromine	2.6	—
Cadmium	1.3	—
Cerium	7.7	11.5
Cesium	0.4	—
Chlorine	207	1000
Chromium	15	10
Cobalt	7	5
Copper	19	15
Dysprosium	2.2	—
Erbium	0.34	0.6
Europium	0.45	0.7
Fluorine	74	—
Gadolinium	0.17	1.6
Gallium	7	7
Germanium	0.71	5
Hafnium	0.60	—
Holmium	0.11	0.3
Iodine	1.10	—
Lanthanum	6.1	10
Lead	16	25
Lithium	20	65
Lutetium	0.08	0.07
Mercury	0.18	0.012
Molybdenum	3	5
Neodymium	37	4.7
Nickel	15	15
Niobium	4.5	—
Praseodynium	2.7	2.2
Rubidium	2.90	100
Samarium	0.42	1.6
Scandium	3	5
Selenium	4.1	3
Silver	0.20	0.50
Strontium	100	500

Table 2 (continued)
AVERAGE CONCENTRATIONS OF ELEMENTS IN COAL

Element	U.S. average	Worldwide average
Tellurium	0.1	—
Terbium	0.1	0.3
Thallium	0.1	—
Thorium	1.9	—
Thulium	0.07	—
Tin	1.6	—
Tungsten	2.5	—
Uranium	1.6	1.0
Vanadium	20	25
Ytterbium	1	0.5
Yttrium	10	10
Zinc	39	50
Zirconium	30	—

From U.S. National Committee for Geochemistry, Trace Element Geochemistry of Coal Resource Development Related to Environmental Quality and Health, National Academy Press, Washington, D.C., 1980.

Table 3
OXIDES AND TRACE ELEMENTS IN ASH (COAL FROM NEBRASKA)

Components	Sample 1	Sample 2	Sample 3
	Concentration (%)		
Ash	71.6	10.7	17.1
SiO_2	50	19	32
Al_2O_3	34	8.4	8.1
CaO	1.2	14	3.5
MgO	.61	1.10	.66
Na_2O	.18	.27	.49
K_2O	.82	.82	1.3
Fe_2O_3	5.4	19	43
MnO	.008	.012	.017
TiO_2	.68	.47	.44
P_2O_5	1.0	.14	.35
SO_3	1.8	24	7.2
Cl	.20	.10	.10
	Concentration (ppm)		
Cd	2	50	1.5
Cu	84	130	290
Li	232	24	12
Pb	50	1750	560
Zn	1360	6640	960
B	200	150	70
Ba	200	500	1000
Be	15	10	—
Ce	500	—	—

Table 3 (continued)
OXIDES AND TRACE ELEMENTS IN ASH (COAL FROM NEBRASKA)

Components	Sample 1	Sample 2	Sample 3
Co	200	10	15
Cr	100	70	50
Ga	50	50	20
Ge	—	150	200
La	150	—	—
Mo	—	70	50
Nb	20	20	20
Nd	300	—	—
Ni	300	50	50
Sc	50	20	15
Sr	100	150	200
V	200	70	70
Y	150	50	20
Yb	15	—	—
Zr	150	100	150

After Burchett, R. R., Coal Resources of Nebraska, Rep. NP 23879, University of Nebraska, Lincoln, 1977.

Table 4
TRACE ELEMENTS — MINERAL CORRELATIONS

Elements	Minerals
As, Be, Cu, Sb	Pyrite
B, Cd, Zn, Hg	Sphalerite
B, Cd, Mn, Se, Mo, V	Calcite
B, Co, Mn, Cd, Mo, Se, V, Zn	Quartz
B, Cu, F, Hg, Sn, V	Clays

Note: Correlation coefficients >0.5.

After Miller, W. G., M.S. thesis, University of Illinois, Urbana-Champaign, 1974.

for a number of trace and minor elements. Three coals, crushed and sized to 3/8 in. by 28 mesh, were separated into six specific gravity fractions by floating them in mixtures of perchloroethylene and naphtha. The heaviest of these six fractions (1.60 sink) was then separated in perchloroethylene and naphtha, but only two fractions were analyzed, one with specific gravity of less than 1.25 and one with specific gravity heavier than 1.60. By use of a technique similar to that of Zubovic, the trace elements determined in these samples are listed in order of decreasing affinity for the clean coal fractions, or decreasing organic affinity (see Table 5). The sequence was determined by comparing ratios of the amount of an element in the lightest float fraction (always less than 1.30 sp gr) to the amount of the element in the 1.60 sink fraction. The numerical values thus determined are not given because they vary with the particle size distribution of the coal, the specific gravity of the liquid used to make the first (lightest) separation, and the size distribution of the mineral fragments in a single coal. How-

Table 5
AFFINITY OF ELEMENTS FOR PURE COAL AND MINERAL MATTER AS DETERMINED FROM FLOAT-SINK DATA

	Davis coal	DeKoven coal	Colchester (No. 2) coal	Herrin (No. 6) coal
Clean coal—lightest specific gravity fraction (elements in "organic combination")	B	Ge	Ge	Ge
	Ge	Ga	B	B
	Be	Be	P	Be
	Ti	Ti	Be	Sb
	Ga	Sb	Sb	V
	P	Co	Ti	Mo
	V	P	Co	Ga
	Cr	Ni	Se	P
	Sb	Cu	Ga	Se
	Se	Se	V	Ni
	Co	Cr	Ni	Cr
	Cu	Mn	Pb	Co
	Ni	Zn	Cu	Cu
	Mn	Zr	Hg	Ti
	Zr	V	Zr	Zr
	Mo	Mo	Cr	Pb
	Cd	Pb	Mn	Mn
Mineral matter—specific gravity greater than 1.60 (elements in "inorganic combination")	Hg	Hg	As	As
	Pb	As	Mo	Cd
	Zn		Cd	Zn
	As		Zn	Hg

After Gluskoter, H. J., Proc. Coal Agglomeration and Conversion Symp., Morgantown, W.Va., 1975.

ever, the sequence given in Table 5 does indicate which elements are primarily in organic combination, which are in inorganic combination, and which are, apparently, both inorganically and organically combined in coals of the Illinois Basin.

Enrichment factors (EF) for an element in coal where EF is defined as the ratio of an element concentration in coal to some reference element (Sc which is relatively constant throughout these coals was used in paper by Lyon et al.[12]) divided by a similar ratio for average crustal material. This is expressed as

$$EF = \frac{\text{elem/Sc (in coal)}}{\text{elem/Sc (crustal average)}} \quad (1)$$

World average crustal values were compiled from published data. Table 6 shows these results and compares them with similar data obtained by the U.S. Geological Survey and the Illinois State Geological Survey. The results of these calculations indicate excellent agreement between EF, for the ORNL and USGS coals. Values for Illinois coal are in general somewhat higher than for the other two sample groups but show the same general trend.

EFs of trace elements in coal with respect to the average composition of the crust of the Earth, is shown in Table 7.[2] The values for the crust of the Earth elemental abundances can be found in several reports, e.g., Taylor[13] Turekian and Wedepohl.[4]

There are many good papers in the literature discussing occurrence and origin of elements in coal.[15-121] We shall mention only some with apologies to the authors whose papers have been omitted. For example, Somerville and Elder[122] have reported a de-

Table 6
ENRICHMENT FACTORS (EF) OF ELEMENTS IN COAL

EF	3-Plant ORNL data	Element (799 USGS Coals)	114 Illinois coals
	Mn, Na	—	—
<1	Al, Ca, Cr, Fe, K, Ni, Ti, V, Zn	Mn, Na, Al, Cr, K, Ti, V, Mg	Mn, Na, Al, Cr, K, Ti, Mg
	Mg	Fe, Ni, Ca	V, Cu, Th
	Br, Cd, Pb, Sb, Cu	Zn	Ga, Fe, Ni
1—10	As, Hg, Se, U	Sb, Cd, Pb, Cu, Hg, As, Se	Br, Pb, Sb, As, Hg, Se, U
>10	S, Cl	S	S, Cl, Cd, Zn

Note: Scandium is used as reference element.

After Lyon, W. S. et al., in *Nuclear Activation Techniques in the Life Science,* International Atomic Energy Agency, 1978.

Table 7
ENRICHMENT FACTORS OF TRACE ELEMENTS IN COAL WITH RESPECT TO THE AVERAGE COMPOSITION OF THE EARTH'S CRUST

Limited enrichment		Moderate enrichment		Marked enrichment	
Element	Enrichment	Element	Enrichment	Element	Enrichment
Hg	2.25	As	8.3	Se	82
Mo	2.00	Cd	6.5		
Pb	1.28	B	5.0		
		Sb	5.5		

After U.S. National Committee for Geochemistry, Trace Element Geochemistry of Coal Resource Development Related to Environmental Quality and Health, National Academy Press, Washington, D.C., 1980.

tailed study of trace element and major constituents in Mercer County and Dunn County (North Dakota) lignite and their ashes. Concentrations of most elements were determined by spark source mass spectrometry (MS), with a detection limit of 0.1 ppm. Some elements were detected using other techniques including: atomic absorption, ion-selective electrode, or gravimetric analysis. Concentration values which are the average of 12 samples are shown in Table 8.

In their report, Lyon et al.[12] have analyzed monthly coal samples from three steam plants by instrumental neutron activation analysis (INAA), isotope dilution spark source MS, and several variations of atomic absorption (AA) spectroscopy. Statistical analyses run on individual steam plant and combined three-plant data included (1) means and standard deviations, (2) interelement correlation coefficients for combined data and data by steam plants, and (3) test for significance of difference in data generated for a given element by two different analytical methods. The elemental composition of coal for a single plant varied considerably from month to month; the variation was even greater when samples from the three plants were compared. Interelement correlation coefficients varied considerably in coal entering different plants, but an overall general pattern was seen. Comparisons were made between data obtained on

Table 8
ELEMENTS IN NORTH DAKOTA LIGNITE

Element	Conc (ppm)	Element	Conc (ppm)	Element	Conc (ppm)	Element	Conc (ppm)
Ag	<0.1	Er	<0.1	Mo	22.2	Se	0.9
Al	6697	Eu	0.3	Na	2395	Si	2395
As	10.1	F	20.8	Nb	3.9	Sm	0.5
Au	<0.1	Fe	7216	Nd	0.9	Sn	5.1
B	63	Ga	4.6	Ni	11.6	Sr	1029
Ba	229.8	Gd	0.2	Os	<0.1	Ta	0.1
Be	0.3	Ge	0.6	P	131	Tb	0.2
Bi	<0.1	Hf	<0.1	Pb	5.4	Te	<0.1
Br	1.7	Hg	0.2	Pd	—0.1	Th	3.6
Ca	16108	Ho	<0.1	Pr	0.9	Ti	301
Cd	0.2	I	0.4	Pt	<0.1	Tl	<0.1
Ce	14.1	Ir	<0.1	Rb	4.1	Tm	<0.1
Cl	46.6	K	462	Re	<0.1	U	3.2
Co	5.0	La	5.8	Rh	<0.1	V	21.9
Cr	65.3	Li	1.2	Ru	<0.1	W	0.6
Cs	0.3	Lu	<0.1	S	13000	Y	23.1
Cu	22.9	Mg	5039	Sb	0.3	Yb	<0.1
Dy	<0.1	Mn	249	Sc	8.0	Zn	10.8
						Zr	68.4

After Somerville, M. H. and Elder, J. L., in Environmental Aspects of Fuel Conversion Technology III, Ayer, G. A. and Massoglia, M. F., Eds., Environmental Protection Agency, Washington, D.C., 1977.

three elements by instrumental neutron activation analysis and isotope dilution spark source MS; significant differences were observed for two of the elements determined. A comparison between spark source and AA for three elements, however, gave excellent agreement.

Table 9 is a combined summary of analytical results for elemental concentrations in coal for all three plants. Where values are lacking, it indicates failure to detect the element or in several instances failure to receive a particular month's sample. Maximum and minimum values and the coefficients of variation show the often wide range of concentrations found.

Although the three plants are all located within 40 km of each other, each received coal from a different source. Bull Run's coal comes from Eastern Kentucky by train. Kingston has been using a mixture of local Tennessee and Kentucky coal and western low-sulfur coal brought in by train. X-12 receives the most homogeneous coal in that it is from the immediate surrounding area.

Table 9 lists the elemental composition for the entire 15-month study for combined data from all three plants. The standard deviations, minimum, and maximum values are shown to indicate the wide range of values obtained. The coefficient of variation (CV) is the ratio: standard deviation × 100/mean, and is an indicator of the overall variation. Thus, Hg and Cd, elements in low concentrations and widely varying, show high CVs whereas most of the more common elements have CVs in the neighborhood of 30.[12]

In order to provide information on the process of metal absorption by lignite under laboratory conditions Ibarra et al.[123] have studied absorption of strontium, lead, uranium, and thorium ions from their solutions by several Spanish lignites. Adsorption isotherms were adequately described by means of the Langmuir equation and could be characterized by two numerical constants: the sorption capacity of lignite, and the

Table 9
ELEMENT COMPOSITION OF COAL FROM COMBINED THREE STEAMPLANT DATE

Element	Mean (ppm)	SD	Minimum value (ppm)	Maximum value (ppm)	CV
Al	2.64%	0.84	1.74%	6.60%	32.1
As	27.35	12.54	5.8	58	45.8
As	36.23	13.27	16	65	36.6
Ba	155.6	42.34	94	300	27.2
Ba	129.1	47.97	40	220	35.6
Br	8.33	2.80	2.5	16.5	34
Ca	1349	396.1	500	2210	29.4
Cd	0.40	0.309	0.10	1	75.8
Ce	29.5	6.66	21	65	22.5
Cl	819.2	279.4	136	1311	34.1
Cr	29.54	6.56	21	59	22.2
Cr	24.91	9.95	7	48	39.9
Co	7.78	1.26	5.3	11.6	16.2
Cs	1.83	0.654	1	5.4	35.7
Cu	20.38	6.85	12	50	33.6
Eu	0.641	0.131	0.44	1.25	20.5
Fe	1.51%	3780	0.43%	2.13%	25.1
Ga	7.13	1.62	4.9	15	22.8
Hf	0.994	0.209	0.54	1.76	21.0
Hg	0.3345	0.5079	0.1	3	151
I	1.32	0.40	0.75	2	30.3
In	0.522	0.322	0.1	1	61.8
La	14.49	3.30	10.6	32	22.8
K	4550	1779	2370	14000	39.1
Mg	3419	1198	2030	9920	35.0
Mn	39.5	23.6	16	137	59.6
Mo	2.14	1.29	0.4	5	60.3
Na	356.1	94.6	244	750	26.6
Ni	17.9	6.32	7	32	35.4
Pb	10.8	3.34	5	21	30.9
Rb	37.1	12.65	25	99	34.1
S	1.95%	5130	0.7%	2.6%	26.3
Sb	1.14	0.434	0.7	2.6	36.4
Sc	5.86	1.15	4.4	12	19.6
Se	4.55	1.4	2.3	8	30.8
Sm	2.86	0.675	2	5.8	22.6
Sr	115.6	31.7	46	185	27.4
Sr	114.9	37.6	44	230	32.7
Ta	0.333	0.087	0.2	0.6	26.1
Th	4.58	1.65	3	13	35.9
Ti	1242	339	866	2960	27.3
Tl	0.958	0.423	0.2	1.7	44.2
U	2.08	1.08	1.29	8.7	49.8
V	45.49	11.64	26	99	25.6
Zn	26.42	8.63	10	47	32.7

After Lyon, W. S., et al., in *Nuclear Activation Techniques in the Life Science,* International Atomic Energy Agency, 1978.

geochemical EF. Very high values of this EF have been found for heavy metals. Humic acids (HA) were extracted from the lignites, and their metal-binding ability was determined as a function of pH. Higher pH favors metal retention, also higher atomic weight and valency. A role of humic substances in the retention and accumulation of metals in coals is suggested.

In their experiments the coals used were a lignite from Utrillas (L-U) and another from Puentes de Garcia Rodriguez (L-P). A lignite from Puentes de Garcia Rodriguez, heated in an oven at 150°C (L-PR), has been also used in order to investigate the influence of HA on the adsorption of metallic cations.

To determine the adsorption isotherms, samples of lignites (0.5 g) were ground to pass a 0.25-mm sieve (576 mesh, UNE 90) and were treated with 50-mℓ solutions containing variable amounts of strontium, lead, uranium, and thorium ions, prepared from their nitrates. The flasks were shaken for 12 hr. After adsorption equilibrium had been attained, the metal concentration was determined in centrifuge aliquots.

The concentrations were fitted to the Langmuir equation

$$N = \frac{N_\infty ac}{1 + ac} \qquad (2)$$

where N denotes the concentration of metal in the substrate, N_∞ the saturation capacity, c the equilibrium concentration of metal in the aqueous phase, and a the ratio between the adsorption and desorption constant.

The adsorption of metallic cations by lignites may now be characterized by two quantities.[124] One of these is N_∞, the saturation capacity; the other is the slope of the tangent of the isotherm at very small concentrations, and it is named the geochemical EF.

The EF values can be calculated from the equations of the adsorption isotherms. They are listed in Table 10. A value of 1×10^4 has been reported by Szalay[124] for uranium sorbed on HA extracted from French peats. The EF values for stronium, lead, and thorium are not described in the literature, but they are generally similar to those reported for transition metals and rare earths by Szalay.

The uranium concentration in the crust of the Earth is about 4 g/ton. The EF obtained here for uranium ($0.2—4.5 \times 10^3$) indicates that uranium concentrations, ranging from 0.8 to 16 kg/ton lignite, can be found in these coals. The values of 0.2 to 12 kgU/ton reported for lignites of Calaf basin by Calvo[125] coincide adequately with these results.

The saturation capacity values (Table 10) increase with the HA content of lignites; so, the lignite L-P has a greater saturation capacity than L-U, although both have the same specific surface area, and the lignite L-PR has the greatest saturation capacity. Thus, the HA seem to be mainly responsible for the sorption process. The HA extracted from lignite show, effectively, a very high retention capacity for metals. In the experiments performed by Calvo,[125] the lignites had not been previously treated with water or hydrochloric acid to wash away adsorbed ions which would cause an increase in sorption capacity. The saturation capacity values obtained were similar to those reported by other workers for uranium, gallium, berillium, titanium, and nickel sorbed on peats, coals, and HA. The above experiments demonstrate that strontium, lead, uranium, and thorium, are huminophilic elements.

Therefore, HA resulting from the peat-forming processes as well as those existing in already formed peats and lignites can exert a strong concentration effect upon heavy metals transported in solution at low concentrations by natural waters.

Coals of different ranks have different elemental composition. This subject is discussed in some detail in the chapter on coal characterization. Here, only some of the published work will be mentioned. We shall start with the results presented in Reference 126, in which the trace element concentrations in mineral matter from coals of different ranks are presented (see Table 11).

Haught[127] has discussed the occurrence and distribution of trace elements in coal. Diverse information explaining the occurrence and range of concentration of both ma-

Table 10
ADSORPTION OF METALLIC CATIONS ON LIGNITES

Lignite	Sr	Pb	U	Th
L-U	[a]0.26	0.29	0.21	0.18
	[b]0.65×10^2	1.07×10^3	2.10×10^2	1.84×10^3
L-P	[a]0.80	1.01	0.40	0.22
	[b]5.85×10^2	6.27×10^3	4.54×10^3	3.51×10^3
L-PR	[a]1.09	1.82	0.70	0.32
	[b]3.45×10^2	6.20×10^3	3.22×10^3	4.46×10^3

[a] Saturation capacity (meq metal/g coal)
[b] Enrichment Factor.

After Ibarra, J. V., Osacar, J., and Gavilian, J. M., Retention of metallic cations by lignites and humic acids, *Fuel*, 58, 827, 1979.

Table 11
TRACE ELEMENT CONC. (>100 ppm) IN MINERAL MATTER FROM COALS OF DIFFERENT RANK

Anthracite		Bituminous						Lig. & Sub. B	
		High-volatile		Medium-volatile		Low-volatile			
Element	ppm	Element	ppm	Element	ppm	Element	ppm	Element	ppm
Sn	795	Sr	1515	Mn	1043	Sr	606	Ba	3248
Ba	715	Ba	956	Ba	652	Ba	548	Sr	3011
Zn	568	B	587	Sr	486	Zr	339	B	659
Cu	335	Zr	313	V	284	Cu	281	Mn	445
Cr	251	Zn	236	Zr	237	Mn	207	Cu	423
Mn	223	Cu	223	Cu	228	V	206	Zr	158
V	205	V	190	Ni	192	Zn	171	Sn	101
Ni	182	Cr	147	B	159	Cr	164		
Sr	146	Pb	140	Zn	142	Co	127		
La	117	Sn	130	Cr	123	Y	113		
		Mn	130	Y	110	Ni	104		
		Ni	117						

After Hamrin, C. E., Calaytic Activity of Coal Mineral Matter, Annu. Rep. FE-2233-3, National Technical Information Service, Springfield, Va., 1977.

jor and minor elements found in coal and coal ash is related to ash constituents found in modern-day plants. Element content in ''intrinsic'' ash (original plant ash) vs. ''extraneous'' ash (clays, etc. washed into the swamp) is used to explain the wide variability in concentration of some elements found in coal. Soluble elements carried into coal-forming swamps were absorbed through plant roots, concentrated in the leaves or plant tissue, and remained during peatification unless redissolved and carried away. Some secondary enrichment of elements like Ge occurred by differential leaching and reprecipitation. Minor elements in coal such as Cu, Zn, B, Co, and Mn are essential in low levels for most green plants, and therefore are found in low levels in the coal unless enriched due to extraneous source material.

In the work by Hawley,[128] concentrations of 15 trace elements were determined in coals from 9 major seams of the Sydney coalfield in Nova Scotia to learn if trace element distribution could be used to identify individual coal seams. Elemental analyses were carried out utilizing emission spectroscopy. Compared to U.S. coal ashes, the Nova Scotia coals contained concentrations of Mn, As, Pb, Sr, and Ba, while having lower amounts of Be, B, V, Cr, Co, Ni, Ge, Mo, and Sn. Levels of Zn were about equal. Individual seams varied significantly in trace element content, especially with respect to Mn, Pb, and As, and the usefulness of this data for correlation was further limited because of the limited number of samples and analysis of several seams. No relationship was found between trace element content and coal lithotypes (fusian, vitrain, clarain, etc.). Chromium, Ge, Ba, Zn, and Sr were concentrated in light and intermediate specific gravity fractions.

Headlee and Hunter[52] have determined concentrations of 38 minor and trace elements in coals sampled from 31 seams in West Virginia. Ashed coals were analyzed for element content by emission spectroscopy. A major portion of the ash from some of the coals was a result of the combustion of pyrite and calcite. These minerals were both secondary in origin and occurred extraneous to the coal substance. The alkali metals, with the exception of Li, were not as abundant in coal ashes as in the crust of the Earth, while Ag, B, and Hg were concentrated 100-fold or more in coal ash. Silicon and Al were the elements most concentrated in the coals and Mg was the lowest. Calcium and Sr exhibited the highest concentration variance while Cr and V were the most uniform in concentration. Germanium was accumulated in the organic rather than the mineral matter of the coals. Variations in concentration of the inorganic elements studied affect catalytic hydrogenation and gasification; e.g., As and Sb would probably act as catalyst poisons. Mineral matter in high-ash West Virginia coals was predominantly extraneous while that of low-ash coals was inherent.

Rao[85] has analyzed selected coals from northern Alaska for 15 minor and trace elements, and float-sink tests were performed to determine elemental distribution between the inorganic and organic fractions of the coals. Samples were collected from freshly exposed coal seams in operating mines, abandoned mines, and seams exposed by natural erosion. Elemental analysis was performed spectrographically and by atomic absorption. Element concentration did not vary significantly between coal ashes from the various coal fields sampled. Concentrations of Pb, Ga, Cu, Be, Ba, Ni, Ti, V, Zr, Co, Cr, Ge, and Sn were lower in the ash of low-rank coals than higher rank coals. Generally the higher ranked coals exhibited greater elemental concentrations in the float ash than in the sink ash. The organic association (affinity) of the elements in the coals studied was ranked as follows: Ge > (V, Be, Co) > Ni > (Cr, Ba, Zr) > Ti > (Ga, Cu) > Pb. Trace element concentration in Alaskan coal was comparable to that in coals of the lower 48 states.

Miller[72] has discussed the relationships between minerals and selected trace elements in some Pennsylvanian Ag coals of Northwestern Illinois. Some trace elements in Illinois coals are associated with mineral species and may be removed by the coal preparation process which removed the minerals from the coal. Fifteen whole coal samples from the Illinois No. 2, 5, and 6 coals, and two sets of washability samples were studied. Mineral identifications were performed on low-temperature ash by X-ray diffraction methods. Proximate, ultimate, and sulfur forms analyses were performed on the whole coals, and two coals were float-sink separated at specific gravities ranging from 1.24 to 2.89. Variables were compared by correlation techniques and cluster analysis. Amounts of illite, kaolinite, mixed-layer clays, quartz, pyrite, sphalerite, calcite, and iron sulfate minerals were compared to 32 elements, resulting in correlations that showed associations between (1) As, Sb, Cu, Ga, Ge, Pb, and Ni and pyrite; (2) V and mixed-layer clays; (3) Hg and sulfide minerals; (4) Mo and calcite; (5) Mn and quartz; (6) Zn, Cd, and sphalerite, and (7) P and F, possibly as fluorapatite. Washability curves showed that 21 elements were associated with mineral matter in the coal, and that B, Be, Ge, Na, P, Mg, Cl, and Br tended to be mainly in organic association in the coal. Areal distributions showed a northwesterly increase in quartz, sphalerite, and mixed-layer clays, suggesting that the source area of the sediments trapped in the coal was to the northwest.

In-depth information on the occurrence, production, composition, and physical and chemical properties of coal were presented by Ctvrnicek et al.[129] Discussion from this data base was directed toward economic analyses, mining techniques, transportation modes, environmental effects of western coal utilization, and the effects of western coal properties on combustion equipment. Information cited included typical minimum average, and maximum values of ten elemental oxides (Si, A, Fe, Ti, P, Ca, Mg, Na, K, S) found in coals from six states in the Northern Great Plains region, one state in the Coastal region, six states in the Interior region, and six states in the Eastern region. The average trace element content in ash of coals of the Eastern region, Interior region, and the Western states was tabulated for 29 elements. Other data included a complete analysis of 15 coals mined in the Western states, 2 coals from Illinois, and 1 from West Virginia, plus the relative combustion fouling potential of each sample. It was found that low-sulfur western coals are low in Fe, compared to eastern coals, though the SO_3 content in ash from western coal is higher than that of eastern coals. Western coals are two to four times higher in average Ca and Mg content and two to eight times lower in average K content than eastern coals. Sodium content follows no particular trend. Elemental types present in eastern and western coals are nearly identical although elemental concentrations vary somewhat. Western coals with higher Na concentrations were found to have greater boiler fouling problems than coal with low Na concentrations.

Cahill et al. [130] have analyzed 70 whole coal samples to approximate the range of concentrations of the elements occurring in the coal. Samples of whole coal, and low- and high-temperature coal ashes were analyzed for trace element concentrations by numerous procedures including neutron absorption. Results of the analytical procedures were compiled into a table containing the mean, minimum, and maximum values of approximately 60 different elements. The element concentrations were then compared to the average abundance in the crust of the Earth, indicating that Se is highly enriched and that Na, Mg, Mn, K, Si, P, and Ti were depleted. A majority of the elements show no trend while the remaining elements were only slightly enriched.

Casagrande and Erchull[83,131] have studied metals in Okefenokee Peat-forming environments and their relationship to consistuents found in coal. Concentrations of 17 trace elements were determined in two peats and their corresponding organic fractions from major vegetational areas of the Okefenokee Swamp, Georgia, in order to establish distribution trends for the elements. Peat was collected from both open marsh and swamp environments. All elemental analyses were carried out by AA. There was no significant variation in concentrations of Ca, Fe, K, Mg, and Na with depth in either of the peat cores sampled. With the exception of Hg and Pb, elemental concentrations were far below average crust values of the Earth. Relative trace metal abundance was as follows: Ba $>$ Cu $>$ Cr $>$ Pb $>$ Zn $>$ Mn $>$ Co $>$ Ni $>$ Hg. The humic fraction generally contained the highest trace element concentrations, while HA had the lower. Vegetational environment, although determining the previtrinitic (humic) constitution of coal, did not significantly influence trace metal distribution in peats. Trace metal concentrations in coals and coal fractions obtained by specific gravity separations were analogous to those found in peats and their organic fractions. Trace element concentration in subtropical peats of the Okefenokee Swamp was not affected by ground or surface water composition nor the highly leachable metal content of nearby mineral veins and bedrock.

Loevblad[132] has analyzed coal samples from Norway, England, U.S., N. Germany, Poland, U.S.S.R., and Australia for number of trace elements. Most of the elements were determined by neutron activation; Hg, Zn, and Cd by radiochemical separation cedure; Be, Pb, and Tl by emission spectroscopy; F by digestion with sulfuric acid; while S was determined by gravimetically. There are several very useful bibliographies on the subject of trace elements in the coal, e.g., Walker.[133]

II. NATURE AND MODE OF OCCURRENCE OF TRACE ELEMENTS

There is a large amount of data on trace elements in coals, although many of the data are suspect because they were obtained on coal ashes using spectrochemical analysis techniques. Losses of volatile elements during ashing of the coals and matrix effects in the spectrochemical procedure account for many of the erroneous values.

Although there is a large body of data on trace elements in coal, there is much less known about the forms in which such trace elements occur and such knowledge is important for an understanding of trace element behavior during coal conversion. Trace elements can occur in coal basically in either organic or inorganic forms, and most trace elements are probably found in both combinations.

Elements such as Fe, Ca, Zn, Mg, Si, etc. may occur predominantly in mineral species of these elements, e.g., FeS_2, $CaCO_3$, ZnS, etc. However, for many elements such as Hg, As, Sb, Pb, Cd, Co, Ni, Se, etc., no specific mineral of the element may be present, but the element may be distributed among several mineral species. Several authors have discussed the organic vs. inorganic occurrence of trace elements in coals. Ruch et al.,[134-136] and Gluskoter,[6] used float-sink specific gravity separations of coal to determine organic vs. inorganic affinity of trace elements in four Illinois coals. Zubovic[138,139] has determined similar affinities for coals.

Some of the general conclusions reached by Gluskoter[137] are described here:

1. Elements that have the largest ranges in concentrations are those that are found in distinct mineral phases in the coals; elements with narrow ranges are often those found in organic combination in coal.
2. Only four elements are, on the average, present in coals in concentrations significantly greater than their average concentration in the crust of the Earth. There are boron, chlorine, selenium, and arsenic. Not all are concentrated in each of the samples analyzed from the three geographic groups (eastern U.S., western U.S., and the Illinois basin).
3. Most of the elemental concentrations in coals are lower than their average concentration in the crust of the Earth.

Generalization from the statistical analyses of the analytical data from five bench (vertical segments of the seam) sets from the Illinois basin include the following:[137]

1. Wide variations in elemental concentrations are present between benches of a single coal sampled.
2. Although elements may be concentrated within any bench of a coal, concentrations are more commonly observed at the top and/or bottom of the coal seam.
3. Most elements occur in significantly higher concentrations in the fine-grained sedimentary rocks associated with the coals (roof shales, underclays, and partings) than is coal.

According to Gluskoter,[137] washability curves and histograms of washability data are effective means of depicting the mode of combination of elements in coal; they indicate whether the elements are associated with the organic or inorganic fractions of the coal. It is useful to quantify the information presented on the washability curves and produce an "organic affinity" index.

The concept of an organic or inorganic affinity for elements in coal, was developed by Goldschmidt,[4] who pioneered modern investigations of trace elements in coals, and identified trace elements in inorganic combination with minerals in coals. He also postulated the occurrence of metal organic complexes in coal; the observed concentrations of vanadium, molybdenum, and nickel were attributed to the presence of such complexes.[4]

Nicholls[8] approached this problem by plotting the analytical data for the concentration of a single element in coal or in coal ash against the ash content of the coal. Diagrams depicting a number of such points for a single coal seam, or for a group of coal seams in a single geographic area, were interpreted for degree of inorganic or organic affinity of the element.

Nicholls[8] concluded that boron is almost entirely associated with the organic fraction in coals. Elements barium, chromium, cobalt, lead, strontium, and vanadium are, in the majority of cases, associated with the inorganic fraction; nickel, gallium, germanium, molybdenum, and copper, may be associated with either or both fractions. The author has subdivided the third group into nickel and copper, which are in inorganic combination when found in large concentrations, and into gallium, germanium, and molybdenum, which are largely in organic combination when found in large concentrations.

Gluskoter[137] has calculated an index or organic affinity of the elements from the washability curves (determined on specific gravity fractions of the washed coals. Elements have been classified as "organic", "intermediate-organic", "intermediate-inorganic", and "inorganic", on the basis of the organic affinity index. The following generalizations are suggested:

1. Germanium, beryllium, boron, and antimony are classified within the organic group in all samples. Germanium has the highest value of organic affinity in each coal.
2. Zinc, cadmium, manganese, arsenic, molybdenum, and iron are within the inorganic group in all four samples.
3. A number of metals including cobalt, nickel, copper, chromium, and selenium have organic affinities within the intermediate categories. This suggests a partial contribution from sulfide minerals in the coal but also suggests the presence of these elements in organometallic compounds as chelated species, or as adsorbed cations.

Swaine[140] indicated that some trace elements in coal may be associated with phenolic, carboxylic, amide, and sulfhydroxy functional groups in the organic fraction of coal. In the inorganic fraction, the trace elements are often associated with clays, silicates, carbonates, sulfides, sulfates, and other minerals. Table 12 shows the association of some trace elements with mineral matter. Layer silicate minerals (clays), quartz, and pyrite are widely distributed in coals. Kaolinite is the most common layered silicate, although montmorillonite and illite are also found. Dolomite, siderite, calcite, and aragonite are the most common carbonate minerals. Pyritic and marcasitic crystalline forms of FeS_2 are by far the dominant sulfide minerals occurring in coals of the U.S. Barite, zircon, and fluorapatite occur sporadically.

The concentration of elements in a given coal deposit depends in part on the conditions that prevailed in the peat swamp in which the coal-forming material was deposited and on the subsequent genetic history of the coal seam. These conditions, especially the hydrological conditions that affected the transport of inorganics into the swamp, varied geographically and temporally within a given peat swamp. Therefore, the concentration of elements, as both mineral matter and organically combined species, varies from one location to another in a coal bed and from one bed to another.

Table 12
MINERAL ASSOCIATIONS OF TRACE ELEMENTS WITH MINERAL MATTER IN COAL

Element	Mineral association
Arsenic	Arsenopyrite (FeAsS)
Barium	Barite ($BaSO_4$)
Boron	Illite and tourmaline (complex aluminium silicates)
Calcium	Gypsium ($CaSO_4 2H_2O$), calcite ($CaCO_3$)
Cadmium	Sphalerite (Zn, Cd)S
Cobalt	Linnaeite (CO_3S_4)
Copper	Chalopyrite ($CuFeS_2$)
Fluorine	Fluorapatite ($Ca_5(PO_4)_3(F,OH)$)
Iron	Pyrite (FeS_2), marcasite (FeS_2), hematite (Fe_2O_3), siderite ($FeCO_3$)
Lead	Galena (PbS)
Manganese	Siderite $(Fe,Mn)CO_3$ and calcite $(Ca,Mn)CO_3$
Mercury	Pyrite (FeS_2)
Molybdenum	Molybdenite (MoS_2)
Nickel	Millerite (NiS)
Phosphorus	Fluorapatite $Ca_5(PO_4)_3(F,OH)$
Silicium	Quartz (SiO_2)
Strontium	Goyazite group (hydrous strontium aluminium phosphates)
Titanium	Oxide (TiO_2)
Zinc	Sphalerite (ZnS)
Zirconium	Zircon ($ZrSiO_4$)

After Swaine, D. J., *Trace Subst. Environ. Health,* 11, 107, 1977.

Using the formulas for the coal minerals and the percentage of these in coals, one can calculate the major element concentrations in mineral matter. These values as reported by Hamrin[126] are given in Table 13 and are in good agreement with the report of Ruch et al.[135] Their procedure assumes that each of the elements is entirely present in the mineral matter which is particularly suspect for the titanium, boron, phosphorous, fluorine, and vanadium.

According to Zubovic[139] the suggestion that organic complexes of various metals exist in coal and other organic sediments is based on indirect evidence. With the exception of V and Ni porphyrins, no other organic complex has ever been isolated or identified in organic sediments. The suggestions by many investigators that metallo-organic complexes exist in coal are based on observations that elements such as Ge and Be were always found in large amounts in the ash of clean coal and ash of pure coal macerals such as vitrinite.

The first attempt to quantify the organic association of metals in coal was made by Horton and Aubrey.[141] Three vitrain samples were separated by sink-float techniques, and the quantities of each element contained in the different specific gravity fractions were determined. In their samples, Ge and V were 100% associated with the organic fraction, whereas Sn was 100% associated with the mineral fraction of coal. The other elements had intermediate associations.

The result of their work is shown in Table 14. Zubovic et al.[9] have reported a dominant organic association for the same six elements (Ge, Be, Ga, Ti, B, V). However in their study Zn showed no organic association (see Figure 1 showing the relation of organic affinity and ionic potential).

Ruch et al.[136] reported on the organic association of 21 elements in 4 coal samples. Ge, Be, and B were found to be organically associated, whereas Hg, Zr, Zn, Cd, As, Pb, Mo, and Mn were largely associated with the mineral matter of the coal. The other elements had varying degrees of organic association.

Table 13
MINERAL MATTER CONCENTRATIONS

Element	Percent
Si	20.5
Fe	10.7
S	12.3
Al	9.0
Ca	3.9
K	1.3
Cl	—
Ti	—
Mg	1.3
Na	0.2
O	39.3
C	1.1
H	0.4
Total	100.0

From Hamrin, C. E., Catalytic Activity of Coal Mineral Matter, Annu. Rep. FE-2233-3, National Technical Information Service, Springfield Va., 1977.

Table 14
AVERAGE ORGANIC AFFINITY OF SOME ELEMENTS (%)

Element	Ref. 141	Ref. 9
Germanium	100	87
Beryllium	75—100	82
Gallium	75—100	79
Titanium	75—100	78
Boron	75—100	77
Vanadium	100	76
Nickel	0—75	59
Chromium	0—100	55
Cobalt	25—50	53
Yttrium		53
Molybdenum	50—75	40
Copper	25—50	34
Tin	0	27
Lanthanum		3
Zinc	50	0

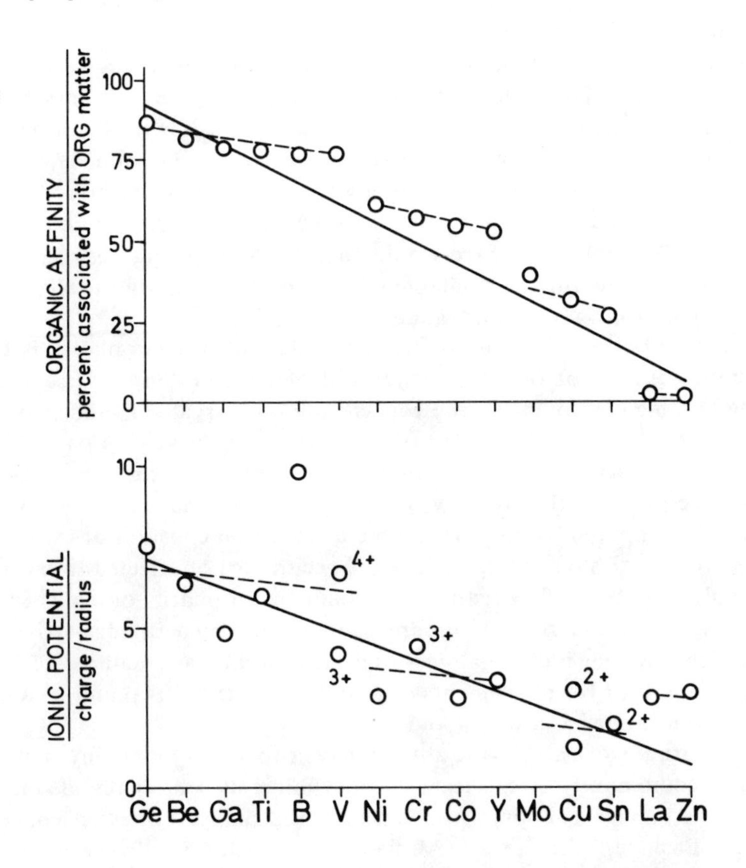

FIGURE 1. Relation of organic affinity and ionic potential of the elements arranged in order of decreasing affinity for organic matter. (After Zubovic.[139])

Zubovic[139] finds additional evidence for the existence of organic complexes (chelates) for some of the elements in coals in the relation of the organic affinity of the elements in coal to the complexing ability of these metals with organic ligands. The stability of chelates of the metals in directly related to the ionic potential of these elements. The organic affinity of the metals in coal is also related to their ionic potential (see Figure 1). This suggests that those metals having high organic affinity in coals are present as chelates.

Compilations of stability constants for a wide range of organic ligands give a stability order of Ge > Y > La for these trivalent metals. This is the same order as is their organic affinity. Other compilations for the bivalent metals suggest that their stability order is Be > Cu > Ni > Co > Zn > Fe. In coal, the organic affinity for some of these elements is Be > Ni > Co > Cu > Zn. Iron, of course, is always dominantly present as the sulfide. With the exception of copper, the two series are identical. The displacement of copper to a lower organic affinity than is indicated by its stability series results from its occasional interaction with sulfide ions, as discussed previously. The evidence indicates that metals that have high organic affinity in coal are present as chelates,[139]

The geographic distribution of the elements in a coal basin appears to be controlled by their chemical properties. Elements possessing a high organic affinity, such as Ge, Ge, B, and Ga have high concentrations in the near source areas and elemental depletions in distal areas. The transition metals, which have a lower organic affinity, are uniformly distributed throughout the basin. Most of the chalcophilic elements behave in the same manner.

Physicochemical processes of trace-element emplacement in coal are described in details by Zubovic.[139] Trace elements in solution entering an area in which organic matter in being deposited are subjected to retention according to the principles of coordination chemistry. It can be assumed that a large variety of organic liquids is present within such an organic environment. This environment would also be conducive to the growth of sulfate-reducing bacteria and consequently to the production of H_2S, which dissociates into HS^- and $S^=$. There would then be competition between the organic ligands and sulfide ions for the available metal ions, particularly the chalcophilic elements such as Cu, Pb, Zn, Cd, Sb, Hg, and Fe.[139]

The partition of these elements between sulfide and organic phases is dependent upon the concentration of organic ligands and of the sulfide ions. Because this is a dominantly organic environment, the concentration of organic ligands would remain relatively uniform.

In addition to a discussion of the availability of the reacting anionic species, consideration must be given to the stability of the compounds that could be formed. This stability can be compared by considering the dissociation constant of the compounds.

Finkelman et al. [142] have studied modes of occurrence of selected trace elements in several Appalachian coals. Their study was conducted primarily on polished blocks of coal using a scanning electron microscope (SEM) with an energy dispersive X-ray detector. With this system individual *in situ* mineral grains as small as 0.5/μm can be observed and analyzed for all elements of atomic number 11 (Na) and greater that are present in concentrations as low as about 0.5 wt %.

From their earlier works,[143,144] the authors have concluded that many elements occur in this coal predominantly as specific accessory minerals. For example zinc occurs as zinc sulfide and copper as copper iron sulfide. SEM analysis of other coals from the Appalachian Basin appears to substantiate this conclusion. Preliminary estimates based on the new data suggest that most of the selenium in these coals and much of the lead occur as 1 to 3/μm particles of leaa selenide, which are often associated with cadmium-bearing sphalerite and chalcopyrite. These fine-grained mineral intergrowths occur exclusively within the organic constituents and in all probability formed in place.

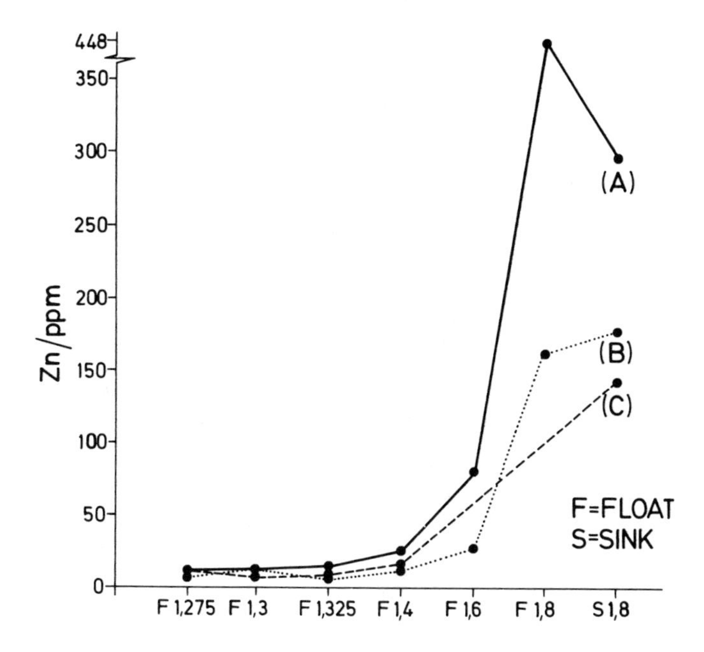

FIGURE 2. Concentration of zinc (whole-coal basis) in size gravity separates of the Upper Freeport coal sample. (As reported by Finkelman, R. B.[142])

Because substantial amounts of Zn, Cu, Pb, Cd, and Se occur as finely dispersed mineral grains in the organic matrix, considerable amounts of these elements can be retained in the lighter specific gravity fractions of cleaned coal.

Figure 2 illustrates the concentration of Zn in a size-gravity separation of the Upper Freeport coal (as reported by Finkelman et al.[142]) Similar results have been obtained for Cd, Cu, and Pb on six samples of this coal. The divergence of these curves in the high specific gravity range is consistent with the observation that these elements occur as fine-grained minerals which are increasingly released from their organic matrix with fine frinding. It is evident from Figure 2 that the concentrations, on a whole coal basis, of Zn (this would apply to Cu, Cd, and Pb as well) are much greater for the higher specific gravity fractions.

The inverse problem of release of metals from coal into water was studied by Akers.[145] The objective of his research was to determine if there is a significant difference in the leaching rates of the metals under consideration and to attempt to approximate the source of any differences found. The metals studied were Al, Ca, Co, Cu, Fe, Mg, Mn, Ni, Pb, and Zn. Two fresh channel samples of the Pittsburgh coal seam were taken about 200 ft apart at the Bud Waal's mine near Easton, W. Va. These samples were crushed, riffled, and ground to minus 200 mesh. After grinding, the samples were each split again and one half of each sample was analyzed by AA spectroscopy, while the other half was analyzed by emission spectroscopy.

Table 15 contains a listing of the average aqueous concentration of the ten elements under consideration as found in water draining from the Bud Waal's Mine immediately below the coal collection point. Also, the average coal concentrations of these metals are again presented. In order to compare coal leaching rates directly (by eliminating the coal concentration variable), each average elemental aqueous concentration was divided by the average elemental concentration in the coal to yield a leaching rate number. This method indicates that the metals with the largest leaching rate numbers are being removed relatively quickly from the coal and those with the lowest, relatively

Table 15
LEACHING RATES OF COAL ASSOCIATED METALS

Metal	Av aq conc (ppm)	Av coal conc (ppm)	Aq conc/coal conc
Aluminium	129.4	85,750	1.51×10^{-3}
Calcium	373.8	11,500	3.25×10^{-2}
Cobalt	1.1	240	4.58×10^{-3}
Copper	1.1	145	7.59×10^{-3}
Total iron	554.5	67,790	8.18×10^{-3}
Magnesium	117.7	3,680	3.20×10^{-2}
Manganese	27.7	125	2.22×10^{-1}
Nickel	2.0	260	7.69×10^{-3}
Lead	1.8	95	1.89×10^{-2}
Zinc	4.5	145	3.10×10^{-2}

After Akers, D. J., Uranium in Coal Carbonaceous Rocks in the United States and Canada, Rep. 157, Coal Research Bureau, West Virginia University, Morgantown, 1978.

slowly. When the metals are listed in order of leaching rate, the following series is produced:

$$Mn>Ca>Mg>Zn>Pb>Fe>Ni>Cu>Co>Al$$

Based on the work of Ruch et al.,[134] these metals can be divided into four groups: calcite metals, organic metals, and clay metals. The calcite metals are of course Ca and Mg, and the clay metal Al. Calcite is very reactive under acid conditions which would explain the rapid leaching rates of Ca and Mg. Clays on the other hand are relatively stable under weathering conditions and are slow to break down to release Al. This appears to hold even under acid weathering conditions.[145]

It is of interest to mention here the work by Wewerka et al.[146,147] They have performed the mineralogical analyses of the bulk refuse samples from the Illinois Basin. Through studies of the statistical correlation of refuse compositional data and direct microprobe observations of refuse micromineralogy, the authors have established structural relationships and associations among the various elements and minerals in the refuse.

To quantify the relationships among the various elements in these refuse fractions, they used a specially designed computer program to calculate Pearson's correlation coefficients for each of the elements studied. These coefficients range on a scale from + 100 among elements that are present in identical amounts in each of the waste fractions to − 100 for elements that exibit totally inverse compositional behaviors. Intermediate values of the coefficients among elements indicate varying degrees of similarities or differences in the manner in which the elements are distributed in the waste fractions.

A rather laborious perusal of these coefficients reveals that the various elements fall into groups representing the classes of major minerals present in the wastes. The authors have chosen to designate these as the silicate group (elements with $R > 75$ for Si and Al), the sulfide group ($R > 75$ for Fe and S) and the organic group ($R > 75$ for C). Although the ordering in the table is based on high positive correlations among the various trace elements and the major elements in the refuse samples, the trace elements in each mineral group also have high positive correlations with one another. These results are shown in Table 16.

Table 16
MINERAL ASSOCIATIONS OF TRACE ELEMENTS IN COAL

Group	Elements
Silicate	Mn, Ni, Hf, Zn, Cu, Dy, Zr, Ca, Eu, Sm, Ce, La, Lu, Sc, Sc, Th, Yb, V, U, Na, Si, Mg, Ti, Al, Y, Li, K, Cr, Be, P, B
Sulfide	Mo, As, Fe, S, Se, Cd
Organic	H, N, C, Cl

After Wewerka, E. M., Williams, J. M., Vanderborgh, N. E., Wagner, P., Wanek, P. L., Olsen, J. D., Trace Element Characterization and Removal/Recovery from Coal and Coal Wastes, National Technical Information Services, Springfield, Va., 1978.

Table 17
ELEMENTS ASSOCIATED WITH LABILE MINERALS IN ILLINOIS BASIN COAL REFUSE

Mineral	Major elements	Minor elements
Pyrite/marcasite	Fe, S	Tl, Mn, As, Cu, Mo, Se, Cd
Carbonates	Ca, Mg, C, O	
Phosphates	P, Ca, Y, F, O	La, Ce, Pr, Nd, Sm, Eu, Gd, Th, Dy, Ho, Er, Tm, Th, U, Ba

After Wewerka, E. M., Williams, J. M., Vandorborgh, N. E., Wagner, P., Wanek, P. L., and Olson J. D., Trace Element Characterization and Removal/Recovery from Coal and Coal Wastes, National Technical Information Service, Springfield, Va., 1978.

Several minor minerals were identified in the refuse samples by Wewerka et al.[146] Particles of rutile (TiO_2) and zircon (ZrO_2) are frequently observed in the clay fractions. Apatite and fluorapatite also were readily observed with the probes as constituents of the clay phases. Interestingly, these phosphate minerals appear to be the main source in the refuse of U, Th, and many of the rare-earth elements. Phosphorus also was present in these wastes in relatively high concentrations ($\sim$0.3 wt%).

Trace elements listed in Table 7 are those identified either by statistical studies or direct microprobe analyses as being present in or associated with labile or environmentally active mineral phases.

III. ASH

Trace elements in coal ash have been studied because of the environmental concerns resulting from increased coal utilization and because of the possibilities for the recovery of some elements from coal ash. The chemical composition of coal ash varies widely but in general is within limits shown in Table 18.

Ash composition has been studied by many authors and valuable information on elemental composition of coal has been obtained. Many of these studies are referred to in other chapters of this book; here we shall mention only some.

Table 18
TYPICAL VALUES FOR OXIDES CONC IN COAL ASH

Constituent	Conc (%)
Silica (SiO_2)	30—60
Aluminium oxide (Al_2O_3)	10—40
Ferric oxide (Fe_2O_3)	5—30
Calcium oxide (CaO)	2—20
Magnesium oxide (MgO)	0.5—4
Titanium oxide (TiO_2)	0.5—3
Alkalies (Na_2O + K_2O)	1—4

Table 19
ANALYSIS OF ASH FROM HIGH-VOLATILE UTAH BITUMINOUS COAL

Component	Conc (%)	Component	Conc (ppm)
Ash	8.9	Li	144
MgO	1.03	Pb	42
Na_2O	4.55	Zn	61
Al_2O_3	17.6	B	920
SO_3	3.9	Ba	940
Cl	<0.20	Be	5.4
CaO	7.3	Co	9
SiO_2	46.0	Cr	88
P_2O_5	1.0	Ga	46
TiO_2	0.93	La	76
MnO	<0.05	Nb	22
Fe_2O_3	4.00	Ni	25
K_2O	0.58	Sc	15
Cd	<1	Sr	140
Cu	64	V	150
		Y	66
		Yb	7
		Zr	220

After Hausen, L. D., Phillips, L. R., Mangelson, N. F., and Lee, M. L., *Fuel*, 80, 323, 1980.

Let us mention a few specific analyses. For example Hausen et al. [148] have reported analysis of ash from high-volatile Utah bituminous coal; their results are shown in Table 19.

Abernethy et al.[17] have presented analyses for 10 elements in 373 U.S. coal samples. Sulfur trioxide retained in the ash was analyzed, as were the oxides of Si, Al, Fe, P, Na, K, Ti, Ca, and Mg by various wet chemical methods. Three main constituents, Si, Al, and Fe oxides, made up 90% of the ash from bituminous coals. Compared to the U.S. average, central U.S. coals were low in Al oxide, and high in Fe oxide. Lignite and some subbituminous coal ashes were high in oxides of Ca and Mg and sulfur trioxide.

In an early work Barret[149] described ash as essentially a four-component system of silica, alumina, basic oxides (CaO, MgO, K_2O, Na_2O), and ferric oxide. Phase equilibrium diagrams are included that describe the relationship between the composition of the ash and the ash-softening temperature. As composition was reported for numerous

U.S. bituminous coals in relation to its effect on slagging properties of ash and to ash clinkering.

The electrical resistivity of 17 commercially produced fly ashes from both eastern and western coals were studied by Bickelhaupt[150] to determine the effects of chemical composition on surface resistivity. The elemental composition of a fly ash influences its electrical resistivity and thus considerably affects its collectibility by electrostatic precipitation. Standard chemical analyses, AA, and colorimetric tests were performed. An air pycnometer and microparticle classifier (for detection of particle size distribution) were used, and a laboratory device was constructed for the measurement of fly ash resistivity under different conditions. A correlation was established between certain elemental constituents of fly ash, especially Li, Na, K, and Fe, and resistivity. A procedure was proposed whereby the resistivity of a particular fly ash might be predicted from measurable chemical constituents, thus allowing for the objective selection of conditioning agents and the anticipation of problems resulting from a particular configuration of plant operating conditions.

Coal-ash composition as related to high-temperature fireside corrosion and sulfur-oxides emission control is discussed by Borio and Hensel.[151] A method by which the analyses of coal samples can be used to evaluate the corrosive potential of coal was proposed, and related corrosion studies were conducted. These included the reduction of high-temperature corrosion by introducing certain additives and the effects of those additives on air pollution and ash deposits. Combustion testing of bituminous coal samples from six geographical areas was performed in a laboratory pulverized combustion furnace (100 lb coal per hour) for approximately 300 hr. Chemical analyses of the samples were done by standard ASTM procedures and soluble alkalies were determined by an acid leach technique. This acid leaching process appeared to be the best method to measure the amounts of Na and K which are available for reaction in forming the trisulfates. Na and K react with Fe and the trisulfates to form alkali-iron-trisulfates at high combustion temperatures. Calcium and Mg were found to have an inhibiting effect on the formation of potassium-iron-trisulfates, and additions of Ca and/or Mg have a large effect on the melting characteristics of coal ash. This could be beneficial or detrimental depending on both the ash composition and the amount of additives used. A nomograph was prepared to determine the corrosive rate of coals based on the K, Na, Fe, Ca and Mg content in the coal. Since the nomograph was based on results of tests on bituminous coal, it may not indicate corrosive rates for other coal types.

Biochemical and geochemical origins of ash-forming ingredients in coal are discussed by Deul and Annell.[193] The biochemical and geochemical origins of ash-forming components in coal were investigated by analyzing the ashes of peat, humic acids (HA) extracted from peat, and the ashes of living plants similar to those forming peats. The composition of coal and weathered coal were compared with the data generated in this research. Semiquantitive analyses were reported for 33 elements in plant ash and peat ash and for 36 elements in HA extracts from peat. The plant specimens tested included 16 HA extracts from peat, 16 samples typical of Carboniferous Age coal-forming plants, and 5 specimens representing Teritary flora. Thirteen peat samples from the Rice Lake Bog near Duluth, Minn. and HA extracts from these peats were analyzed. The plant ashes were found to be rich in K, Ca, Mg, Na, Li, Ba, and Si. These elements were notably depleted in the peat ashes, probably because they are relatively soluble and lost during degradation of early peatification stages. Elements concentrated in peat ash which are retained from the original mineral content of the coal plants are Zn and B. Humic acids in peat have high concentrations of Fe, Cu, Mo, Pb, Cr, Ni, Co, Sn, and Zn. The elements found in HA ashes but not in plant or peat ashes include Nb, Bi, Ge, and Au. During coal weathering U, Mo, and possibly Fe and Ge are combined with the organic matter in coal. Minerals added to coal during the peatification stage are detrital clay minerals.

Davison et al.[39] have studied trace elements in fly ash — dependence of concentration on particle size. The concentrations of 25 trace elements were measured in fly ash from a coal-fired power plant in order to develop a theory for the dependence of trace element concentration on particle size. Airborne fly ash and fly ash retained in the plant collection system were sampled. Elemental analyses were performed by AA, emission and MS, and X-ray fluorescence. Elements studied were classified into three groups. One group exhibited a concentration increase with decreasing particle size (e.g., As Se, Zn), while a second group showed a nonuniform concentration dependence on particle size (e.g., Fe, Mn, Si). Finally, a third group exhibited no concentration trends as a particle size function (e.g., Bi, Sn, Cu). Elements were concentrated in the smallest fly ash particles by volatilization in a high-temperature combustion zone followed by recondensation and adsorption onto the greater surface area provided by the finer particle.

Demeter and Bienstock[154] have studied sulfur retention in anthracite ash. Sulfur retention properties of anthracite ash from four producing regions in Pennsylvania were checked by both ashing in the laboratory and burning on a small-scale chain-grate stoker. The effect of temperature on S retention in the ash was determined by ashing samples in the laboratory at 750 and 1200°C. In addition several tests were made in which anthracite was mixed with dolomite before firing in the chain-grate stoker. Flue gas and corresponding ash samples were taken during combustion in the small-scale stoker and analyzed by standard ASTM methods. Results of the tests and analyses were (1) when laboratory ashing of anthracite was done at 750°C, the percentage of S retained in the ash was related to the S and Ca content of the coal, (2) when the ashing temperature was 1200°C (the decomposition temperature of calcium sulfate) very little S was retained in the ash (3) dolomite did not improve S retention due to the high temperature in the fuel bed, and (4) the percentage of S retained is almost directly proportional to the percentage of original coal C in the ash.

Bryers and Taylor[155] have examined the relationship between ash chemistry and ash fusion temperatures in various coal size and gravity fractions. Mineral matter distribution was determined in specific gravity fractions of four Wyoming subbituminous coals, and ash chemistry was statistically related to ash-softening temperature to establish the effect of mineral matter on their fireside fouling and slagging potential. Mineralogical analysis was determined by X-ray diffraction, IR spectroscopy, and SEM. Inherent ash existent primarily in fractions of sp gr 1.40 or less, was concentrated with respect to K and Si, indicative of feldspars (e.g., orthoclase) or clays (e.g., biotite or muscovite). Calcite, illite, pyrite, gypsum, and quartz were also identified. Calcium and Mg were predominant in the extraneous ash. Mineral matter distribution in the coals as well as their ash chemistry significantly affected their fouling characteristics.

Composition of the ash of Illinois coal is discussed by Rees.[86] The effects of preparation (sizing and washing) on elemental composition of the ashes of Illinois coals were discussed. Coals analyzed included face channel samples, run-of-mine coal, and both sized and washed coals. Elemental determinations were made utilizing wet chemical procedures. Ash analyses of 20 Illinois coals were presented in tabular form showing ash-softening temperatures, percent ash, percent S, and percents (as the oxide) of Al, Ti, Fe, Mg, Ca, Na, K, and Si. Principal minerals in the coals studied were clays (aluminum silicates), pyrite and marsacite, and carbonates (primarily calcite). During ashing, clays lost the water of hydration, carbonates lost CO_2, and pyrite was converted to iron and sulfur oxides, fractions of which were retained in the ash as sulfate. Percentages of Fe and Si were highest in the ash of large-sized coals, while Ca was highest in the ash of small-sized coals. Aluminum concentrations were unaffected by the size of original coals. Clays were concentrated in the light specific gravity fractions (1.30 F) as evidence by high percentages of Si and Al (20 to 50%). Iron and Ca percentages were the greatest in the heavy specific gravity fractions (1.70 S), indicative of the concentration of pyrite and calcite.

Thiessen et al.,[156] in an early work, examined the composition of several coal samples from Illinois and Pennsylvania to determine a relationship between the ash-softening temperature and the tendency to clinker. Samples from the Herrin No. 6 (Ill.), Upper Freeport, and Pittsburgh (Pa.) coal seams were submitted for chemical and petrographic analysis. Results of the petrographic analysis on the coal ash-forming material indicated that better than 95% was composed of detrital clay, kaolinite, calcite, and pyrite. Analysis of the ash from combusted coal indicated it was composed of oxides of Al, Si, Ca, and Fe. Testing the effect of the four common oxides in the coal ash on the ash-softening temperature and on clinker formation was done by using synthetic mixtures and standard ASTM procedures. Specific results of these tests can be found in tables in the report. Some conclusions reached by the authors were that (1) a correlation exists between the ash-softening temperatures of the natural coal ashes and of the synthetic mixtures containing proportional amounts of the four basic oxides (Al, Si, Ca, and Fe), (2) the initial deformation temperature of the synthetic mixture is related to the mixture with the lower melting point, (3) the temperature at which clinker formation starts is dependent upon the fusion temperature of that portion of the ash with the lowest softening temperature, (4) localized concentrations of more readily fusible material increases the tendency towards clinker formation, and (5) iron oxides formed during the combustion of pyrite have an accelerating influence on clinker formation.

Relation of sulfur content to ash fusion characteristics was described by Reiter.[157] Ash fusion properties of coal were discussed in reference to a study made by a large, commercial utility. Several relationships between coal composition and ash fusion behavior were detailed. The ash fusion temperature depends greatly upon the percentage of metallic oxides present in the ash. Oxides of Fe, Ca, Mg, Na, K and similar metals form silicates of low melting points. As the percentage of SiO_2 in the ash decreases the ash fusion temperature also decreases. The percentage of S in the ash may also be used as an approximate indicator of ash fusing behavior. Ash fusion temperatures generally decrease as the S content of the coal increases.

Rees et al.[158] have studied sulfur retention in bituminous coal ash. Tests were performed on samples of Illinois No. 5 coal and Illinois No. 6 coal to determine some of the factors affecting S retention in coal ash. Total S, CaO, and Fe_2O_3 were found to affect the amount of S retained in the coal ash. Samples of the two coals were ashed at 800°C and were analyzed for S and CaO content by gravimetric methods. A relationship was found to exist between S retention and the total S and CaO content of the coal; that is, S retention in the ash was greater in samples with higher amounts of S and CaO. However, because the CaO retains the S in the ash in the form of $CaSO_4$, this reaction would not be applicable to industrial uses where combustion temperatures exceed the decomposition temperature of $CaSO_4$. Iron oxide, formed from pyrite in the coal, was found to have the ability to catalyze the transformation of SO_2 to SO_3. The SO_3 more readily reacts with CaO to form $CaSO_4$, which is retained in the coal ash.

Sage and McIlroy[92] have discussed the relationship of coal ash viscosity to chemical composition. Laboratory analyses of coal slags were used to predict the behavior of slag in power boilers. An attempt to relate slag composition (amount SiO_2, Al_2O_3, Fe_2O_3, TiO_2, CaO, MgO, Na_2O, and K_2O) to the slag viscosity was made. A previous method cited was developed by the Bureau of Mines and describes the relationship of SiO_2 content of the slag and the slag viscosity over a given temperature range. Designed as the "Equivalent Silica Percentage", the relationship is described by the following equation:

$$\frac{SiO_2 \times 100}{SiO_2 + Fe_2O_3 + CaO + MgO} = \text{Equivalent silica (\%)} \qquad (3)$$

This correlation holds reasonably well for slags with SiO_2 to Al_2O_3 ratios not in excess of 1 to 4. The authors found that the point at which the slag viscosity increases at a significantly higher rate per unit decrease in temperature was described by the relationship:

$$\frac{Fe_2O_3}{Fe_2O_3 + 1.11\ FeO + 1.43\ Fe} \times 100 \qquad (4)$$

This relationship was termed the "Ferric Percentage". At or above a "Ferric Percentage" of 20 the temperature at which the viscosity increases more rapidly becomes largely dependent on the content of Fe in the slag. A final method of estimating slag viscosity, the base to acid ratio, was determined in the equation:

$$\text{Base/acid} = \frac{Fe_2O_3 + CaO + MgO + Na_2O + K_2O}{SiO_2 + Al_2O_3 + TiO_2} \qquad (5)$$

As this ratio increases to 1, the viscosity of the slag decreases.

Dutcher et al.[159] have used electron probe to study ash properties. Samples of several coal seams were examined by an electron microprobe X-ray scanner to assay the lithotype-ash-element associations that commonly characterize coals. By using the "sweep" mode on the electron probe, an area as large as 360 × 360 μm can be scanned by the electron beam in such a way that a two-dimensional image of the ash distribution can be obtained. A summary of the results of the study are as follows:

1. Ash-forming material did not occur with the inert fractions (macerals from the micrinoid and fusionid group).
2. Material occurring in the lumina of remnant cells was often composed largely of Si, Al, and pyrite.
3. High S values were always associated with high Fe concentrations; however S also occurred without the presence of Fe, and high concentrations of Fe did not indicate the presence of high values of S.
4. Fusionids exhibiting bogen structure may have very high concentrations of admixed mineral matter.
5. Spore exine material was very low in ash-forming elements.
6. The mid-line material of a coalfied spore often may represent inorganic substances that migrated into the cell after deposition of the spore.
7. Organic matter also may enter into the cell cavity at the same time as the inorganic substances.

Fisher et al.[160] have discussed size dependence of the physical and chemical properties of coal fly ash. Some morphologic and chemical properties of fly ash escaping a power plant stack are dependent on the size of the particulate. Size-classified fly ash was collected in the stack beyond the electrostatic precipitator. The mean size of each of the four fractions was measured by optical sizing, centrifugal sedimentation, and Coulter analyses to produce a count mean diameter, mass mean diameter, and a volume mean diameter for each fraction. Particles in each size fraction were quantitatively classified into 11 morphologic types by light microscopy, and 35 elemental concentrations were measured in each by instrumental neutron activation analyses, (INAA), Major, and 17 trace element concentrations were independent of particle size, with the exception of Si which decreased with particle size. Elements which increased in concentration with decreased particle size were Cd, Zn, Se, As, Sb, W, Mo, Ga, Pb, V, U, Cr, Ba, Cu, Be, and Mn, nine of which are volatile elements. Refractory elements included in the above group were thought to be the result of mineral decomposition and/or original elemental distributions in the coal.

Gronhovd et al.[161-163] have studied factors affecting ash deposition from lignite and other coals. The fouling potential of 31 U.S. coals was evaluated and mechanisms were proposed to explain ash deposition. During the course of the investigation the mode of occurrence of six minor elements in the coals and ashes were determined. Coals studied were primarily western in origin and ranged in rank from lignite to bituminous. Elements and minerals were identified by X-ray fluorescence, diffraction, and electron-microprobe analysis. Sodium was organically bound and did not exist in a mineral matrix. Calcium, Fe, Al, and Mg were all organically associated in the lignites to some extent. Sodium, K, and S were vaporized during combustion, but a small percentage of Na and S were retained in the ash as sodium sulfate. The less volatile organically bound elements (Ca, Mg, Al, and Fe) were agglomerated in larger ash particles. Extraneous mineral matter including kaolinite, quartz, and pyrite remained largely unreacted during combustion.

Characterization of ash from coal-fired power plants is discussed in paper by Ray and Parker.[164] A summary of existing data on the chemical and physical characteristics of coal ashes and effluent produced by the burning of coal in steam-electric generating plants was presented. Tabulations show the concentrations of 26 different trace elements from 8 separate power plant locations representing a cross-section of typical North American steam coals. Analyses were performed upon the feed coal, bottom ash superheater ash, inlet fly ash, dust collector fly ash, outlet fly ash, and gas effluent. Analytical techniques included absorption spectroscopy, gas chromatography (GC), INAA, ion selective electrodes, optical emission spectroscopy, photon activation analysis, plasma emission spectroscopy, spark source MS, wet chemistry, and X-ray fluorescence spectroscopy. The elements found concentrated equally in the bottom ash and fly ash were Al, Ba, Ca, Ce, Co, Eu, Fe, Hf, K, La, Mg, Mn, Rb, Sc, Si, Sm, Sr, Ta, Tn, Ti; those elements tending to be discharged to the atmosphere were Hg, Cl, and Br.

Hendrickson[165,166] has compiled information on U.S. coal ash. Analyses of coal ash from locations in 12 different states were tabulated for S, Si, Al, Fe, Ti, P, Ca, Mg, Na, and K. Average trace element analyses for the eastern region, the interior region, and the western region were also tabulated and included values for Ba, Be, B, Cr, Co, Cu, Ga, Ge, La, Pb, Li, Mn, Mo, Ni, Sc, Sa, Sn, V, Yb, Zn, Zr, As, Bi, Ce, Nd, Nb, Rb, and Tl. Other information compiled included the variations in ash composition with coal rank, the range of trace elements in coal ash, ash fusion temperature data, and a brief section on fly ash resistivity.

IV. SUMMARY OF DATA ON ELEMENT CONCENTRATIONS IN COAL

In this chapter information on elemental concentrations in coal are summarized. The attention is focused on trace elements, i.e., elements which do not form the bulk of coal structure. However, for the sake of completeness some information is given about all other elements. This information is often needed by researchers involved in coal analysis. Additional data about elements in coal can be found in Volume 2.

Table 20
HYDROGEN CONTENT OF SOME U.S. COALS

Arithmetic mean (%)	Geometric mean (%)	Range (%)	SD	Number of samples	U.S. area
4.7	4.6	3.8—5.8	0.48	29	Western
4.9	4.9	4.0—6.0	0.44	22	Eastern
5.0	5.0	4.2—6.0	0.31	110	Illinois basin

After Gluskoter, H. J., Trace Elements in Coal, Circ. 499, Illinois State Geological Survey, Urbana, 1977.

Table 21
HYDROGEN CONTENT OF COALS

Appalachian bituminous	Interior bituminous	N Great Plains lignite	N Great Plains subbituminous	Rocky Mountain subbituminous	Rocky Mountain bituminous
4.9	5.1	6.7	6.4	5.4	5.4

After Zubovic, P., Hatch, J. R., and Medlin, J. H., Proc. U.N. Symp. World Coal Prospects, Katowice, 15—23. 10, 1979.

A. Elements

1. Hydrogen (H)

An extensive compilation of data for hydrogen content of U.S. coals is reported by Gluskoter.[137] Data reported in this work include data from following reports: Ruch et al.,[88,134] Gluskoter and Lindhal,[167] Gluskoter,[6] Frost et al.,[168] Kuhn et al.,[169] and Dreher and Schleicher.[170] Reported results are shown on Table 20, where geometrical means were calculated by taking the logarithm of each analytical value, summing the logarithms, dividing the sum by the total number of values, and obtaining the antilogarithm of the result.

Let us mention some other reported numbers for hydrogen content of coal. Debelak et al.[171] have reported following figures for Kentucky coals, No. 11 (5.5%), No. 4 (4.8%), No. 9 (5.2%), No. 3 (5.5%).

Schultz et al.[172] have found hydrogen values in the range 4.4 to 5.1%, all reported on moisture-free basis. This figure is slightly lower when reported for wet coal (as received by analyst). For example Illinois No. 6 coal has 4.18% of hydrogen as received, while the value for dry coal is 4.58%.

Zubovic et al.[173] have recently reported mean values of hydrogen content of coals from different regions of U.S. Their results are shown in Table 21.

Petrakis and Grandy[174] have reported chemical analysis of two subbituminous coals, two high-volatile coals, and one low-volatile coal. Some of the analytical data presented were from Pennsylvania State University coal data base. Hydrogen concentrations of coals investigated are shown in Table 22.

Recent interest in coal utilization problems, such as processing, underground gasification, and combating the methane danger, leads to the need for techniques to determine rapidly the hydrogen content in coal seams. Neutron methods seem to be particularly useful for this purpose. In order to test their usefulness, neutron parameters of coal vs. its hydrogen content were calculated by Morstin and Wozniak.[153] This is described in more detail in Chapter 5, where the applications of nuclear methods to coal analysis are presented.

Vypirakhina and Aronov[175] presented results of the development of a method of

Table 22
HYDROGEN CONCENTRATION OF COALS

Coal	Hydrogen (wt % daf)
Hagel, N.D.	4.9
Wyodak Gillete, Wyo.	5.5
Deitz, Wyo.	5.5
Kentucky No. 11	5.4
Illinois No. 6	5.4
Lower Dekoven, Ill.	5.8
Pacahoutas No. 3, W. Va.	3.9
Lower Kittanning, Pa.	4.8

After Petrakis, L. and Grandy, D. W., *Fuel,* 59, 227, 1980.

Table 23
LITHIUM IN U.S. COALS

Region	Coal	Conc range (ppm)	Arithmetic mean (ppm)	Geometric mean (ppm)
Appalachia	Bituminous	0.70—350	21	14
Interior province	Bituminous	0.44—210	10	6.7
N Great Plains	Lignite	0.37—33	4.4	3.2
N Great Plains	Subbituminous	0.22—61	5.8	3.7
Rocky Mountain	Subbituminous	0.52—87	12	7.6
Rocky Mountain	Bituminous	0.58—70	12	8.3
All U.S.	Different	0.17—350	15	8.3

After Zubovic, P., Hatch, J. R., and Medlin, J. H., Proc. U.N. Symp. World Coal Prospects, Katowice, 15—23. 10, 1979.

determining the amount of hydrogen in coal that is capable of being replaced by deuterium at various temperatures. The method consists of treatment of the coal with gaseous deuterium under pressure at 250 to 400°C, with the subsequent combustion of the treated coal and the determination of the amount of deuterium in it from the density of the water of combustion measured by the flotation method.

Hydrogen in coal is also studied by the techniques of NMR, which allows a nondestructive determination of hydrogen (e.g., see Alemany et al.,[176] Gerstein and Pembleton[177]

2. Lithium (Li)

Data on lithium concentrations in coals were summarized in the work by Abernethy and Gibson.[178] They reported that lithium content of representative West Virginia coal beds calculated as percentage of lithium in the ash, averaged 0.035 with a range of 0.01 to 0.065%. The concentration in ash from various parts of the beds ranged from a minimum of less than 0.002 to a maximum of 0.31%. The ash of Pennsylvania anthracites contained from a trace to 0.01% lithium.

In most of the western coal ashes tested, the lithium was not present in percentages high enough to be detected. The maximum concentration was in the range of 0.01 to 0.1%.

Butler[179] found the lithium content of ash of Svalbard coal was highly variable with a maximum concentration of 0.45%. De Brito[180] detected lithium in the ash of Portuguese anthracite. Zubovic et al.[173] have reported values for lithium concentrations in U.S. coals; their results are shown in Table 23.

Table 24
LITHIUM CONCENTRATION IN U.S. COALS

Coal	Conc (ppm)
Anthracite	33
Bituminous	23
Subbituminous	7
Lignite	19
U.S. av	20
World av	65

After Swanson, V. E., Medlin, J. H., Hatch, J. R., Coleman, S. L., Wood, G. H., Jr., Woodruff, S. D., and Hildebrand, R. T., Open File Rep. 76-468, U.S. Geological Survey, Reston, Va., 1976.

The average concentration of lithium in 12 lignite samples from North Dakota was reported to be 1.24 ppm.[122] Hansen et al.[181] have reported lithium concentration in the ash of high-volatile bituminous coal (8.9% ash) from Utah to be 144 ppm.

Wewerka et al.[182] have reported lithium concentration in raw coals from the Illinois Basin to be in the range 14 to 46 ppm, with mean value 30 ± 13 ppm. Lithium concentration in eight coal samples from Nebraska was in the range 2.1 to 166 ppm.[3]

Lithium concentration in different U.S. coals is shown in Table 24. The U.S. average is obtained from the analysis of 799 samples of different coals (see Reference 2) while the estimated worldwide average is obtained from Bertine and Goldberg.[184]

3. Beryllium (Be)

Old works on beryllium concentrations in coal are summarized in the report by Gibson and Selvig.[185] They quoted measurements done for some Russian and German coals. Zilbermintz et al.[186] reported that a few Russian coals from the Donetz basin contain beryllium. Using the same coal samples that were examined for vanadium and germanium content, beryllium was determined spectrographically by heating the ash (containing 1% added platinum) in a pure carbon arc. Visual estimations were made by comparing certain beryllium lines with platinum lines. Of 604 coal ashes examined, none showed over 0.1% beryllium, only 38 showed over 0.001%, and over half showed none. Nazarenko[187] examined 21 other ashes from Russian coals and reported that 4 samples contained a very small quantity of beryllium.

Stadnichenko et al.[188] studied the geochemistry of beryllium in American coals and the results for maximum and minimum averages of beryllium content of coal beds are given in Table 25. The conclusions of their investigations were

1. The study of 1385 samples of coal from most of the coal-producing regions of the U.S. shows a wide variation in beryllium content. The areal distribution of the sample localities shows beryllium-rich and beryllium-poor regions. The rich and the poor distribution of beryllium in coal beds depends upon the availability of beryllium to the swamps at the time of deposition of the coal. This availability was dependent upon the type of rock being eroded in the surrounding borderlands.
2. There is also a pattern in the distribution of beryllium in coal in a basin. Coal

Table 25
BERYLLIUM IN U.S. COAL

Region	Coal	Range in ash (%)	Range in coal (ppm)	No. of samples
Eastern province	Northern part	0.0041—0.0016	4.1—1.6	104
	Lower Kittanning bed	0.0041—0.0016	4.1—1.6	104
	Middle Kittanning bed	0.014 —0.0016	4.2—1.5	46
Appalachian region	Southern part			
	Alabama	0.015 —0.0011	4.6—0.5	68
	Eastern Kentucky	0.11 —0.001	31.0—0.4	87
	Tennessee	0.0046—0.0005	11.0—0.1	29
	Virginia and W. Virginia	0.0072—0.0007	3.6—0.2	30
Interior province	Eastern region			
	Illinois (all samples)	0.011 —0.0004	6.3—0.7	253
	Bed 5	0.011 —0.0011	3.2—1	90
	Bed 6	0.0044—0.0004	4.0—0.7	74
	Indiana (all samples)	0.017 —0.0032	12.0—1.5	83
	Western Kentucky (all samples)	0.0093—0.0004	9.5—0.5	96
	Bed 9	0.0057—0.0013	4.3—1.7	47
	Western region			
	McAlester Basin	0.003 —0.0001	2.9—0.1	44
	All other	0.011 —0.0003	5.1—0.5	110
Northern Great Plains province	Jurassic and Cretaceous	0.0094—0.0015	5.8—2.7	58
	Paleocene and Eocene	0.013 —0.0001	9.1—0.1	131
Rocky Mountain province	Sweetwater County, Wyo.	0.038 —0.0003	13.0—0.1	17
	All other	0.062 —0.0001	31.0—0.1	174

After Stadnichenko, T., Zubovic, P., and Sheffey, N. B., Beryllium Content of American Coal, Bull. 1084-K, U.S. Geological Survey, Reston, Va., 1961.

sampled near the edge of a basin has a higher beryllium content than that sampled in the center. Coal that was deposited near eroding rocks rich in beryllium or that had access to water from such areas is also high in beryllium.

3. In sink-float experiments, beryllium is consistently associated with the lighter organic-rich fractions. The beryllium content of the sink fractions, particularly those with a high percentage of ash, consistently is below the average for that of the crust of the Earth. The analysis of petrographic constituents of coal shows that beryllium is most often associated with vitrain and least with fusain.
4. The accumulation of beryllium in coal is concluded to be a syngenetic process. The beryllium now present in the coal probably is a result of accumulation by plants and (or) by the adsorption from solution by the organic matter in the coal-forming swamps to form metallo-organic complexes. Although some of the original beryllium may have been lost, there is no possible way to ascertain this. Further studies should be made of coal beds which contain large amounts of beryllium. The economic aspects of the beryllium in coal are not predictable.

Goldschmidt and Peters[189] found beryllium in the ashes of 12 English and German coals. Three samples had 0.1 to 1.0% Be, four had 0.1%, two had 0.01 to 0.1% one had 0.01% and two had 0.001%. Grillot[190] discussed the importance of beryllium and its occurrence in coal. Other data on beryllium in ash are shown in Table 26.

Beryllium in coal was also studied by Albernethy and Gibson[178] and Abernethy and Hattman;[191] analyses were performed on 34 samples from various coalfields of the U.S. along with the float-sink fractions of 3 coals. The prescribed method requires only wet chemical, analytical apparatus of the type owned by most small laboratories.

Table 26
BERYLLIUM IN ASH

Source of coal	Percentage of Be in ash	Ref.
United States:		
Pa., anthracite	0.001—0.009	192
Texas, Colorado, North and South Dakota	0.1—1.0 max.	193
West Virginia	0.0007—0.0108	194
Belgium	0.002—0.0500	195
Germany, Newrode	0.001	196
Germany	0.0013	197
England, Newcastle	0.036	196
England, vitrain	0.005—0.01	141
Nova Scotia	0.0014	198
Portugal, anthracite	Up to 0.01	180
Russia	0.01—0.1 (3 samples)	186
	0.001—0.01 (36 samples)	
	0.001 (255 samples)	
	0 (308 samples)	

Table 27
BERYLLIUM IN U.S. COALS

Region	Coal	Range	Arithmetic mean (ppm)	Geometric mean (ppm)
Appalachian	Bituminous	0.23—25	2.4	1.9
Interior province	Bituminous	0.05—18	2.4	1.9
N Great Plains	Lignite	0.08—14	0.97	0.61
N Great Plains	Subbituminous	0.05—13	0.97	0.61
Rocky Mountain	Bituminous	0.08—6.2	1.2	0.88
All U.S.	Different	0.05—330	2	1.4

After Zubovic, P., Hatch, J. R., and Medlin, J. H., Proc. U.N. Symp. World Coal Prospects, Katowice, 15-23.10, 1979.

Beryllium content on the ashes ranged from 0.0005 to 0.020% in the 34 coals, and analyses on float-sink fractions showed that Be content of the coal increased as the percent ash decreased, indicating that Be is associated with the organic fraction of the coal.

Zubovic et al.[9,173] have also studied beryllium in U.S. coals. Their results are shown in Table 27. Gluskoter[137] has studied beryllium in U.S. coals. His results are shown in Figure 3, where the distribution of beryllium in coals is presented, and in Table 28.

The average concentrations of beryllium in 12 lignite samples from North Dakota was reported to be 0.31 ppm.[122] Hansen et al.[181] have reported beryllium concentration in the ash of high-volatile bituminous coal (8.9% ash) from Utah to be 5.4 ppm.

Beryllium concentration in eight coal samples from Nebraska is found to be in the range 1 to 10 ppm.[3] Its concentration in raw coals from the Illinois Basin was found to be in the range 1.5 to 3.3 ppm with mean value 2.2 ± 0.8 ppm.[182,199] Values up to 15 ppm are reported in some coals from Australia.[132]

Table 29 shows beryllium concentrations in different U.S. coals as reported by Swanson et al.[183] Their findings are reported by U.S. average value (from Reference 2) and worldwide average.[184] The geochemistry of beryllium in coals is described in the work by Hak and Babcan.[200] Beryllium content of coals is also discussed in papers by Danchev et al.,[201] Drever et al.,[202] Jedwab,[195] Odor,[203] Phillips,[204] and others. Losev

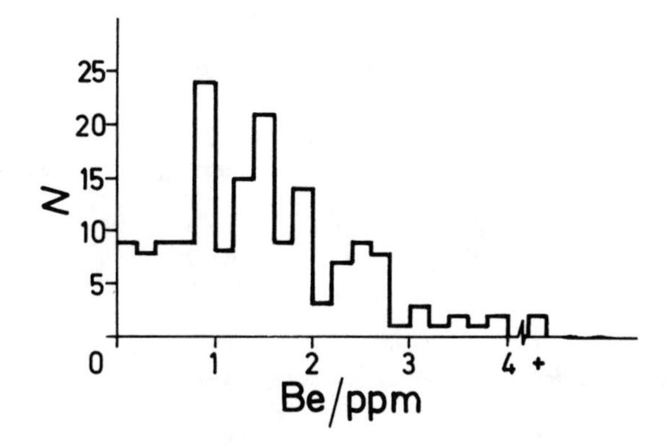

FIGURE 3. Distribution of values for beryllium concentration in U.S. coals. (As reported by Gluskoter.[137])

Table 28
BERYLLIUM IN SOME U.S. COALS

Region	Range	Arithmetic mean (ppm)	Geometric mean (ppm)	Number of samples
Illinois Basin	0.5—4.0	1.7	1.6	113
Eastern U.S.	0.23—2.6	1.3	1.1	23
Western U.S.	0.10—1.4	0.46	0.35	29

After Gluskoter, H. J., Trace Elements in Coal: Occurrence and Distribution, Circ. 499, Illinois State Geological Survey, Urbana, 1977.

Table 29
BERYLLIUM IN U.S. COALS

Coal	Conc (ppm)
Anthracite	1.5
Bituminous	2.0
Subbituminous	0.7
Lignite	2.0
U.S. av	2.0
Worldwide av	3

Data from Swanson, U. E., Medlin, J. H., Hatch, J. R., Coloman, S. L., Wood, G. H., Jr., Woodruff, S. D., and Hildebrand, R. T., Open file Rep. 76-468, U.S. Geological Survey, Reston, Va., 1976.

et al.[205] have discussed distribution of Be in pyrolysis products after treating coals with organic acids.

The origin of Be in coal by studying adsorption of Be on peat and coals is discussed by Eskenazi.[206] Beryllium content of coal and its emission during coal combustion is discussed in many papers, and these aspects will be discussed in detail. Here we shall mention work by Gladney and Owens[207] and Bencko et al.,[208] who studied Be emissions from a coal-fired power plant and its effects on people exposed to Be due to combustion of coal.

Table 30
BORON IN U.S. COALS

Area	Range	Arithmetic mean (ppm)	Geometric mean (ppm)	Number of samples
Western U.S.	16—140	56	48	27
Eastern U.S.	5.0—120	42	28	23
Illinois Basin	12—230	110	98	99

After Gluskoter, H. J., Trace Elements in Coal: Occurrence and Distribution, Circ. 499, Illinois State Geological Survey, Urbana, 1977.

Methods of Be determination in coal are discussed by Bekyarova and Rushev.[209] Spectrochemical determination of Be in coal ash is described in detail in work by Kul'skaya and Vdovenko.[210] Determination of beryllium by AA spectroscopy is described by Owens and Gladney[211] and Gladney.[212] Tamura[213] has described the accuracy in the nondestructive neutron activation analysis of coals for beryllium.

4. Boron (B)

Early studies of boron in coal[185] showed a marked concentration of boron in coal ash. Values up to 1.0% were reported.

Deul and Annell[193] also noted that the boron contents of the coals they examined were high, averaging more than 0.1% boron in the ash which indicates considerable enrichment over that in the crust of the Earth. They suggested that boron is perhaps the only element in coal ash which is directly attributable to the plants from which the coal was formed. Goldschmidt[4] noted some enrichment of boron in coal ash, and Kear and Ross[214] studied the distribution of boron in New Zealand coal. Concentrations of boron in coal reported by several other investigators are given in Table 30.

Roga et al.[215] noted that analyses of Polish coals for boron gave a random distribution that showed no enrichment or deficiency of boron according to geographical location or depth of coal bed. Hutcheon[216] and Kahn and Sen[217] studied boron in the coal and petroleum products used as raw material for graphite.

Abernethy and Gibson[178] have summarized the reported results for both ash and coal. This is shown in Table 31.

Bohor and Gluskoter[218] have discussed boron in illite as an indicator of paleosalinity of Illinois coals. Trace amounts of B locked in the illite lattice reflects the palesalinity of the water in which the Herrin (No. 6) coal of Illinois was deposited. Though 31 samples were too few to be statistically significant, a modified method of boron-in-illite measurement was presented in detail. Additional samples of the Springfield-Harrisburg (No. 5), Danville (No. 7), Indiana Seelyville Coal Member (III), and the Survant Coal Member (IV) were also analyzed. The samples were low-temperature-ashed, sized to less than 1 μm, and leached in successive acids and bases to remove all nonillite minerals. The resulting pure illite was digested and analyzed colorimetrically for B content. Accepted ranges of B concentration indicating levels of paleosalinity were applied to the boron measurements. Areas of the Herrin (No. 6) coal with B values less than 125 ppm, indicating relatively fresh water, coincided with low-sulfur areas of the coal. Highest paleosalinity values (higher B concentration) were in coal furthest from freshwater channel systems, and on the west side of the system, reflecting the seaward direction. Data for boron concentrations in U.S. coals presented by Gluskoter[137] are shown in Figure 4.

Boron is concentrated only in coals of the Illinois Basin. Possibly the presence of boron represents a greater marine influence during and immediately following the time

Table 31
BORON IN COAL ASH

Source in coal	Boron in ash (%)	Ref.
U.S.		
N Great Plains	0.005—0.65	10
West Virginia	0.008—0.096	194
North Dakota	0.21	374
Nova Scotia	0.0052—0.0220	128
Germany, Newrode coal	0.09	196
England, Newcastle coal	0.31	11
England, vitrain	0.02—0.3	141
Spitzbergen	0.1—2.0	364
Portugal, anthracite	0.001—1.0	180
New Zealand	1.51	178
Austria	0.46	178
U.S.		
N Great Plains	116	9
Eastern interior	96	11
Appalachian	25	11
England	2 to 140	178
South Africa	11 to 60	229

After Abernethy, R. F. and Gibson, F. H., Rare Elements in Coal, Rep. BM-IC-8163, U.S. Bureau of Mines, Washington, D.C., 1963.

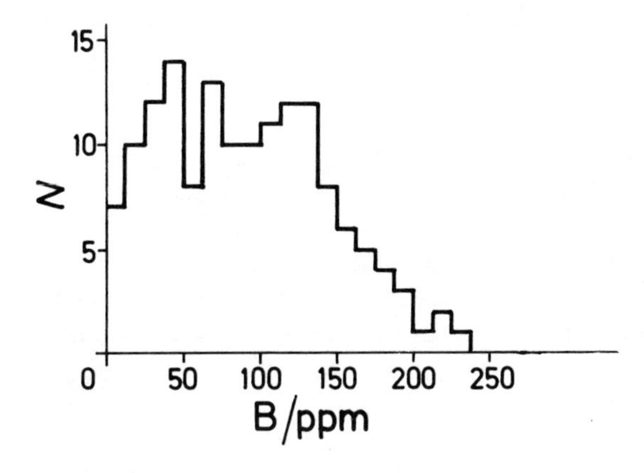

FIGURE 4. Distributions of boron concentration values for U.S. coals. (As reported by Gluskoter.[137])

of the coal swamp in the basin. Since the mean value of B concentrations in Illinois Basin coal is 95 ppm the EF with respect to earth, rocks, and soils is 9.5.

Somerville and Elder[122] have reported the concentration of boron in lignite of North Dakota to be 63 ppm. Hansen et al.[181] have reported boron concentration in the ash of high-volatile bituminous coal (8.9% ash) from Utah to be 920 ppm.

Boron concentration in raw coals from three Illinois basin preparation plants was in the range 53 to 73 ppm, with mean value 63 ± 8 ppm.[182] Burchett[3] has reported values from 10 to 200 ppm for eight coal samples from Nebraska. Duel and Annell[219] (1956) have studied boron concentration in the low-rank coals. They have concluded that boron was highly enriched in the majority of coals, being concentrated from the soil by living plants and therefore not lost during plant tissue decomposition in the early stages of coalification.

Table 32
CARBON IN U.S. COALS

Area	Range (%)	Arithmetic mean (%)	Geometric mean (ppm)
Eastern U.S.	63—80	72	72
Western U.S.	58—74	67	67
Illinois Basin	62—80	70	70

After Gluskoter H. J., Trace Elements in Coal: Occurrence and Distribution, Circ. 499, Illinois State Geological Survey, 1977.

Swanson et al.[183] have reported boron concentrations for U.S. coals to be anthracite, 10 ppm; bituminous coal, 50 ppm; subbituminous coal, 70 ppm; lignite, 100 ppm. The average value of 799 coal samples from the U.S. (Reference 2) is reported to be 50 ppm, which is only slightly lower than the worldwide average value 75 ppm as reported by Bertine and Goldberg.[184] Boron contents of coal is also discussed in the papers by Nazarenko,[187] Rafter,[220] Konieczynki,[221] Kryukova et al.,[222] and others.

The occurrence of boron in South African coals and its behavior during combustion is described in the paper by Kunstmann et al.[223] Allan et al.[224] have described boron content of some Scottish coals and distribution of the boron between various density fractions. Some other aspects of boron compounds and coal are discussed in papers by Nakumara and Kitamura[225] and Lenz and Köster.[226]

Methods of boron determination in coals are described in papers by Millet,[227] Skalska and Held,[228] Kunstmann and Harris,[229] Konieczynski,[230] and others.

5. Carbon (C)

Let us show some example of carbon content of different coals. Table 32 shows data as reported by Gluskoter.[137] Table 33 shows data reported by Zubovic et al.[173] Table 34 shows data reported by Petrakis and Grandy.[174]

Several methods for carbon content determination of coals are developed. New methods are being developed (for example, NMR; see Alemany et al.[176]

H/C ratio is of interest for many uses of coals, especially on generation of oil from coal.[231]

6. Nitrogen (N)

For nitrogen determination, very often the Kjeldahl method is employed, consisting of an attack and digestion of the coal powder with a strongly oxidizing, boiling solution which is actually a mixture of concentrated sulfuric acid, potassium sulfate, and metallic mercury (with some added chromic oxide (CrO_3) in the analysis of coke); this treatment converts the nitrogen to ammonia which is distilled into a measured sulfuric acid solution after making the diluted solution basic with NaOH, and titrated back by standard ammonia solution.

There are many data for nitrogen contents of coals. Let us mention some for the illustration. Gluskoter[137] has reported nitrogen data as shown in Table 35. Table 36 shows geometric mean values for nitrogen content in som U.S. coals as reported by Zubovic et al.[173] Additional information about nitrogen content of coal are presented in Table 37 where data reported by Debelak et al.[171] and Petrakis and Grandy[174] are shown.

The variability of nitrogen as a function of the rank and degree of reduction of the coals is considered by Dobronravov.[232] A mathematical analysis of the results of chem-

Table 33
GEOMETRIC MEAN VALUES FOR CARBON CONTENT OF U.S. COALS

Region	Coal	Fixed carbon (%)	Carbon (%)
Appalachian	Bituminous	53.3	71.1
Interior	Bituminous	46.1	63.0
N Great Plains	Lignite	28.1	38.7
N Great Plains	Subbituminous	33.1	45.5
Rocky Mountain	Subbituminous	35.9	41.4
Rocky Mountain	Bituminous	49.46	58.3

After Zubovic, P., Hatch, J. R., and Medlin, J. H., Proc. U.N. Symp. World Coal Prospects, Katowice, 15-23.10, 1979.

Table 34
CARBON CONTENT OF SOME COALS

Coal	C (wt % daf)
Hagel, N.D.	71.0
Wyodak Gillete, Wyo.	72.2
Deitz, Wyo.	73.9
Kentucky No. 11	76.4
Illinois No. 6	82.0
Lower Dekoven, Ill.	80.6
Pocahontas No. 3, W. Va.	86.4
Lower Kittanning, Pa.	89.1

From Petrakis, L. and Grandy, D. W., *Fuel,* 59, 227, 1980.

Table 35
NITROGEN IN U.S. COALS

Area	Range (%)	Arithmetic mean (%)	Geometric mean (%)	Number of samples
Illinois Basin	0.93—1.8	1.3	1.3	110
Eastern U.S.	0.94—1.8	1.3	1.3	22
Western U.S.	0.59—1.5	1.0	1.0	29

After Gluskoter, H. J., Trace Elements in Coal: Occurrence and Distribution, Circ. 499, Illinois State Geological Survey, Urbana, 1977.

ical and petrographic investigations of the coals of the Kuzbass has been performed, and formulas are given for calculating the nitrogen content on the fusible components and on coal containing definite proportions of them. In the work by Karr et al.[233] the presence of nitrate mineral species in Montana and North Dakota lignites was confirmed through IR and X-ray diffraction analyses of the lignites, their low-temperature ash, and water-derived extracts. IR analysis could not be performed on the whole coal, and nitrates detected in the low-temperature ash could originate partially or entirely in the asher. For the above reasons, finely ground whole coal samples were treated with stepwise water, extractions in which samples held in paper thimbles with glass wool were submerged in water with a temperature of 90°C for 2 hr. The water was

Table 36
NITROGEN CONTENT IN U.S. COALS

Region	Coal	Nitrogen (%)
Appalachian	Bituminous	1.25
Interior	Bituminous	1.19
N Great Plains	Lignite	0.57
N Great Plains	Subbituminous	0.78
Rocky Mountain	Subbituminous	0.96
Rocky Mountain	Bituminous	1.13

After Zubovic, P., Hatch, J. R., and Medlin, J. H., Proc. U.N. Symp. World Coal Prospects, Katowice, 15-23.10, 1979.

Table 37
NITROGEN CONTENT OF SOME COALS[171,174]

Coal	Nitrogen (%)
Hagel, N.D.	1.6
Wyodak Gillette, Wyo.	1.0
Deitz, Wyo.	1.7
Kentucky No. 11	1.2
Kentucky No. 4	1.3
Kentucky No. 9	1.2
Kentucky No. 3	1.2
Illinois No. 6	1.7
Lower Dekoven, Ill.	1.8
Pocahontas No. 3, W. Va.	1.2
Lower Kittanning, Pa.	0.9

filtrated and evaporated, and the dried residue analyzed by IR for nitrates. Five to ten successive leaches of each sample were analyzed. Sodium sulfate, alkaline earth nitrates, calcium sulfate, kaolinite, and Na oxalate were identified in the extracts. X-ray powder diffraction patterns of the original coals and the extracts verified the presence of Ca and Na nitrates as well as quartz, kaolinite, and gypsum. The separation of nitrogen-containing constituents from coal is described by Ito.[234]

Nitrogen in coal can be conveniently determined by neutron activation analysis (NAA). This is discussed in details by Hamrin et al.[235]

7. Oxygen (O)

Oxygen analysis is often related to determination of moisture. Therefore it is important to investigate whether the moisture evolving at 105°C from a coal is all H_2O (88.81% O) or contains significant quantities of CH_4, C_2H_6, N_2, H_2, etc. This can be done by analyzing the moist "as received" coal for total oxygen, expelling the water at a desired temperature, and reanalyzing the same dry coal for oxygen. In order to avoid oxidation this should be done in vacuum or an inert atmosphere of nitrogen or argon.[236]

The procedure used for oxygen determination is described in detail by Volborth et al.[236] In the ultimate coal analysis total hydrogen and carbon are determined by burning the coal placed in a porcelain boat in a combustion tube in an oxygen atmosphere by slowly increasing the temperature to 850°C. The water and carbon dioxide so formed are collected in an absorption tube and weighed. Since these determinations are performed as "air dried", 35°C, (ad) samples but reported usually on "as received" (ar) basis, the result for carbon can be recalculated back to coal on an "as received" basis by multiplying with 100 minus the "air drying loss" (ad-loss), and dividing by a hundred:

$$\%C(ar) = \%C(ad) \times \frac{(100-\text{ad-loss})}{100} \qquad (6)$$

and in the case of hydrogen:

$$\%H(ar) = \%H(ad) \times \frac{(100-\text{ad-loss})}{100} + 0.1119 \times \text{ad-loss} \qquad (7)$$

because only about 1/9 of the air-drying loss is hydrogen.

For oxygen, if calculated "by difference" (diff) on "air dried" basis according to formula:

$$O(\text{diff,ad}) = 100 - \text{ash(ad)} - C(ad) - H(ad) - N(ad) - S(ad) \qquad (8)$$

where the values for ash, carbon, hydrogen, nitrogen, and sulfur reported on "air dried" basis are substracted from 100. The recalculation to "as received" basis is

$$\%O(ar) = \%O(ad) \times \frac{(100-\text{ad-loss})}{100} + 0.8881 \times \text{ad-loss} \qquad (9)$$

In the above calculations the values for hydrogen and oxygen include the hydrogen and the oxygen of the moisture. In case it is desired to report the content of these elements excluding the contribution due to moisture (88.81% 0, 11.19% H), one drops the respective addition terms.

Volborth et al.[237] have reported data on oxygen determinations of six Wyoming coals, and this is shown in Table 38.

Accurate oxygen determination for coal ash and LT-ash (or mineral matter) is important for calculation of data in the ultimate analysis of coal as such. Knowledge is required for recalculation of the data on a dry and dry-ash free basis. The routinely used "oxygen by difference" values are inadequate for accurate work. In order to determine the organic oxygen in coal one also has to correct for oxygen in mineral matter and oxygen in the water removed as moisture. The Parr formula and other methods of empirical estimation are inadequate and may be replaced in some cases by the oxygen determination. The complete data provide a quantitative basis for stoichiometric interpretation of coal analysis.

Fast neutron activation analysis (FNAA) is successfully used for oxygen determination in coals. Volborth et al.[236] present figures for oxygen determinations by FNAA and by difference; this is shown in Table 39.

According to Volborth et al.[237] knowing oxygen in both ashes as well as other elemental data permits a rough estimate of the nature of the mineral matter in coal in terms of whether it is predominantly kaolinitic and clayey, and whether it has much or little sulfide, sulfur, iron, and calcium. It is probable that one could roughly estimate the nature of the mineral matter simply from a knowledge only of the oxygen content in the different types of ash. The sum of the cations, the SO_4, S, Cl, F, etc. in ash is also best estimated by the accurate analysis of oxygen. This value may be as reliable as the value derived from the summation of the results of the chemical silicate

Table 38
OXYGEN IN WYOMING COALS

Sample no. and origin	Oxygen determined: O (%) as rec'd	Oxygen determined: O (%) oven dried 105°C	Oxygen calc: O (%) calc dry	Oxygen calc: O (%) calc as rec'd daf
5-74; K-46566, USBM	41.6 ± 0.1	23.1 ± 0.1	23.8	21.0
Wyodak Strip Mine	42.8 ± 0.1	23.2 ± 0.1	24.1	21.6
22-74; K-46430, USBM	42.1 ± 0.17	25.2 ± 0.1	24.6	23.1
Wyoda, Upper Bench				
6-74; K-46565, USBM	43.6 ± 0.23	22.9 ± 0.2	23.2	21.3
29-74; K-46216, USBM	36.3 ± 0.1	21.1 ± 0.13	22.1	20.5
30-74; K-46217, USBM	35.9 ± 0.2	22.1 ± 0.1	21.9	19.9
31-74; K-46218, USBM	35.9 ± 0.1	21.9 ± 0.1	21.9	20.0

After Volborth, A., Miller, G. E., Garner, C. K., and Jerabek, P. A., *Fuel,* 57, 49, 1978.

Table 39
OXYGEN DETERMINATION BY FAST NEUTRON ACTIVATION ANALYSIS (FNAA)

Oxygen (%)	HVA bituminous coal, WV, PSOC-121: As rec'd	Dry	Daf	Lignite, Texas, PSOC-140: As rec'd	Dry	Daf
By difference	7.75	6.69	7.15	42.58	18.80	21.10
By FNAA as is	9.18 ± 0.04	8.14	7.15	47.34 ± 0.15	26.01	23.95
By FNAA dried	9.17	8.13 ± 0.07	7.14	48.03	27.06 ± 0.13	25.10

After Volborth, A., Miller, G. E., Garner C. K., and Jerabek, P. A., *Fuel,* 57, 49, 1978.

analyses. The material balances for coal and coke can be estimated better if accurate oxygen data on ash and coal are available rather than using oxygen "by difference" as in coal or oxygen based on assumption of strict stoichiometry in ash when Cl, S, and F are not known. Data obtained by Volborth et al.[237,238] indicate that in approximate recalculations of whole coal analyses where the estimated oxygen in ash and the estimated oxygen due to the crystalline water of the mineral matter are used to calculate the "organic" oxygen, one may use a factor of 0.46 ± 0.04 for oxygen in high-temperature ash (HTA), the factor for low-temperature ash (LTA), however, may vary too much for a "universal" factor to be meaningful. It appears from these limited data that the variation is at least 0.40 to 0.54. Further work on a larger number of coal ash matched with proximate and ultimate analyses on the same coal is required in order to explore and interpret fully the stoichiometry of coal ash and coal. Such work may permit considerable simplifications of practical coal analysis and reduce the empirical content of the overall interpretation of coal analysis.

The physical and chemical characteristics of lignites and their upgraded products are greatly influenced by the presence and nature of various oxygen-containing functions. Analytical methods which can be used to determine the number and type of O-functions before and after upgrading are therefore of interest. The analytical procedures that have been previously employed are time consuming and require relatively elaborate apparatus. Furthermore, the reproducibility of the results obtained often leaves much to be desired.[226] Coal samples studied by Lenz and Köster[226] are characterized as shown in Table 40.

Table 40
COAL SAMPLE CHARACTERIZATION[a]

#[b]	Coal rank	C	H	N	O	S	Atomic ratio
1a	Lignite (air contact)	67.2	5.4	0.73	25.7	0.73	$C_{100} H_{96.5} N_{0.93} O_{28.9} S_{0.41}$
1b	Lignite (no air contact)	66.9	5.45	0.73	27.15	0.99	$C_{100} H_{97.5} N_{0.9} O_{30.5} S_{0.30}$
1c	Lignite (air contact, demineralized)	66.6	5.3	0.50	27.0	0.54	$C_{100} H_{95.0} N_{0.6} O_{30.4} S_{0.30}$
2	Lignite (air contact, demineralized)	72.4	4.95	—	22.45	0.43	$C_{100} H_{81.8} O_{23.3} S_{0.20}$
3	High-volatile bituminous coal (air contact)	83.8	5.4	1.61	7.8	1.09	$C_{100} H_{78} N_{1.65} O_{7.0} S_{0.79}$
4	Medium-volatile bituminous coal (air contact)	87.3	4.5	1.56	5.5	1.09	$C_{100} H_{62} N_{1.60} O_{4.7} S_{0.47}$
5	Anthracite (air contact)	91.5	3.8	—	2.65	0.88	$C_{100} H_{99.5} O_{2.17} S_{0.36}$

[a] All values are wt %, normally daf.
[b] Sample number.

After Lenz, U. and Koster, R., *Fuel*, 57, 489, 1978.

Table 41
OXYGEN CONTENT OF SOME COALS

Coal	Rank	0 (wt % daf)
Hagel, N.D.	Lignite	21.8
Wyodak Gillete, Wyo.	Subbituminous C	21.0
Deitz, Wyo.	Subbituminous B	18.4
Kentucky No. 11	HVB	10.0
Illinois No. 6	HVB	9.5
Lower Dekoven, Ill.	HVA	4.8
Pocahontas No. 3, W. Va.	LV	6.8
Lower Kitanning, Pa.	LV	3.9

From Petrakis, L. and Grandy, D. W., *Fuel*, 59, 227, 1980.

Oxygen content decrease with coal rank is also demonstrated by Petrakis and Grandy.[174] Their numbers are shown in Table 41.

Lenz and Köster[226] have determined fraction of hydrogen and oxygen in hydroxil groups by O-diethylborylation and by O-acetylation.

Table 42 shows how the hydroxyl group content (% H_{OH} and % O_{OH} as determined by O-diethylborylation) of various lignite samples increases during demineralization and storage in air. The various lignite samples were found to have 11 to 20% H_{OH} and 36 to 70% O_{OH}. For the bituminous and anthracitic coals investigated, the % H_{OH} was found to decrease with increasing degree of coalification from about 5.0% H_{OH} to 2.2% H_{OH}, whereas the observed % O_{OH} is nearly independent of the degree of coalification. Overestimation of % H_{OH} can occur with bituminous and anthracitic coals when small but not negligible amounts of amino or thio groups are present, as these also react with triethylborane with evolution of ethane.

By the O-diethylborylation of the low-rank, demineralized lignite sample (lc) it was possible to detect more hydroxyl groups than were observed by O-acetylation: 15% vs. 11% H_{OH} and 46% vs. 33%. The corresponding difference of four hydroxyl groups per C_{100}-units is presumably due to incomplete O-acetylation of carboxylic acid groups. For the demineralized lignite sample (2) the values obtained using activated triethylborane are in good agreement with those determined by the O-acetylation method: 21% H_{OH} and 74% O_{OH} are found by both methods.

Table 42
FRACTION OF HYDROGEN AND OXYGEN IN HYDROXYL GROUPS AS DETERMINED BY O-DIETHYLBORYLATION

Sample no.	OH (mmol/g)	H_{OH} (wt %)	H_{OH} (%)	H_{OH}/C_{100}	O_{OH} (wt %)	O_{OH} (%)	O_{OH}/C_{100}
1a	6.74	0.67	12	12	10.8	42	12
1b	6.20	0.62	11	11	9.9	36	11
1c	7.83	0.78	15	14	12.5	46	14
2	9.98	1.00	20	17	16.0	71	17
3	2.75	0.28	5	4	4.4	56	4
4	1.57	0.16	3.5	2	2.5	46	2
5	0.79	0.08	2.2	1	1.3	51	1

Note: Wt % H_{OH} (wt % O_{OH}) means weight of hydrogen (oxygen) in the form of hydroxyl groups as a percentage of the total weight of the coal sample; % H_{OH} (% O_{OH}) means weight of hydrogen (oxygen) in the form of hydroxyl groups as a percentage of the total weight of hydrogen (oxygen) in the coal sample; H_{OH}/C_{100} (O_{OH}/C_{100}) means hydrogen (oxygen) atoms in the form of hydroxyl groups per 100 carbon atoms in the coal sample.

After Lenz, U. and Köster, R., *Fuel,* 57, 489, 1978.

By contrast, the O-acetylation method gives 14% H_{OH} and 48% O_{OH} for the sample (la), whereas the O-diethylborylation method detects only 12% H_{OH} and 42% O_{OH}. This difference of two hydroxyl groups per C_{100}-units is probably due in this case to an *in situ* demineralization during the O-acetylation, whereby inorganic acetates are formed leading to higher values.

Oxygen functional groups in coal are discussed by several authors (see, e.g., Kharitonov and Zamai,[239] Wasilewski and Kobel-Najzarek[240]).

Schylyer and co-workers[241] have shown that it has been possible, using the technique of changed particle activation analysis, to follow the time course of the oxidation of coal exposed to air. The kinetics have been studied and seem to be consistent with a rapid initial uptake of oxygen containing molecules followed by slow diffusion into the surface of the coal particles. In this latter regard a study has been undertaken to study the depth profile of the oxygen into the coal particle surface. The depth of penetration of the activating particle is determined by the incident energy and therefore, by comparison to the appropriate standards, the depth profile may be determined either by varying the incident energy or by varying the particle size. Both approaches have been used and give consistent results. The depth to which a significant amount of oxygen penetrates varies from about 3 μm for very high-rank coals to about 20 μm for low-rank coals. This diffusion depth seems to be related to the porosity of the coals. A model for the low-temperature air oxidation of coal has been developed to explain the results from the above-mentioned experiments.

The curves derived from the kinetic experiments are shown in Figure 5. The three coal types all follow the same general behavior, a rapid uptake of oxygen followed by a more gradual uptake which continues for a considerable amount of time in the case of the low-rank coal. The lower curve in each case is the uptake of oxygen after the sample has been placed in the vacuum oven at 65°C for 3 hr and then irradiated. In all cases this uptake is slower and in the case of anthracite the oxygen to carbon ratio remains nearly constant after about 15 min. In the case of lignite, the irreversible uptake (after heat and pumping) appears to increase for a long period of time. This may be due to the chemical oxidation of the coal itself. Determination of oxygen in coat by NAA is described by Hamrin et al.[235]

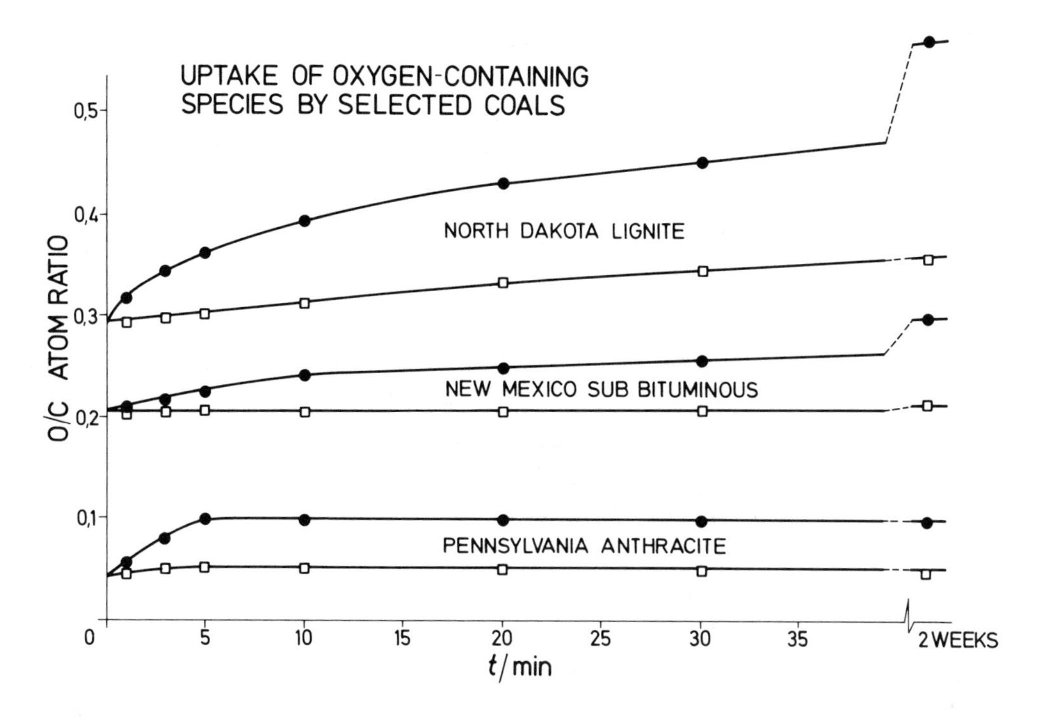

FIGURE 5. Plot of oxygen adsorption vs. time for three ranks of coal. Squares represent the oxygen adsorbed on coal after heating under vacuum for 3 hr at 65°C. Solid circles represent total oxygen. (After Schylyer et al.[241])

8. Fluorine (F)

Coal normally contains trace amounts of fluorine ranging up to about 0.02% percent. These small quantities of fluorine generally are volatilized when coal is burned in boiler furnaces, but they may be a transient factor in some types of corrosion and occasionally in deposit formation, and may contribute to atmospheric pollution.

A standard method is needed for the determination of fluorine in coal, and there is little published information on the occurrence of fluorine in coals. Churchill et al.[242] reported 85 to 295 ppm fluorine in eight coals they analyzed, and Bradford[243] found 40 to 132 ppm in six western coals.

Fluorine determinations were made on raw coal samples by first ashing the sample of coal in a bomb calorimeter and then distilling the product in perchloric acid; coal ash analysis for F was done by fusing the ashed sample with NaOH in a nickel crucible and then distilling the fused product in perchloric acid. Examination of the analyses indicated that the amount of F retained in the ash varies with different burning temperatures. Generally F content decreased as the combustion temperature increased; however, not all of the F was eliminated at the high combustion temperatures. Samples with high F content also generally had high P content, but no correlation was found between low F and low P content.

Churchill et al.[242] also reported fluorine determinations of six coal samples taken in Vancouver, Wash. Four of the coals were from Utah, and the source was not given for two samples; the range of fluorine was 145 to 295 ppm.

Lessing[244] found that fluorine in hot ammoniacal liquor caused corrosion of porcelain tower fillings in a gas works in England. He showed that the coal consisting of a mixture of Midland end West County coals contained small quantities of fluorine that occurred mainly as calcium fluoride. Natural coal dust, containing most of the fusain, was higher in fluorine content than dust-free coal.

Table 43
FLUORINE IN COAL

Coal origin	Specific gravity fraction	% coal in fraction	Fluorine conc (ppm)
Lower Kittanning, Pa.	Head coal	100	137
	FL. 1.30	20.1	30
	FL. 1.30—1.40	30.3	56
	FL. 1.40—1.60	24.0	123
	Sink 1.60	25.6	270
Uncorrelated coal bed, Mahoska County, Iowa	Head coal	100	100
	FL. 1.30	20.5	65
	FL. 1.30—1.40	30.3	85
	FL. 1.40—1.60	22.0	114
	Sink 1.60	27.2	110

After Schultz, H., Hattman, E. A., and Booher, W. B., *Prepr. Pap. Natl. Div. Environ. Chem. Am. Chem. Soc.*, 15, 196, 1975.

Kokubu[245] reported the fluorine content of Japanese coals ranged from 100 to 480 ppm and that the fluorine increased in going from coal to coaly shale.

In the extensive study of fluorine in coal, Crossley[246] discussed the geographical distribution of fluorine in British coals, the occurrence of fluorine in minerals associated with the coal, the relation between the amount of fluorine and phosphorus, the nature of the fluorine compounds, and fluorine in malting coals. The fluorine probably is combined mainly with phosphate as fluoropatite. The fluorine content ranged from 0 to 175 ppm, but generally was less than 80 ppm. Coals responsible for etching of glass in an annealing furnace contained 85 to 130 ppm, and coals causing corrosion in porcelain scrubber fillings had 120 to 140 ppm.

McGowan[247] developed a spectrophotometric method of determining fluorine in coal using oxygen bomb washings. Methods of fluorine determination in coal are discussed by Abernethy and Gibson.[248] Kunstmann et al.[223] investigated fluorine and boron in South African coals, and Durie and Schafer[249] studied fluorine and phosphorus in Australian coals.

Schultz et al.[172] have reported data on fluorine concentration in coal from two locations in the U.S. Their results are shown in Table 43 for different specific gravity fractions. Reduction in fluorine concentration on removal of sink 1.60 fraction was 42% for coal from the Lower Kittanning coalbed, Center County, Pa., and only 6% for coal from an uncorrelated coalbed in Mahoska County, Iowa.

Fluorine concentrations reported by Gluskoter[137] for U.S. coals are shown in Figure 6 and Table 44 according to area of origin. Additional information on fluorine concentration in U.S. coals can be obtained from Table 45 which shows data reported by Zubovic et al.[173]

Fluorine concentrations in North Dakota lignite are reported to be 20.8 ppm (average value for 12 samples). Burchett[3] has reported fluorine concentration in the range 55 to 200 ppm for eight coal samples from Nebraska. Its concentration in raw coals from three Illinois Basin preparation plants was in the range 126 to 243 ppm, with mean value 190 ± 40 ppm.[199]

Loevblad[132] has measured fluorine in coal samples from seven countries and founded in concentration range 17 to 83 ppm, with mean value of 53 ppm.

Fluorine for U.S. coals is also reported by Swanson et al.[183] as follows: anthracite, 61 ppm; bituminous, 77 ppm; subbituminous, 63 ppm; and lignite, 94 ppm.

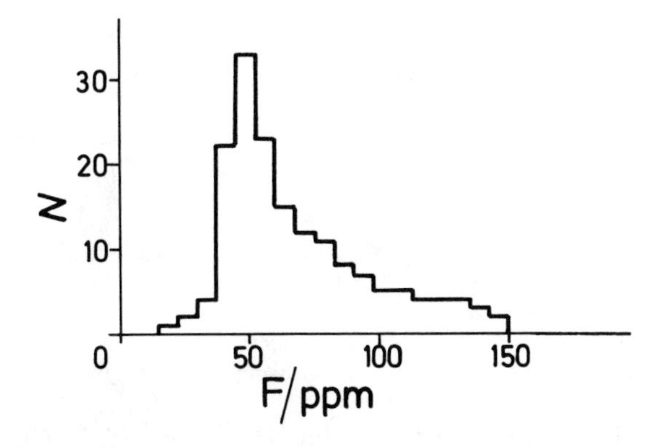

FIGURE 6. Distribution of fluorine concentration values for U.S. coals. (As reported by Gluskoter.[137])

Table 44
FLUORINE IN SOME U.S. COALS

Area	Conc range (ppm)	Arithmetic mean (ppm)	Geometric mean (ppm)	Number of samples
Western U.S.	19—140	62	57	29
Eastern U.S.	50—150	89	84	23
Illinois Basin	29—140	67	63	113

After Gluskoter, H. J., Trace Elements in Coal: Occurrence and Distribution, Circ. 499, Illinois State Geological Survey, Urbana, 1977.

Table 45
FLUORINE IN SOME U.S. COALS

Region	Coal	Conc range (ppm)	Arithmetic mean (ppm)	Geometric mean (ppm)
Applachian	Bituminous	20—1900	93	69
Interior province	Bituminous	18—630	71	59
N Great Plains	Lignite	15—1300	130	22
N Great Plains	Subbituminous	20—1400	79	54
Rocky Mountain	Subbituminous	2—900	100	74
Rocky Mountain	Bituminous	15—940	110	77
All U.S.	Different	0.45—1900	86	64

After Zubovic, P., Hatch, J. R., and Medlin, J. H., Proc. U.N. Symp. World Coal Prospects, Katowice, 15—23. 10, 1979.

9. Sodium (Na)

Interest in sodium concentration determination has increased recently because sodium can be a source of corrosion in a combined-cycle power system, particularly corrosion of the blades of a high-temperature, gas turbine.

From the studies of the material balances it appears that Na is retained by the particulate matter during combustion. A small amount of sodium is present as NaCl, whereas a large quantity is present as Na_2SO_4. Comparison of ash fouling tendencies of high- and low-sodium lignites is discussed by Gronhovd et al.[161] Results for sodium determination in U.S. coals as summarized by Gluskoter[137] are shown in Table 46.

Table 46
SODIUM CONC IN SOME U.S. COALS

Area	Conc range (%)	Arithmetic mean (%)	Geometric mean (%)	Number of samples
Illinois Basin	0—0.2	0.05	0.03	113
Eastern U.S.	0.01—0.08	0.04	0.03	23
Western U.S.	0.01—0.60	0.14	0.06	29

After Gluskoter, H. J., Trace Elements in Coal: Occurrence and Distribution, Circ. 499, Illinois State Geological Survey, Urbana, 1977.

Abernethy et al.[250] have determined Na values (together with P, Cl, and K) for coals from eastern, central, and western portions of the U.S. These values represent tests performed on 62 seams from 11 states (Alabama, Illinois, Indiana, Kentucky, Ohio, Pennsylvania, Tennessee, Utah, Virginia, and West Virginia) plus 1 sample from each of 4 counties in Colorado. Na concentrations were determined using a flame photometer. Data are reported on an ash basis for Na. Mean, high, and low values are reported. No attempt was made to correlate the amounts of P, Cl, Na, and K to other properties of coal. Extensive data (30 pages of tables) for P, Cl, Na, and K content of individual samples with reference to location, bed, mining type, sample type, coal size, coal rank, and mode of handling is included in this report.

Average value for 12 samples of North Dakota lignite was reported to be 2395 ppm by Somerville and Elder.[122]

Burchett[3] has measured sodium concentration in eight coal samples from Nebraska to be in the range 0.021 to 0.098%. Sodium concentration in raw coals from the Illinois Basin was reported to be within the range 0.04 to 0.18% with mean value 0.10 ± 0.06%.[199]

Seven countries included in the study by Loevblad[132] had sodium in coal within the concentration range 0.006 to 0.182%, with the mean value of 760 ppm.

Sondreal et al.[251] have studied lignite ash. The authors have calculated that sodium, which is involved in boiler fouling, increased in concentration with lower elevation of the coal bed. Chlorine, which in Europe is a predictor of Na content, was not found to be a predictor in the coals from nine mines in North Dakota end Montana.

Swanson et al.[183] have reported highest Na concentration in lignite (0.21%), followed by subbituminous coal (0.10%). Anthracite and bituminous coal were reported to have similar concentrations, 0.05 and 0.09% respectively.

Methods for the detection of sodium in coal have been discussed by several authors. For example, Anderson and Beatty[252] have described spectrographic determination of sodium in coal ash. Gluskoter and Ruch[253] have described the use of NAA, while Muter and Cockrell[254] have described analysis of sodium in coal ash by AA spectrometry.

10. Magnesium (Mg)

Table 47 shows data for magnesium concentration as reported by Gluskoter.[137]

Magnesium concentration in North Dakota lignite is reported by Somerville and Elder[122] to be 5039 ppm.

Burchett[3] has reported magnesium concentration in eight coal samples from Nebraska to be in the range 0.047 to 0.574%. Raw coals from three Illinois Basin preparation plants had somewhat lower values; ranging from 0.08 to 0.27%, with mean value 0.14 ± 0.07%.[182,199]

Cooley and Ellman[255] studied lignite and subbituminous coals from eastern Montana. They found that Mg is in higher concentrations in lignite samples than in subbituminous coal samples. Mg content generally decreased in the samples of subbituminous coal in a westerly direction.

Table 47
MAGNESIUM CONC. IN U.S. COALS

Area	Conc range (%)	Arithmetic mean (%)	Geometric mean (%)
Western U.S.	0.03—0.39	0.14	0.12
Eastern U.S.	0.02—0.15	0.06	0.05
Illinois Basin	0.01—0.17	0.05	0.05

After Gluskoter, H. J., Trace Elements in Coal: Occurrence and Distribution, Circ. 499, Illinois State Geological Survey, Urbana, 1977.

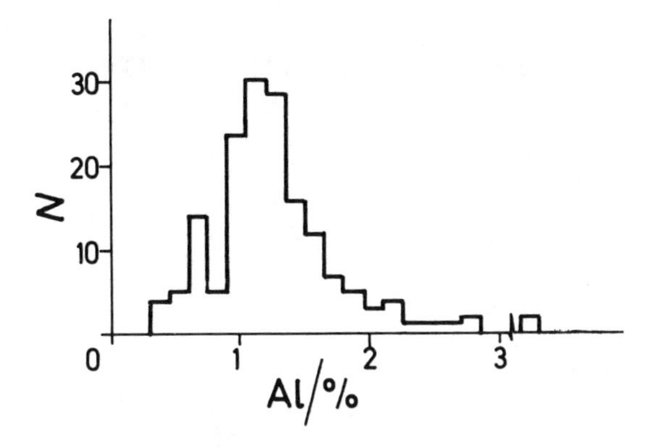

FIGURE 7. Distribution of aluminium concentration values for U.S. coals. (After Gluskoter.[137])

Magnesium concentration in U.S. coals (as reported by Swanson et al.[183]) is highest for lignites (0.31%), followed by subbituminous coal (0.18%) and bituminous coal (0.08%).

Methods of magnesium determination in coal and ashes were discussed by several research groups (for illustration see Muter and Cockrell,[254] Savranskaya and Khanina[256]).

11. Aluminum (Al)

Data reported by Gluskoter[137] is shown in Figure 7 and Table 48.

Aluminum in North Dakota lignite was found to be present in concentration of 63400 ppm.[122]

Aluminum concentration in coal samples from Nebraska is found to be within the range 0.4 to 13%.[3] Raw coals from the Illinois Basin had mean aluminum concentration 3.1 ± 0.6%,[199] while the range was 2.29 to 4.00%.

Cooley and Ellman[255] in the study of coals from eastern Montana found that Al content was higher in subbituminous coal samples than in lignite.

Table 49 shows Al concentrations in U.S. coals, U.S. average as reported in Reference 2, and worldwide average based on paper by Bertine and Goldberg.[184] Aluminum in coal and its ashes is also discussed by Kranz et al.,[257] Panin and Glushnev,[258,259] and others. Aluminum presence in coal mine wastes is discussed by Berg.[260] Methods of aluminum determination are discussed by Hitchen and Zechanowitsch[261] and Navalikhin et al.[262]

12. Silicon (Si)

Figure 8 and Table 50 show data reported by Gluskoter.[137] Silicon concentration in 12 samples of North Dakota lignite is reported to be 11,000 ppm.[122]

Table 48
ALUMINUM IN U.S. COALS

Area	Conc range (%)	Arithmetic mean (%)	Geometric mean (%)
Illinois Basin	0.43—3.0	1.2	1.2
Eastern U.S.	1.1—3.1	1.7	1.6
Western U.S.	0.31—2.2	1.0	0.86

After Gluskoter, H. J., Trace Elements in Coal: Occurrence and Distribution, Circ. 499, Illinois State Geological Survey, Urbana, 1977.

Table 49
ALUMINUM CONCENTRATION IN U.S. COALS

Coal	Conc (%)
Anthracite	2.0
Bituminous	1.4
Subbituminous	1.0
Lignite	1.6
U.S. av	1.4
Worldwide av	1.0

From Swanson, et al., Open File Rep. 76—468, U.S. Geological Survey, Reston, Va., 1976.

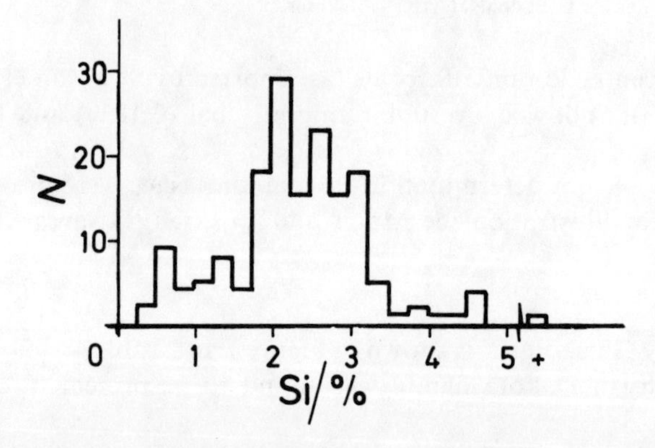

FIGURE 8. Distribution of silicon concentration values for U.S. coals. (As reported by Gluskoter.[137])

Silicon concentration in coal samples from Nebraska is found to be within range 0.6 to 17%.[3] The concentration range for Illinois coals is reported to be 4 to 8% with the mean value around 6%.[199] Silicon is found[255] to occur in greater concentrations in subbituminous samples than samples of lignite and was found to be extremely variable within the samples of coals from eastern Montana.

Silicon concentration in U.S. lignite is higher (4.9%) then in bituminous and subbituminous coal (2.0 to 2.7%).[183]

Silicon distribution in coal ash is discussed by Panin and Shpirt.[259] Schultz et al.[263] have discussed silicon minerals occurring in a coal dust.

Several methods for determining silicon contents in coals are described in literature.

Table 50
SILICA IN U.S. COALS

Area	Conc range (%)	Arithmetic mean (%)	Geometric mean (%)
Western U.S.	0.38—4.7	1.7	1.3
Eastern U.S.	1.0—6.3	2.8	2.6
Illinois Basin	0.58—4.7	2.4	2.3

After Gluskoter, H. J., Trace Elements in Coal: Occurrence and Distribution, Circ. 499, Illinois State Geological Survey, Urbana, 1977.

Table 51
PHOSPHORUS IN SOME U.S. COALS

State	County	Bed	Phosphorus, % of coal as rec'd
Alaska	Low Matanuska District	Upper Shaw	0.143
Alabama	Tuscaloosa	Milldale	0.034
Illinois	Christian	—	0.016
Illinois	Williamson	No. 6	0.002—0.007
Kentucky	Harlan	High Splint	0.002
Maryland	Allegany	Big Vein	0.017
North Dakota	Mercer	—	0.005
Ohio	Belmont	Pittsburgh No. 8	0.006
Ohio	Jefferson	Pittsburgh No. 8	0.009
Ohio	Meigs	Redstone	0.009
Pennsylvania	Allegheny	Pittsburgh	0.006—0.010
Pennsylvania	Cambria	Lower Kittanning	0.003
Pennsylvania	Clearfield	Lower Kittanning	0.005
Pennsylvania	Fayette	Pittsburgh	0.012—0.020
Pennsylvania	Somerset	Lower Kittanning	0.012—0.122
Pennsylvania	Somerset	Upper Kittanning	0.004
Pennsylvania	Westmoreland	Pittsburgh	0.007—0.018
Texas	Milam	—	0.004
Utah	Carbon	—	0.003—0.015
Washington	Pierce	Wilkeson	0.052
West Virginia	—	New River coal	0.004
West Virginia	Raleigh	Beckley	0.006
West Virginia	—	Pocahontas coal	0.002

After Gibson, F. H. and Selvig, W. A., Rare and Uncommon Chemical Elements in Coal, Tech. Pap. 669, U.S. Bureau of Mines, Washington, D.C., 1944.

For example Bernas[264] has used AA spectrometry, while Navalikhin et al.[262] have applied FNAA.

13. Phosphorus (P)

The first analysis of coal for phosphorus was done by Campbell.[265] Interest in phosphorus concentration in coal has increased when coals low in phosphorus were required in making metallurgical coke. The first review article on this subject was by Gibson and Selvig[185] in which they cover early works and present the phosphorus content of a number of coals. The coals analyzed were all U.S. coals and phosphorus content showed wide variation (0.002 to 0.143%) as seen in Table 51.

Another review of phosphorus occurrence in coal is published by Abernethy and

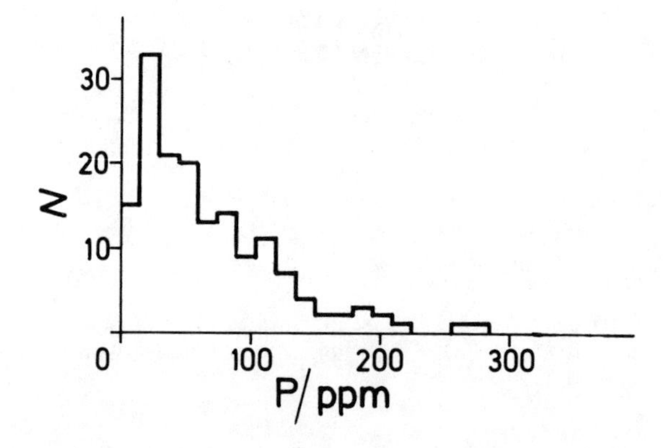

FIGURE 9. Distribution of phosphorus in U.S. coals. (As reported by Gluskoter.[137])

Gibson.[178] They also cover older works which include Geer et al.,[266] who investigated the manner in which phosphorus occurs in coking coals of Washington and found that the phosphorus content of coal from two beds could be reduced greatly by coal-washing methods. Coal from the west No. 3 bed at Wilkeson, Pierce County, containing 32.6% ash and 0.132% phosphorus could be washed to produce a product containing 12.1% ash and 0.043% phosphorus. This result was accomplished with a full-sized coal-washing table, a refuse product containing 59.3% ash, and 0.358 phosphorus being removed. Coal from the east No. 2 bed at Wilkeson, containing 22.6% ash and 0.206% phosphorus, was difficult to wash because it contained a large amount of intermediate-density material. By discarding a middling as well as a refuse product, a washed coal was produced that contained 12.0% ash and 0.068% phosphorus. The phosphorus in coal from five other beds examined could not be reduced appreciably by washing, because it was associated with the clean coal rather than with impurities.

Two phosphorus minerals, evansite ($3\ Al_2O_3 \times P_2O_5 \times 18\ H_2O$) and wavellite ($3\ Al_2O_3 \times 2\ P_2O_5 \times 13\ H_2O$) were identified in shale from the middle parting of the Roslyn bed by a petrographic examination. About 12% of the total phosphorus in the shale occurred in the form of these minerals, but the form in which the remaining 88% of the phosphorus occurred could not be determined petrographically. Selvig and Seaman[267] investigated the distribution of ash-forming mineral matter in coal from the Pittsburgh bed, Fayette County, Pa. They reported that the top 10 in. of coal contained relatively large amounts of phosphorus compared to the other benches of the bed. A recalculation of their data to the percentage of phosphorus in the coal shows that bed samples from four locations in a mine contained 0.012 to 0.020% phosphorus. In comparison, four samples of the top 10 in. of coal contained 0.024 to 0.092% phosphorus.

Abernethy et al.[250] determined phosphorus content (together with Cl, Na, and K) in coals from eastern, central, and western portions of the U.S. These values represent tests performed on 62 seams from 11 states (Alabama, Illinois, Indiana, Kentucky, Ohio, Pennsylvania, Tennessee, Utah, Virginia, and West Virginia) plus 1 sample from each of 4 countries in Colorado. Phosphorus determinations were done by colorimetric methods. Data are reported on a dry coal basis; mean, high, and low values are reported. No attempt was made to correlate the amounts of P, Cl, Na, and K to other properties of coal. Extensive data for P, Cl, Na, and K content of individual samples with reference to location, bed, mining type, sample type, coal size, coal rank, and mode of handling is included in this report.

Data reported by Gluskoter[137] are shown in Figure 9 and Table 52. Phosphorus concentration in North Dakota lignite is reported to be 131 ppm (average of 12 samples) by Somerville and Elder.[122]

Table 52
PHOSPHORUS IN U.S. COALS

Area	Conc range (ppm)	Arithmetic mean (ppm)	Geometric mean (ppm)
Western U.S.	10—510	130	82
Eastern U.S.	15—1500	150	81
Illinois Basin	10—340	64	45

After Gluskoter, H. J., Trace Elements in Coal: Occurrence and Distribution, Circ. 499, Illinois State Geological Survey, Urbana, 1977.

Burchett[3] has reported phosphorus concentration in coal from Nebraska to be in the range 0.1 to 3100 ppm. Coal from Illinois basin had much narrower concentration range: 320 to 680 ppm, with mean value 480 ± 150 ppm.[199] Durie and Schafer[249] have investigated phosphorus concentration in Australian coals. Estimated worldwide average for phosphorus in coal is 500 ppm.[184]

14. Sulfur (S)

Because of its widespread environmental significance, sulfur is probably the most widely publicized and one of the most studied elements associated with coal. It may occur in three forms: in organic combination with the coal material, as the mineral pyrite or marcasite, or as sulfate.[268] At sulfur content above 0.6%, pyritic and organic forms are often roughly equal, whereas below 0.6% about 70 to 100% is in the organic form.[269] Although sulfur is usually encountered in the three forms mentioned above, elemental sulfur concentrations as high as 15% have been found by some workers.

The Interior region coals and most Appalachian coals were formed under the influence of high-sulfate ion concentrations in saline water. As a result, much pyritic sulfur is included in these coals. Coals of other provinces formed under freshwater conditions contain lower sulfur concentrations. Large deposits of low-sulfur coals are located in the West, whereas ample reserves of medium- to high-sulfur bituminous coals are located in Illinois, Indiana, Missouri, Kansas, and West Kentucky[270] (see also Ensminger[271]).

The coal-forming process is a geochemical continuum that includes: peat- lignite-bituminous coal- anthracite coal. Contemporary coal progenitors can be found in peat-forming systems of the Okefenokee Swamp, Georgia and the Everglades Swamp, Florida. Each peat-forming system is a dynamic community composed of plants, surface litter, water, minerals, microorganisms, and peat; in the case of the Okefenokee, its present vegetation is similar to that which gave rise to the huge Braunköhle accumulations in Germany and some of the lignites in the Dakotas.[272] Virtually all the sulfur in some coals can be accounted for in the peat-forming stage of coal formation. The forms of sulfur that have been found in peat include pyrite, sulfate, hydrogen sulfide, elemental sulfur, carbon bonded (C−S) sulfur, and ester sulfate (C−O−S). Organic sulfur is the major variety of sulfur in virtually all peats studied: it represents over 60% of the total sulfur. One origin for the organic sulfur in peat is the sulfur-containing amino acids such as methionine, cystine, and cysteine that are contributed to the sediment by decaying plants and associated microflora and microfauna, but less than 40% of the total organic sulfur can be accounted for as sulfur associated with amino acids. As a function of depth, fewer sulfur-containing amino acids were observed in peat from a marine environment, even though total organic sulfur increased substantially with depth.

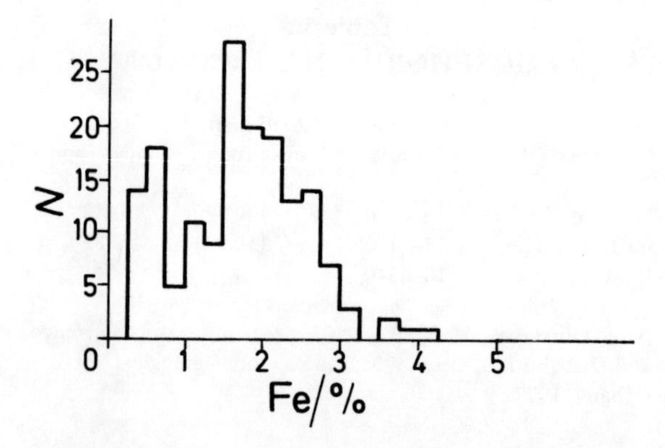

FIGURE 10. Distribution of iron concentration values for U.S. coals. (As reported by Gluskoter.[137])

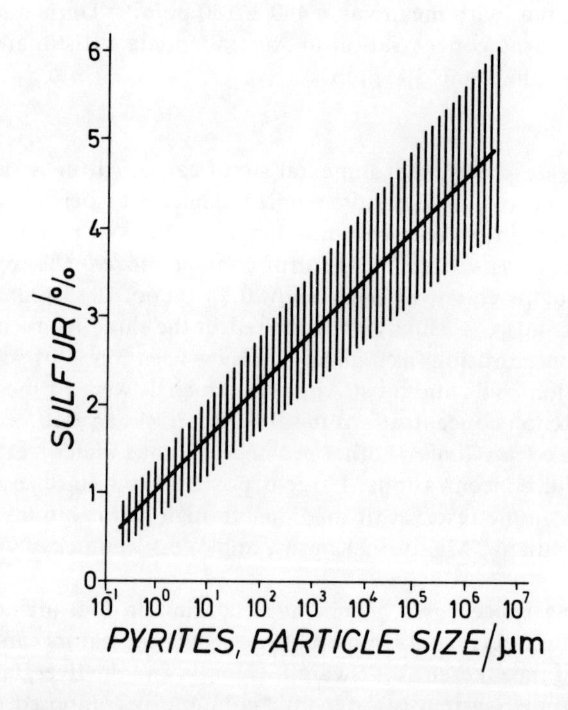

FIGURE 11. Typical distribution of pyrite in coal. (After Meyers.[273])

Distribution of total sulfur, sulfate sulfur, organic sulfur, and pyritic sulfur for 150 samples from the western and eastern U.S. as well as from the Illinois Basin is shown in Figure 10.[137] Figure 11 shows a typical distribution of pyrite in coal.[273]

According to Zubovic et al.[173] the forms-of-sulfur concentration in the coals from the different areas show an interesting relationship. In the high-sulfur coals, pyritic sulfur is the dominant form. This is evident from Table 53, in which sulfur forms in selected bituminous coals is shown.

From Table 53 it can be seen that the total sulfur content of these coal samples varies from 0.38% to a high of 5.32%. This is essentially the range of sulfur content which is normally found among coal samples on either a worldwide or regional basis. The purite sulfur content of these selected coals varies from a low of 0.09% to a high

Table 53
SULFUR FORMS IN SELECTED BITUMINOUS COALS

Region and country	Location or mine	Sulfur (%) Total	Pyritic[a]	Organic	Ratio pyritic to organic sulfur
Asia					
U.S.S.R.	Shakhtersky	0.38	0.09	0.29	0.031
China (Mainland)	Taitung	1.19	0.87	0.32	2.7
India	Tipong	3.63	1.59	2.04	0.78
Japan	Miike	2.61	0.81	1.80	0.45
Malaysia	Sarawak	5.32	3.97	1.35	2.9
North America	W. Virginia	1.20	0.27	0.93	0.29
U.S.	Eagle No. 2	4.29	2.68	1.61	1.7
Canada	Fernie	0.60	0.03	0.57	0.053
Europe					
Germany	—	1.78	0.92	0.76	1.2
United Kingdom	Derbyshire	2.61	1.55	0.87	1.8
Poland	—	0.81	0.30	0.51	0.59
Africa					
S. Africa	Transvaal	1.39	0.59	0.70	0.84
Australia	Lower Newcastle	0.94	0.15	0.79	0.19
South America					
Brazil	Santa Caterina	1.32	0.80	0.53	1.5

[a] Pyrite + sulfate reported as pyrite.

After Meyers, R. A., *Coal Desulfurization,* Marcel Dekker, New York, 1977.

of 3.97%, while the organic sulfur content varies from a low of 0.29% to a high of 2.04%. Generally speaking, organic sulfur levels much greater than 2% or much less than 0.3% are almost never encountered, and pyritic sulfur levels greater than 4% are also uncommon. However, the pyrite content of a few coals can approach zero when there is both little inherent pyrite and when careful mining operations (such as manual labor) prevent the mining of pyrite-containing formations adjacent to the coal seam. The ratio of pyritic to organic sulfur can vary over 2 to 3 orders of magnitude.[273]

The sulfur content and sulfur forms distribution of U.S. coals have been more extensively reported than those of other countries. Still, no set of data is available which fully and statistically describes both the sulfur content and sulfur distribution for U.S. coals. However, let us mention work by Hamersma et al.[747] Their results for the distribution of sulfur forms (dry moisture-free basis) in run-of-mine U.S. coals is shown in Figure 12.

In the low-sulfur coals the organic sulfur content is generally less than 0.5% whereas in the high-sulfur coals the mean concentration is more than 1%. This would appear to indicate that the organic sulfur content of coals, which was derived from the original plant tissues, constitutes less than 0.5% of the coal. As additional sulfur becomes available, probably from bacteriological processes which produce H_2S, to form pyrite, some of the H_2S reacted with the organic matter and produced the higher organic sulfur contents of pyritic coals.

In the standard procedure for determining the forms of sulfur in coals, total sulfur is determined by the Eschka method, pyrite is determined from the concentration of iron in a nitric acid extract (corrected for HCl-soluble iron), sulfate is determined gravimetrically in an HCl extract, and organic sulfur is calculated by difference. The

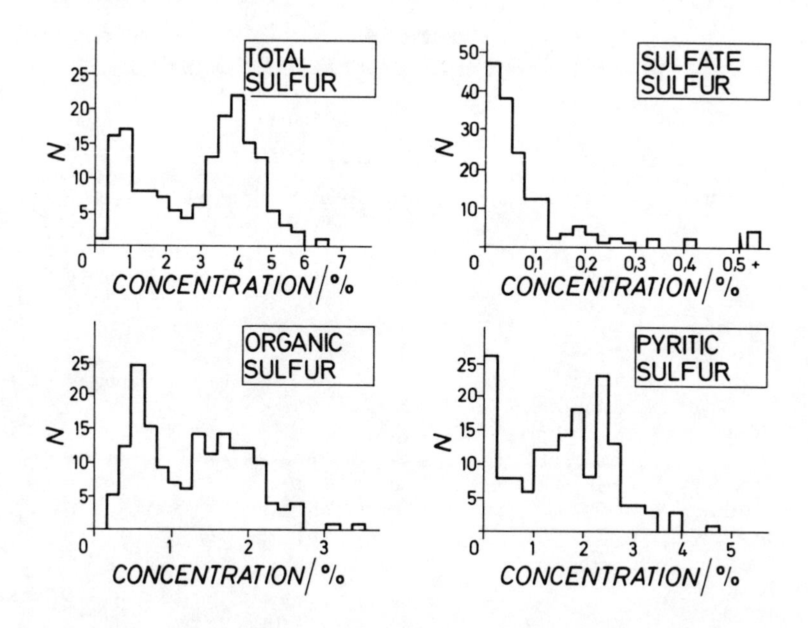

FIGURE 12. Distribution of sulfur forms in run-of-mine U.S. coals. (As reported by Hamersma et al.[748])

Eschka method consists of a fusion of the coal powder with a mixture of ignited magnesium oxide (2 parts) and sodium carbonate (1 part) in a muffle furnace at 800°C, digestion with hot water, filtration, and precipitation of the sulfate ion by barium. Some pyrite can still be detected petrographically in the nitric-acid-insoluble residue from bituminous coals of relatively high rank, and the amount was estimated to be equivalent to 0.1 to 0.2% pyritic S in dry coal. Very small grains of pyrite embedded in organic matter appeared to be inaccessible to the nitric acid, thus leading to analytical error. It has been suggested that the problem can be overcome by performing the nitric acid extraction on −300 mesh coal instead of the usual −72 mesh.[269]

Let us mention some of the reports on the study of sulfur in coal. Casagrande et al.[274] have described an investigation into the origin of the organic sulfur in the coal-forming sequence. Because of the chemical heterogeneity and associated chemical complexity of working with either bituminous or anthracite coals, the earliest stage of coal formation, that is the peat-forming stage, was studied. Modern progenitors of coal from the Okefenokee Swamp were studied, as these peat-forming systems closely approximate ancient systems that eventually gave rise to huge coal deposits. While amino acid sulfur from source plants is an important progenitor of organic sulfur in peat, it was found that H_2S can react with the organic matter in peat to produce organic sulfur, a source of organic sulfur in coal that has not been previously discussed.

Mukherjee and Chowdhury[275] have analyzed sulfur forms in the coals from India; their results are shown in Table 54.

Boateng and Phillips[276] have investigated coal surfaces. Examination of cut surfaces of several U.S. and Canadian bituminous coals with the optical microscope and a SEM equipped with an energy-dispersive-X-ray-analyzer revealed a heterogeneous distribution of Fe and S. Uniformly distributed S content was attributed to organic S in the coal matrix; fluctuations in S content appeared to be due to inorganic S, mainly pyrite. High S coals were high in pyritic S.

Gluskoter and Simon[277] have determined the occurrence and distribution of S forms in selected Illinois coals (see also Gluskoter and Hopkins,[278] Gluskoter,[279,280]). All data

Table 54
SULFUR DISTRIBUTION IN COALS FROM INDIA

Sample	Ash (%)	Volatile matter (%)	Fixed carbon (%)	Total sulfur (%)	Sulfur distribution (% of total)		
					Organic	Pyritic	Sulfate
Overall	4.4	41.5	52.0	4.46	83.8	11.3	4.9
Gravity fractions							
1.30 (float)	1.7	42.7	53.7	4.06	88.4	3.2	8.4
1.30—1.50	9.5	39.6	49.2	6.23	70.5	13.5	16.0
1.50—1.80	27.7	32.5	37.3	7.37	40.0	37.6	22.4
1.80 (sink)	64.0	19.9	14.3	5.10	16.1	58.4	25.5

After Mukherjee, D. K. and Chowdhury, P. B., *Fuel*, 55, 4, 1976.

were obtained from microscopic chemical and X-ray diffraction analyses of face-channel coal samples. Sulfate S in Illinois coals was contained primarily in gypsum with the following sulfate minerals also being identified: rozenite, melanterite, coquimbite, roemerite, and jarosite. Pyrite was the dominant sulfide mineral occurring in Illinois coal. The coals studied demonstrated significant positive correlation between organic and pyritic S. Pyritic S concentrations were generally highest in fusain. Organic S content in fusain was always less than 1%, and vitrain and clarain had higher organic S to pyritic S ratios than did fusain. Organic S concentration in vitrain was generally lower than that in clarain of the same coal. Organic S exhibited more uniform vertical distribution within an individual seam than pyritic S, and large vertical variations in pyritic S were commonly observed. Total S showed a tendency to be concentrated near the upper and lower portions of the seam.

Gluskoter[280] has discussed minerals constituting inorganic S in coal in textbook fashion, employing scanning-electron photomicrographs of sulfur-bearing coal minerals. Discussion included Fe sulfides (pyrite and marcasite), rarc sulfides (galena, chalcopyrite, arsenopyrite, and sphalerite), and sulfates (gypsum, barite, and ferrous and ferric sulfates). Limited attention is given specific gravity S removal techniques.

Sulfur in lignite, its forms and transformations on thermal treatment, are discussed by Fowkes and Hoeppner.[281] Their study examines the changes in S inducted by the gasification process and determines the distribution of S forms for several North Dakota lignites and gasification chars. All samples obtained from the testing procedures were analyzed for sulfate, pyritic, and nonpyritic sulfide S by an appropriate acid-digestion technique. Analysis of the lignite coal samples found no sulfide S present; however, the pyritic and sulfate sulfur that was present was readily reduced to sulfide S upon heat treatment. During normal ashing of lignite and lignitic gasification char, a major percentage of the S retained in the ash was in the form of sulfate S. It was noted that little or no effect on the amount of S retained in the ash occurred during normal ashing temperatures; however, above 700°C, S evolution increased while a concurrent decrease in the total amount of ash occurred.

Hidalgo[282] has determined clay-sulfur relationship in the Upper Freeport coal of West Virginia. Mineralogical analysis was carried out utilizing the combined techniques of infrared and emission spectroscopy, and X-ray diffraction. Excluding clay partings, intermediate clay content correlated with higher sulfur values. Neither illite nor kaolinite were preferentially present with either high or low sulfur, although there was an apparent increase in illite and a corresponding decrease in kaolinite in samples having greater than 1% sulfur. Pyrite formation was more dependent upon sulfide availability than on iron availability. Much more extensive sampling of several coal seams is needed to further clarify clay-sulfur relationships in coal.

Casagrande and Seifert[283] have studied the origin of sulfur in coal. Organic S in selected peats was characterized and was determined to be a significant factor in influencing the quantity and types of S to be found in coal. Peats were sampled from freshwater and marine environments in Georgia and Florida. Organic S content was determined by wet chemical procedures. "Ester sulfate" containing C–O–S linkage was the predominant organic S form in the peats studied, and total S of the marine peats was higher than that of the freshwater peats. Ester sulfate concentration in the marine peat was a function of sediment depth increasing with depth. Total S and ester sulfate content were closely related as a combined function of depth. High ester sulfate concentrations in peat were significant because once incorporated in the platforming stage, they could then remain throughout the successive stages of coalification and thus contribute to the total S content of the coal.

In the work of Gomez et al.,[493] distribution of sulfur and ash in part of the Pittsburgh seam was studied. Within a small area of southwestern Pennsylvania the areal distributions for ash, total S, and S forms were mapped for the top 12 in., the middle section, the bottom 8 in., and the total thickness of the Pittsburgh seam to facilitate understanding of the inherent vertical and longitudinal variation in S and ash, which are important factors in the mining and utilization of this seam. The information used in this study is that which is typically available in the exploration phase of mine planning. The 108 drill core samples were sectioned into the top 12 in., the bottom 8 in., and the remaining middle portion, and each portion was analyzed for ash, total S, pyritic S, and organic S. These values were contour mapped on the grid pattern of core locations for each section and the total seam thickness. Also placed on the maps were locations of "want" or sandstone "cut-out" areas revealed through coring or mining. High-low ash or S values in the top layer did not correspond with high-low ash or S values in the bottom layer, suggesting that both ash and S in this seam were syngenetic and that environmental conditions changed in the time interval between the deposition of the two layers. Overall, preswamp topography controlled many aspects of the seam such as thickness, "want" area location, S and ash concentrations in the seam as a whole, and the presence or absence of the higher ash and S top and bottom layers of the seam. Preswamp paleotopography and postdepositional erosion influenced the amount of ash and S in the top 12 in. and the bottom 8 in., whether or not these layers are present in the seam today. This is important in mine planning because these are the layers not only highest in S and ash, but also the most variable in content and presence.

Nature and distribution of pyrite in Iowa coals have been studied by Greer.[284] Some of the physical characteristics of pyrite that occur in the high-sulfur reserves of Iowa coals are briefly discussed in this paper. The general characteristics of pyrite that were noted were as follows: it could occur in concentrated form such as nodules (up to 150 mm thick), as narrow seams that could thin out very quickly (in less than 1 mm), or in a finely disseminated form; the size of individual pyrite crystals can vary from 1 to 40 μm in diameter with a majority having a diameter of 1 to 2 μm; spherical assemblages of pyrite, called framboids, can also occur; the average size of these spherical crystals groups is approximately 10 and 20 μm in diameter with some varying from a few to several hundred μm; a majority of the pyrite concentration appears in the dull bands of the coal rather than in the vitrinite. In addition, it was noted that the organic S (since it was not chemically determined) could be shown to be mostly micron size pyrite and small framboids rather than true organically bound S.

Pyritic, sulfate, and organic S content of 31 West Virginia coals and their associated floor and roof rocks were determined in the paper by Headlee and Hoskins:[748] in addition, the mode of occurrence and distribution of inorganic S forms in the coals were explained. Analyses were made utilizing wet chemical techniques and included the Eschka and Powell and Parr methods. Inorganic S was found to occur in both sulfate and sulfide forms. Calcium, iron, and barium sulfates were identified and pyrite was the dominant sulfide mineral in the coals. Sulfate S generally comprised less than 0.1% of the coal. Roof and floor clays had the highest sulfate content (always greater than 0.1%). Shale contained more Ba and Sr than the coals. Pyrite distribution was random and independent of the other mineral matter in the coal. While southern West Virginia coals contained almost no pyrite, there was a steady increase in pyritic S in progressively northern and northwestern West Virginia coals.

Detailed studies of distribution and forms of sulfur in a high-volatile Pittsburgh seam coal are reported by Gray et al.[285] Inorganic and organic sulfur were related to coal entities (macerals) in various size and specific gravity fractions of a sample of the Pittsburgh seam coal to aid in S reduction in coke making. A 3½ ton raw coal sample was crushed to −3 in. top size then screened to 7 fractions. Each size fraction, except the −200 mesh coal, was fractionated by heavy liquids at 13 specific gravities ranging from 1.280 to 1.700. Ash, volatile matter, S forms, and petrographic analyses were performed on each of the 88 resultant samples. Eight size and specific gravity fractions were carbonized in a 30-lb test oven. Relationships between S forms and specific coal entities were examined through statistical analysis to establish the degree of correlation. Macroscopic pyrite (larger than 100 μm) accounted for 84% by weight of the pyrite, and microscopic pyrite the remaining 16%. Macroscopic pyrite forms a coating on cleat faces, fills desiccation cracks, and occurs in fusain layers or as lens-shaped masses. This large pyrite constituted little or no coal cleaning problem. Microscopic pyrite (generally less than 80 μm in size), was associated with all entities except micrinoids. Pyrite in vitrinoids varied greatly in size, degree of dissemination and form. In semifusinoids and fusinoids, pyrite filled voids. Marcasite was present as irregular forms. In all of the size fractions, middling-type coal contained the greatest percentages of the unliberated microscopic pyrite, and also the greatest proportion of the organic S. Depending upon the size fraction, pyritic S correlated to varying degrees (either positively or negatively) with the organic S content of the coal. Total S content of the coal had the greatest effect on the S in coke. Some pyritic S combined with carbon to add to the organic S content of the coke.

These are only some of the published reports on sulfur in coal. Some others of interest to the reader of this text may be: Mansfield and Spackman,[286] Neavel,[287] Rees et al.,[158] Reidenouer,[288] Demeter and Bienstock,[154] Nazarova and Berman,[289] Cheek,[290] Hopkins and Nance,[291] Eddy,[292] Schultz and Proctor,[293] Smith and Batts,[294] Mc Millan,[295] Von Demfange and Warner,[296] Ctvrnicek et al.,[129] Chadwick et al.,[297,298] Levene and Hand,[299] Gladfelter and Dickerhoof,[300] Lloyd and Francis,[301] Avgushevich et al.,[302] Shimp et al.,[303] Soloman and Manzione,[304] Cooper et al.,[305] Attar and Corcoran,[306] Paris,[307] Attar et al.,[308] Antonijevic et al.,[309] Neavel and Keller,[310] Myers,[311] and many others.

This chapter will not be complete without mentioning the pioneering works in this field including Ashley[326] Thiessen,[312,313] Yancey and Fraser,[314] Yancey and Parr,[315] Newhouse,[361] Cady and Leighton,[317] White,[318] and Selvig and Seaman.[267] An estimated worldwide average of total sulfur in coal is 2.0%.[184]

15. Chlorine (Cl)

The first review of chlorine in coal is presented by Gibson and Selvig.[185] Some 20 years later the same subject is covered by Abernethy and Gibson.[178]

Parr and Wheeler[319] determined chlorine in 49 samples of Illinois coal by digesting pulverized coal with water and titrating with standard silver nitrate solution. The chlorine contents of 32 samples ranged from 0.03 to 0.56% and none was detected in 17 samples. The early work on chlorine in coal includes also the report by De Weale.[320]

Selvig and Gibson[321] determined total halogen content by igniting the coal in an oxygen bomb and titrating the chlorine in the bomb washings. The chlorine in 21 coals from various states ranged from 0.01 to 0.46%; no chlorine was detected in 3 coals from western States. From 0 to 0.19% chlorine was reported in Polish coals by Nielecki.[749]

Work in Great Britain on the occurrence of chlorine in coal (Daybell and Pringle[322]), its state of combination (Edgcombe[323]), and its behavior during combustion and carbonization was reviewed by Kear and Menzies,[324] who included a bibliography of 57 references. Many coals contain water-soluble sodium equivalent to only one third or one half of the total chlorine content, and part of the chlorine is believed to be present as chloride ions attached to the coal substance.

Das Gupta and Chakrabarti[325] found that up to 35% of the total chlorine in Indian coals is present in organic combination rather than as inorganic chlorides.

According to Crossley,[246] the chlorine content of English coals ranges from about 0.01 to 1.0% and varies with the area from which it is mined. The U.S. National Coal Board showed that coals with a high chlorine content are found mainly in four divisions, northeastern, northwestern, east midlands, and west midlands, and that mining developments will move into areas with increasing chlorine content. Wandless[327] pointed out that coals containing less than 0.3 percent chlorine are not likely to give rise to the formation of troublesome boiler deposits.

In the report by Gluskoter and Rees[328] a variety of coals from Illinois and Indiana and their associated groundwaters were studied to determine the mode of occurrence of Cl in these coals. Both channel and drill core samples were utilized in addition to uncontaminated groundwater. Chlorine was determined by the bomb combustion method, and values for one seam were regionally mapped. There was a general increase in Cl with seam depth in coals of the Illinois Basin. Within the same mine Cl content of the coal was controlled by the Cl concentration of the associated groundwater and not seam depth. The majority of Cl in Illinois coals occurred as inorganic chlorides, primarily NaCl. There was no evidence for organically combined Cl. Inorganically bound Cl was slightly volatilized by combusion with other materials, probably sulfur-containing compounds. The inorganic chlorides were slightly water-soluble and their removal during preparation was dependent on the size of the coal, Cl content of the original coal and wash water, and the duration of washing. Fresh water flushing of the coal was generally effective for Cl reduction.

In the work by Gluskoter[329] the Cl content of a selected midwestern coal was regionally mapped to determine its distribution, and both bench-scale and field testing in a coal preparation plant were conducted to establish the extent of removal of water-soluble Cl from coal. Analyses were performed on face-channel samples and associated ground waters. Chlorine concentration in the coal was controlled by the composition of the associated groundwater, and increased with coal depth. An apparent equilibrium of the coal with the groundwater composition provided a depth-chlorine correlation. In a natural weathering test Cl content in the coal significantly decreased over an 8-week period. Field testing in a coal preparation plant indicated removal of a portion of the Cl only if the coal was washed with fresh and nonrecycled water.

Gluskoter and Ruch[253] have investigated chlorine and sodium in Illinois coals. Samples from 27 mines in the Illinois Basin were analyzed for soluble and insoluble Na, Cl, and K to determine their mode of occurrence. Analytical techniques used to determine three elements were neutron activation, bomb combustion, flame photometry, and aqueous extraction. Results of these analytical procedures indicated a poor correlation existed between the Cl and Na content in the same coal sample; the stoichiometric concentration of Na and Cl in halite was more closely approximated by the soluble Na and Cl fraction than by the insoluble fraction; the poor correlation between Na

Table 55
CHLORINE DISTRIBUTION IN U.S. COALS

Area	Conc range (%)	Arithmetic mean (%)	Geometric mean (%)
Illinois Basin	0.01—0.54	0.14	0.08
Eastern U.S.	0.01—0.80	0.17	0.10
Western U.S.	0.01—0.13	0.03	0.02

After Gluskoter, H. J., Trace Elements in Coal: Occurrence and Distribution, Circ. 499, Illinois State Geological Survey, Urbana, 1977.

and Cl became even worse when soluble Na and Cl were removed; and values of K in the coal samples analyzed ranged from 0.11 to 0.27% with water-soluble K ranging from trace to a maximum of 0.01%. With low correlations between Na and Cl concentrations, the presence of halite (NaCl) can only account for a portion of the Cl content of the coal. Also, because of the very small amounts of water-soluble K, the presence of sylvite (KCl) is very unlikely. It was generally concluded that all K in Illinois coal is probably in silicate minerals and that there is a strong possibility of an organic association with Cl.

The behavior of chlorine in coal combustion was investigated by Iapulacci et al.[330] It was concluded that the majority of Cl was discharged as hydrogen chloride in the stock gas.

In the work by Abernethy et al.[250] quantitative values for Cl were determined for coals from eastern central, and western portions of the U.S. These values represent tests performed on 62 seams from 11 states (Alabama, Illinois, Indiana, Kentucky, Ohio, Pennsylvania, Tennessee, Utah, Virginia, and West Virginia) plus 1 sample from each of 4 countries in Colorado. Cl determinations were done by potentiometric titration. Data are reported on a dry coal basis. Mean, high, and low values are reported. Extensive data for Cl content of individual samples with reference to location, bed, mining type, sample type, coal size, coal rank, and mode of handling is included in this report.

Gluskoter[137] has presented chlorine data for U.S. coals (see Table 55). Chlorine is concentrated in coals from the Illinois Basin and from the eastern states, but not in coals from western U.S. The EF for coals from the Illinois Basin is 6.0 while for coals from eastern U.S. it is 7.7. The observed correlation of chlorine and depth to the coal bed is not a primary correlation, but it is the result of an increase in salinity of groundwater with greater depth. Chlorine concentration in North Dakota lignite is reported to be 46.6 ppm.[122]

Burchett[3] reported chlorine in coals from Nebraska to be in a wide concentration range of 60 to 9500 ppm. Raw coals from three Illinois Basin preparation plants had chlorine concentration within the range 39 to 460 ppm, with mean value 240 ± 200 ppm.

Chlorine in the coal samples from seven countries studied by Loevblad[132] was in the concentration range 30 to 3680 ppm (mean value 940 ppm). An estimated worldwide average for chlorine concentration in coal is about 1000 ppm,[184] while the U.S. average is quite lower (207 ppm) as reported in Reference 2.

Chlorine in coal, its behavior during coal combustion, coal gasification, coal liquefaction, and as a source of atmospheric chlorine is discussed by many authors (e.g., Iapulacci et al.,[330] Palmer,[331] Vasyutinskii,[332] and others).

16. Potassium (K)

Data reported for potassium concentration in U.S. coals by Gluskoter[137] are shown in Figure 13 and Table 56.

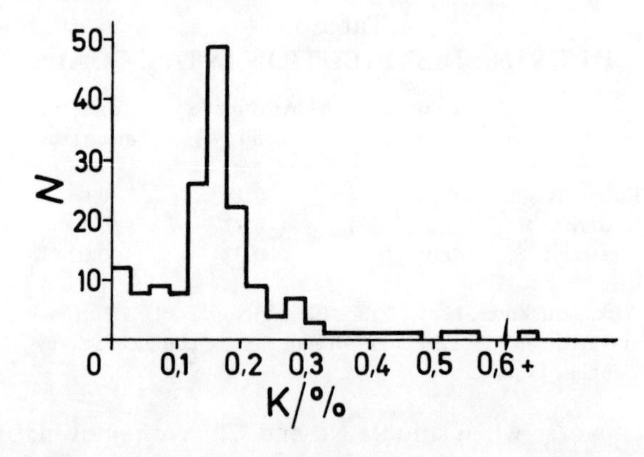

FIGURE 13. Distribution of potassium concentration values for U.S. coals. (As reported by Gluskoter.[137])

Table 56
POTASSIUM IN U.S. COALS

Area	Conc range (%)	Arithmetic mean (%)	Geometric mean (%)
Illinois Basin	0.04—0.56	0.17	0.16
Eastern U.S.	0.06—0.68	0.25	0.21
Western U.S.	0.01—0.32	0.05	0.03

Data from Gluskoter, H. J., Trace Elements in Coal: Occurrence and Distribution, Circ. 499, Illinois State Geological Survey, Urbana, 1977.

Abernathy et al.[250] have determined values for K concentrations in coals from eastern, central, and western portions of the U.S. These values represent tests performed on 62 seams from 11 states (Alabama, Illinois, Indiana, Kentucky, Ohio, Pennsylvania, Tennessee, Utah, Virginia, and West Virginia) plus 1 sample from each of 4 counties in Colorado; K determinations were done using a flame photometer. Data are reported on an ash basis. Mean, high, and low values are reported. No attempt was made to correlate the amounts of K to other properties of coal. Extensive data for P, Cl, Na, and K content of individual samples with reference to location, bed, mining type, coal size, coal rank, and mode of handling is included in this report.

Average concentration of potassium in 12 samples of North Dakota lignite is reported to be 462 ppm.[122] Potassium in eight coal samples from Nebraska is reported to be within the concentration range 0.07 to 1.4%.[3] Raw coals from three Illinois Basin preparation plants had potassium within the range 0.34 to 0.79%, with mean value of 0.50%.[199] Table 57 is based on data reported in Reference 2; potassium concentrations in U.S. coals, and U.S. and worldwide averages are presented.

17. Calcium (Ca)

Calcium in coal occurs in concentrations arround 1% or smaller. For example, Ca concentrations of Dietz No. 2 coal seam (Bighorn Mine,[333]) are in the range 0.24 to 2.78%.

The average concentration of calcium in North Dakota lignite is reported to be 1.6%.[122] Calcium in coal samples from Nebraska is reported to be within 0.25 to 2.4% range.[3] Raw coals from the Illinois Basin preparation plants had calcium in the range 0.12 to 1.79%.[199]

Table 57
POTASSIUM CONCENTRATION IN COAL

Coal	Conc (%)
Anthracite	0.24
Bituminous	0.21
Subbituminous	0.06
Lignite	0.20
U.S. average	0.18
Worldwide average	0.01

From U.S. National Committee for Geochemistry, Trace Element Geochemistry of Coal Resource Development Related to Environmental Quality and Health, National Academy Press, Washington, D.C., 1980.

Table 58
CALCIUM CONCENTRATION IN SOME U.S. COALS

Area	Conc range (%)	Arithmetic mean (%)	Geometric mean (%)
Western U.S.	0.44—3.8	1.7	1.5
Eastern U.S.	0.09—2.6	0.47	0.34
Illinois Basin	0.01—2.7	0.67	0.51

After Gluskoter, H. J., Trace Elements in Coal: Occurrence and Distribution, Circ. 499, Illinois State Geological Survey, Urbana, 1977.

Calcium in U.S. coals was reported by Swanson et al.[183] to be anthracite, 0.07%; bituminous coal, 0.33%; subbituminous coal 0.78%; lignite, 1.2%. The average for U.S. coals determined from analysis of 799 coal samples is found to be 0.54% (see Reference 2). A worldwide average is estimated to be 1.0%. Values reported for U.S. coals by Glusketer[137] are shown in Table 58.

Genetic characteristics of accumulation of calcium in coals are described by Kler.[334] Kler[335] has also discussed the formation of the calcium mineralization of coals.

High-calcium content coals and inferior grades of coal are discussed by Balakhnin and Merentsova,[336] and Corbaty and Taunton.[337] Effects of calcium on coal properties and its behavior during combustion and its usage are discussed by many authors. Muralidhara and Sears[338] have discussed effect of calcium in coal on gasification. Corbaty and Taunton[337] have discussed liquefaction of calcium-containing subbituminous coal.

Otto et al.[339] have discussed effects of calcium as catalyst and sulfur scavenger in the steam gasification of coal char. Calcium present in coal ash has been discussed by several authors. Distribution of calcium compounds in fly ash is discussed by Panin and Shpirt;[259] analysis of calcium in siliceous coal ash of AA spectroscopy is described by Muter and Cockrell.[254]

Production of calcium hydrosilicates and white Portland cement during complex processing of coal ashes is discussed by Martirosyan and Safaryan.[340] Burek and Palica[341] have described the effects of variations in the calcium content on ash determination by X-ray absorption.

18. Scandium (Sc)

Data for some U.S. coals obtained by Gluskoter[137] are shown in Figure 14 and Table

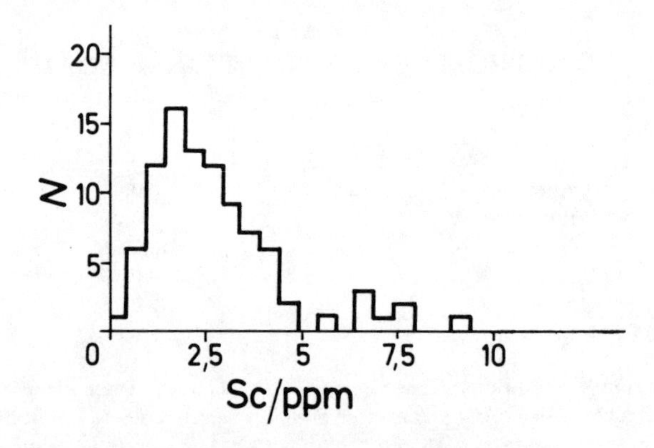

FIGURE 14. Distribution of scandium concentration values for U.S. coals. (As reported by Gluskoter.[137])

Table 59
SCANDIUM IN SOME U.S. COALS

Area	Conc range (ppm)	Arithmetic mean (ppm)	Geometric mean (ppm)
Western U.S.	0.50—4.5	1.8	1.5
Eastern U.S.	1.6—9.3	5.1	4.5
Illinois Basin	1.2—7.7	2.7	2.5

After Gluskoter, H. J., Trace Elements in Coal: Occurrence and Distribution, Circ. 499, Illinois State Geological Survey, Urbana, 1977.

59. Scandium concentration (average value for 12 samples) in North Dakota lignite is reported to be 8 ppm.[122] Masursky[342] has suggested that scandium was probably deposited in coal by water passing through coal beds.

Hansen et al.[181] have reported scandium concentration in the ash of high volatile bituminous coal (8.9% ash) from Utah to be 15 ppm. Scandium in coal samples from Nebraska is reported to be within the range 2 to 30 ppm.[3] Raw coals from the Illinois Basin preparation plants had scandium concentration within the range 5.3 to 8.8 ppm. Scandium average concentration from U.S. coals is reported to be 3 ppm (see Reference 2) while an estimated worldwide average is 5 ppm. Komissarova et al.[343] have discussed the distribution of scandium according to fractions of the gravitation separation of brown coal.

19. Titanium (Ti)

The best known review articles on titanium occurrence in coal are those by Gibson and Selvig[185] and Abernethy and Gibson.[178] It was Wait[344] who first reported the occurrrence of titanium in the ash of some vegetable matters and in coal ash. He found from 0.006% titanium in the ash of cow peas to 0.19% in the ash of oak wood and from 0.41 to 1.55% in five coal ashes. Baskerville[345] reported analyses of three samples of ash from North Carolina peat ranging from 0.20 to 0.29% titanium.

Analyses of ash from U.S. coals published by the Bureau of Mines[321] included the determination of titanium in 116 samples. All of the results calculated as titanium were within the range of 0.3 to 1.5% except for five samples that had a slightly higher content ranging up to 2.2%. Headlee and Hunter[52] obtained similar values for titanium in ash of West Virginia coals, the average being 0.91% with a range of 0.60 to 1.38%.

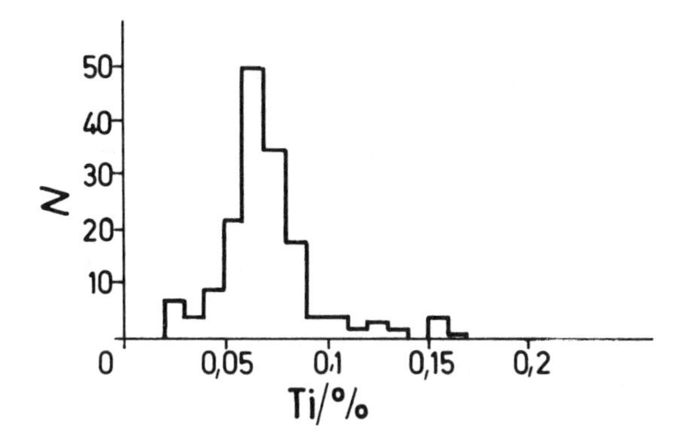

FIGURE 15. Distribution of titanium concentration values for U.S. coals. (As reported by Gluskoter.[137])

Table 60
DISTRIBUTION OF TITANIUM IN U.S. COALS

Area	Conc range (%)	Arithmetic mean (%)	Geometric mean (%)
Illinois Basin	0.02—0.15	0.06	0.06
Eastern U.S.	0.05—0.16	0.09	0.09
Western U.S.	0.02—0.13	0.05	0.05

After Gluskoter, H. J., Trace Elements in Coal: Occurrence and Distribution, Circ. 499, Illinois State Geological Survey, Urbana, 1977.

Analyses of ash of cubes from column samples of coal showed a wider range, the minimum being 0.10% and the maximum 3.44%. Duel and Annell[193] found titanium present in the ash of western coals in about the same abundance as in the crust of the Earth, the maximum concentration being from 0.1 to 1.0% Ti. Zubovic et al.[10] reported from 0.15 to 2.6% titanium in ash of coals from the Northern Great Plains. Jones and Miller[346] reported 0.58 to 1.09% titanium in anthracite ash from Pennsylvania. Nunn et al.[192] found 0.9 to 1.2% titanium in ash from Pennsylvania anthracites.

Jones and Miller[346] found 4.2 to 14.6% titanium in the ash of 15 special samples obtained from deposits called cauldrons and horsebacks in Northumberland and Durham, England. These unusually pure vitrains were float portions separated on a solution of 1.35 sp gr; they contained only 0.1 to 1.0% ash. Eight similar vitrains from Kent coal separated on 1.30 sp gr contained 0.7 to 2.9% ash and 2.0 to 5.9% titanium in the ash. Horton and Aubrey[141] reported values of 0.3 to 0.8% titanium in the ash of three other English vitrains. Reynolds[347] reported 3.2 to 17.7% titanium in ash of special vitrain samples from North Staffordshire and North Wales.

According to King and Crossley[348] the titanium content of ash from English coal varies between 0 and 1.8% titanium. Ash from 45 samples of South African coal was reported to contain from 0.46 to 1.18% titanium.[229,349]

The report in Reference 333 showed a good agreement in titanium values obtained by X-ray fluorescence and by colorimetric determination. The titanium values were in the range 36 to 5000 ppm. For the description of spectrographic determination of titanium, see Valeska and Havlova.[350]

Titanium concentrations in U.S. coals as reported by Gluskoter[137] are shown in Figure 15 and Table 60. Titanium concentration in North Dakota lignite is reported to be

300 ppm.[122] Titanium concentration in coal samples from Nebraska was reported to be within the range 300 to 2900 ppm.[3] Raw coals from three Illinois Basin preparation plants had a titanium mean value of 0.18 ± 0.04%.[199]

Swanson et al.[183] have reported titanium concentrations in U.S. coals to be anthracite, 0.15%; bituminous coal, 0.08% subbituminous coal, 0.05%; lignite, 0.12%. The average value for all U.S. coals is estimated from the analysis of some 800 samples (see Reference 185) to be 0.08%, while the worldwide average is supposed to be 0.05%.

20. Vanadium (V)

Vanadium is ubiquitous in nature. All types of igneous rocks and soils contain an average of 100 ppm vanadium, although its concentration in sea water is much lower, about 0.002 ppm. All living organisms contain vanadium from 0.15 to 2 ppm; certain land plants such as *Amanita muscaria* and marine animals such as *Pleurobranchas plumula* contain vanadium in concentrations as high as 0.65% on a dry weight basis.[351]

Early work on vanadium in coal is compiled in works by Gibson and Selvig[185] and Abernethy and Gibson.[178]

Kyle[352] announced the discovery of vanadium in a deposit, which he called lignite, from San Rafael, Province of Mendoza, Argentina. He reported that the deposit contained 0.63% of ash with 38.22% of V_2O_5 in the ash (21,41% V). In another sample, probably from the same region, Mourlot[353] obtained 38.5% V_2O_5 (21.6% V) from the ash. Torrico and Meca (U.S. Geological Survey, 1896) reported 38% (21.3% V) in the ash of a deposit from Yuali, Peru. Vanadium was reported also in certain Australian "coals", one analysis showing a sample with 1.7% ash that contained 25.1% vanadium in the ash.

Kyle's discovery aroused great interest, and samples of the vanadium-bearing ash from Argentina, and vanadium compounds from it were shown at the Columbian Exposition held at Chicago in 1893.[178] Hewett, after examining the vanadium deposits in Peru, calculated that the hydrocarbon materials containing vanadium were probably asphaltites. Hess[178] suggested that the material analyzed by Kyle was not lignite, but was an asphaltite, and added that much of the "coal" quoted in the literature as carrying vanadium is asphaltite and not coal.

Vanadium was found in some samples of ash from Alberta[178] coals (Research Council of Alberta, 1943) but only in trace amounts of no commercial significance.[178] A study of Hungarian[354] coals indicated that the vanadium was bound to the organic matter in the coal. Zilbermintz[355] and Zilbermintz and Bezrukov[356] described some clarain-vitrain-type coal samples from the eastern slope of the Urals that contained from 0.31 to 4.92% vanadium in the ash. Most of the vanadium content was ascribed to infiltration of vanadium-bearing material during the laying down of the coal beds. Reynolds[347] also found notable quantities of vanadium in special vitrain samples from England, and Horton and Aubrey[141] found as much as 0.5% vanadium in ash of vitrain. The occurrence of vanadium in special anthracite samples in central Europe[178] in coal from Belgium[357] and from Russia,[187] and the geochemistry of vanadium[358] were discussed. Work was reported on the possible harmful effects of vanadium oxide in boiler firing. Headlee and Hunter[359] noted that vanadium in West Virginia coals is remarkably constant in the main body of coal, but found evidence that thin sections of coal outside the main coal bed between shale partings frequently have high vanadium content. (For additional reports on vanadium presence in coal see Vorob'ev,[360] Abbolito,[361] Idzikowski and Sachanbinski.[362]) Data on vanadium content in coal ashes are reported by various investigations. Abernethy and Gibson[178] have summarized these reports, and their work is shown in Table 61. Detailed information on vanadium in U.S. coals is available in References 173 and 137. Table 62 shows data on vanadium concentrations in U.S. coals. It is of interest to note that vanadium concentration can

Table 61
VANADIUM CONCENTRATIONS IN COAL ASHES

Source of coal	Vanadium conc (%)	Ref.
U.S.		
Northern Great Plains	0.001—0.058	10
Texas, Colorado, North and South Dakota	0.01—0.1	219
West Virginia	0.018—0.039	52
Pennsylvania, anthracite	0.01—0.02	192
North Carolina, peat	0.0006—0.0017	345
Australia, Collie	0.012	363
England, vitrain	0.04—0.5	141
England, vitrain	0.7—7.9	347
England, Newcastle	0.05	196
Germany, Newrode	0.12	
Germany	0.1	197
Nova Scotia	0.0061—0.0244	128
Spitzbergen	0.01—0.1	364
Japan		
Fukui	0.007—0.020	365
Heijo	0.062	361
Italy	0.16—0.60	
Russia	0.01—0.17	366
Russia	0.011—0.034	367
Russia		
Ural	0.10—2.85	356
Alai and Turkestan	0.006—0.73	360

From Abernethy, R. F. and Gibson, F. H., Rare Elements in Coal, Rep. BM-IC-8163, U.S. Bureau of Mines, Washington, D.C., 1963.

Table 62
VANADIUM IN SOME U.S. COALS[137,173]

Region	Coal	Conc range (ppm)	Arithmetic mean (ppm)	Geometric mean (ppm)
Appalachian	Bituminous	1.5—150	21	16
Interior province	Bituminous	1.1—350	23	17
N. Great Plains	Lignite	0.90—110	9.0	5.7
N. Great Plains	Subbituminous	0.91—370	19	12
Rocky Mountain	Subbituminous	0.14—330	23	16
Rocky Mountain	Bituminous	1.9—120	15	11
All U.S.	Different	0.03—51000	96	14
Western U.S.		4.8—43	14	12
Eastern U.S.		14—73	38	35
Illinois Basin		11—90	32	29

be as low as 0.03 ppm and as high as 51,000 ppm. Distribution of vanadium concentration in U.S. coals as analyzed by Gluskoter et al. is also shown in Figure 16.

Vanadium concentration in North Dakota lignite is reported by Somerville and Elder[122] to be 22 ppm. Coal samples from Nebraska had vanadium within the concentration range 7 to 150 ppm.[3] Coals from the Illinois Basin studied by Wewerka et al.[199] had only 34 to 100 ppm of vanadium. Even lower values are reported by Loevblad,[132] who studied coal samples from seven countries (5.5 to 39.5 ppm with mean value of 19 ppm). The average vanadium concentration in U.S. coals is reported to be 20 ppm (see Reference 2) which is very close to the estimated worldwide average of 25 ppm.[184]

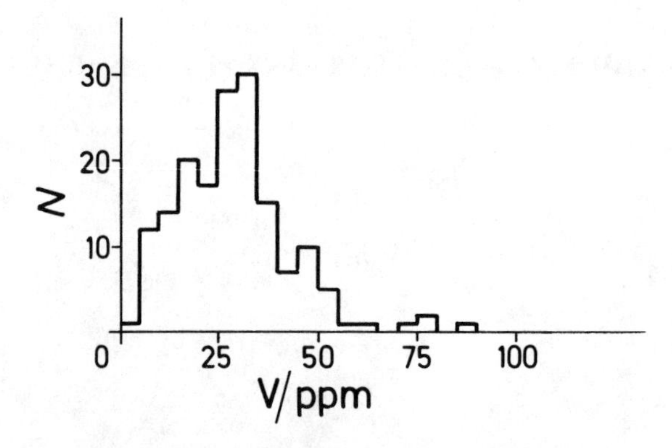

FIGURE 16. Distribution of vanadium concentration values for U.S. coals. (As reported by Gluskoter.[137])

Table 63
CHROMIUM IN COAL ASH

Source of coal	Chromium conc (%)	Ref.
U.S.		
N. Great Plains	<0.0001—0.03	10
North Carolina, peat	0.019—0.025	345
Pennsylvania, anthracite	0.001—0.01	192
Pennsylvania		
Cambria County	0.027	185
Washington County	0.013	185
Texas, Colorado, North and South Dakota	0.01—0.1	219
West Virginia	0.011—0.02	52
England	0.0114—0.0177	178
England, vitrain	0.01—0.1	141
England, Newcastle	0.03	196
England, vitrain	<1.5	347
Germany, Newrode	0.014	196
Nova Scotia	0.0018—0.0079	128
Portugal, anthracite	0.01—0.1	180
Spitzbergen	0.01—0.1	364

After Abernethy, R. F. and Gibson, F. H., Rare Elements in Coal, Rep. BM-IC-8163, U.S. Bureau of Mines, Washington, D.C., 1963.

21. Chromium (Cr)

Chromium was determined by the Bureau of Mines in the ash of two Pennsylvania bituminous coals. One coal from Washington County contained 0.019% Cr_2O_3 in the ash, and the other sample, from Cambria County, contained 0.039% Cr_2O_3 in the ash.[185] Analyses of ash from three samples of North Carolina peat[345] ranged from 0.028 to 0.036% Cr_2O_3. It was also reported that ashes from the Newrode coal in Germany and the Newcastle coal in England contained 0.02 and 0.05% Cr_2O_3, respectively. Other early works on chromium content of coals include reports by Jorissen,[357] Iwasaki and Ukimoto,[368] and Reynolds.[347]

Washability data on chromium in a sample from the Blue Creek coal in Alabama (reported by Gluskoter[137]) give a washability curve with a positive slope. Such a curve shows that the element is concentrated in the inorganic (mineral matter) portion of coal.

Abernethy and Gibson[178] have summarized the report about chromium contents of coal ashes as reported by various investigators. This is shown in Table 63.

Table 64
CHROMIUM CONCENTRATION IN U.S. COALS[137,173]

Region	Coal	Conc range (ppm)	Arithmetic mean (ppm)	Geometric mean (ppm)
Appalachian	Bituminous	1.5—220	18	14
Interior province	Bituminous	2.0—200	20	14
N. Great Plains	Lignite	0.25—43	4.2	2.8
N. Great Plains	Subbituminous	0.54—66	6.7	4.4
Rocky Mountain	Subbituminous	0.54—70	11	6.9
Rocky Mountain	Bituminous	0.81—75	7.9	5.7
All U.S.	Different	0.05—300	15	9.4
Illinois Basin		4.0—60	18	16
Eastern U.S.		10—90	20	18
Western U.S.		2.4—20	9.0	8.1

Data for chromium in U.S. coals are shown in Table 64.[137,173] Chromium concentration in North Dakota lignite is found to be 65 ppm.[122] Hansen et al.[181] (1980) have reported chromium concentration in the ash of high-volatile bituminous coal (8.9% ash) from Utah to be 88 ppm.

Burchett[3] has reported chromium concentration in coal samples from Nebraska to be within the range 7 to 70 ppm. The chromium concentration range for raw coals from Illinois Basin preparation plants is reported to be 30 to 50 ppm; mean value 38 ± 8 ppm.[199]

The concentration range for chromium in coal samples from seven countries studied by Loevblad[132] is 1.8 to 17.4 ppm, with a mean value of 9.1 ppm.

Swanson et al.[183] have reported chromium concentration in U.S. coals to be anthracite, 20 ppm; bituminous coal, 15 ppm; subbituminous coal, 7 ppm; lignite, 20 ppm. The U.S. average is reported to be 15 ppm (see Reference 2), very close to estimated worldwide average of 10 ppm.

22. Manganese (Mn)

Manganese is a relatively common element with crustal abundance of 0.10%. Small quantities of manganese occur in all coal ashes. Early work on manganese determination in coal and coal ashes is summarized by Gibson and Selvig[185] and later by Abernethy and Gibson.[178] Table 65 shows manganese in coal ash.[178]

Methods for determination of manganese in coal have been discussed by several authors including Kessler and Dockalova,[369] Lustigova,[370] Ueda et al.,[371] and others. Mineralogy and geochemistry of manganese in coal-bearing formations during diagenesis was studied by Zaritskii.[372]

Manganese in U.S. coals has been studied by Zubovic et al.[173] and Gluskoter.[137] Their results are shown in Table 66.

Manganese in North Dakota lignite is reported to be 249 ppm.[122] Manganese in coal samples from Nebraska was found to be within the range 8.5 to 90 ppm.[3] Coals from the Illinois Basin had manganese reported within concentration range 34 to 180 ppm.[199] Values of manganese concentration in coal samples from seven countries studied by Loevblad[132] were within the range 10 to 224 ppm, with mean value 72 ppm. The average value of manganese concentration in U.S. coals is estimated at 100 ppm (see Reference 2).

23. Iron (Fe)

Iron in coal has been studied in connection with coal combustion properties. Special attention is paid to the study of behavior of iron compounds during coal

Table 65
MANGANESE IN COAL ASH

Source of coal	Manganese conc (%)	Ref.
U.S.		
West Virginia	0.012—0.18	52
Pennsylvania, anthracite	0.005—0.006	192
Pennsylvania, anthracite	0.02—0.09	373
North Dakota	0.15	374
Montana	0.33	374
Alabama, 2 analyses	0.04—0.05	185
Texas, Colorado, North and South Dakota	0.1—1.0	219
England	0.08—0.18	
Germany, Newrode	0.05	196
England, Newcastle	0.01	196
Germany, Ruhr	0.6	348
Germany, brown coal	0.35	46
Germany	<2.2	78
Portugal, anthracite	0.001—1.0	180
Nova Scotia	0.0165—0.220	128
Spetzbergen	0.01—0.1	364
Russia, peat	0.098—0.109	376
Russia	0.0205	376

From Abernethy, R. F. and Gibson, F. H., Rare Elements in Coal, Rep. BM-IC-8163, U.S. Bureau of Mines, Washington, D.C., 1963.

Table 66
MANGANESE IN U.S. COALS[137,173]

Region	Coal	Conc range (ppm)	Arithmetic mean (ppm)	Geometric mean (ppm)
Appalachian	Bituminous	0.75—1400	27	15
Interior province	Bituminous	1.4—1100	82	42
N. Great Plains	Lignite	7.3—660	75	55
N. Great Plains	Subbituminous	1.4—450	56	29
Rocky Mountain	Subbituminous	1.4—3500	57	28
Rocky Mountain	Bituminous	0.9—590	30	14
All U.S.	Different	0.01—4500	52	25
Illinois Basin		6.0—210	53	40
Eastern U.S.		2.4—61	18	12
Western U.S.		1.4—220	49	28

combustion.[377] There are number of methods applied for iron determination in coal.[378] Atomic absorption is often used for iron determination (e.g., Gladfelter and Dickerhoof[379]). Neutron-irradiation is also sometimes used.[380] Recently Mössbauer spectroscopy was used for the study of iron in coal.[381,382]

Isolation of iron organometallic compounds from brown coal was discussed by Aleksandrov et al.[383] Detection and content of various organic compounds in coal containing iron was discussed by Kamneva et al.[384,385] and Harry and Schafer.[386] Doughty and Dwiggins[387] have discussed the characterization of the valence state of iron in coal dust.

Early works on the effects of iron on coal combustion are reviewed by Moody.[435] Areas briefly discussed include: mineral forms, the relationship of Fe and S in coal, pyrite distribution in coal, changes occurring in coal ash during combustion, the effect of the ash type on the stoker, pulverized-coal, and slag-top furnaces and furnace design. Additional areas of discussion include the relationship of the ash fusion temperature to iron oxide in the coal ash, and the general characteristics of the viscosity of coal ash slag.

Ferrous iron associated with carboxyl groups in a brown coal or in a methacrylate resin is readily oxidized, the oxidation of the iron on the brown coal being more rapid. It would seem that unless brown coal could be sampled from the coal seam under conditions of complete exclusion of air, it is impossible to determine the form in which iron occurs naturally because of the ease of oxidation, particularly on air drying. However, on the basis of other evidence the naturally occurring form of iron organically bound to brown coal would be more likely to be ferrous.[388]

Various techniques may be used to enhance the paramagnetic properties of pyrite to a level that would allow coal and pyrite to be magnetically separated effectively. In the report by Ergun and Bean[389] samples of coal from Ohio, Illinois, Pennsylvania, and West Virginia were used to investigate variables that affect the magnetic susceptibility of pyrite within coal. Tests on the effect of weathering of coal indicate that this process apparently increases the paramagnetic properties of pyrite to levels that would make magnetic separation possible. It was found that during the period of weathering, sulfate sulfur levels as well as magnetic susceptibility increased, and because iron sulfate is more paramagnetic than pyrite, increases in pyrite magnetic susceptibility were attributed to the effects of weathering. The effects of various pulverization techniques as they would influence the paramagnetic properties of pyrite were also studied. These tests indicate that none of the pulverizers tested increase the magnetic susceptibility of pyrite to the magnitude necessary for magnetic separation. It was also found that heat treatment at temperatures below 400°C improves separation only to a limited extent. Initial experiments suggest that if pyrite could be selectively heated in the Mhz and Ghz regions the susceptibility of pyrite could be increased to sufficient levels for magnetic separation.

Characterization of iron minerals in coal by low-frequency IR spectroscopy is described by Estep et al.[375] IR spectra were presented and interpreted to differentiate and identify some of the more commonly occurring Fe minerals in both their original and altered forms in bituminous coal. Low-temperature ashing prior to analysis yielded essentially unaltered mineral residues. Samples were pelletized in a cesium iodide matrix and scanned on an IR grating spectrophotometer. Observed spectral variations in terms of absorption bands between major groups of Fe minerals (including sulfides, carbonates, oxides, hydrous oxide, and sulfates) were sufficient to allow their identification. Similar study was also reported by Belly and Brock.[390] High iron content of coal is of importance for the study of ash-forming processes.[391,392]

Distribution of iron in U.S. coals is shown in Figure 10. Concentration ranges, arithmetic and geometric means for coal samples from different parts of U.S.A. are shown in Table 67. Average iron concentration of Dietz No. 2 seam in Bighorn mines was found to be 0.240% with a concentration range of 0.048 to 0.850%.[333]

Iron in North Dakota lignite is reported to be 0.72%.[122] Iron in eight coal samples from Nebraska studied by Burchett[3] was within the concentration range 1.1 to 5.1%. Illinois Basin coals had iron concentration from 2.4 to 4.3%,[199] with mean value 3.4 ± 1.3%. Coal samples from seven countries studied by Loevblad[132] had an average iron concentration of 0.82% (range: 0.2 to 2.62%).

24. Cobalt (Co)

Early works on cobalt occurrence in coal ash are summarized by Abernethy and Gibson.[178] This is shown in Table 68.

Table 67
IRON IN SOME U.S. COALS

Area	Conc range (%)	Arithmetic mean (%)	Geometric mean (%)
Western U.S.	0.30—1.2	0.53	0.49
Eastern U.S.	0.50—2.6	1.5	1.3
Illinois Basin	0.45—4.1	2.0	1.9

After Gluskoter, H. J., Trace Elements in Coal: Occurrence and Distribution, Circ. 499, Illinois State Geological Survey, Urbana, 1977.

Table 68
COBALT IN COAL ASH

Source of coal	Conc (%)	Ref.
U.S.		
N. Great Plains	<0.0005—0.009	10
Pennsylvania, anthracite	0.001—0.009	192
Texas, Colorado, North and South Dakota	0.01—1.0	219
West Virginia	0.005—0.018	52
England, vitrain	0.01—0.03	141
Finland	<0.001—0.03	393
Nova Scotia	0.0026—0.0196	128
Portugal, anthracite	0.01—0.1	180
Spitzbergen	0.001—0.01	364

From Abernethy, R. F. and Gibson, F. H., Rare Elements in Coal, Rep. BM-IC-8163, U.S. Bureau of Mines, Washington, D.C., 1963.

Table 69
COBALT IN U.S. COALS[137,173]

Region	Coal	Conc range (ppm)	Arithmetic mean (ppm)	Geometric mean (ppm)
Appalachian	Bituminous	0.70—930	7.9	5.5
Interior province	Bituminous	0.20—500	13	6.0
N. Great Plains	Lignite	0.05—20	1.3	0.98
N. Great Plains	Subbituminous	0.25—20	2.2	1.5
Rocky Mountain	Subbituminous	0.06—70	3.8	2.2
Rocky Mountain	Bituminous	0.24—21	2.3	1.7
All U.S.	Different	0.05—930	6.7	3.5
Western U.S.		0.60—7.0	1.8	1.5
Eastern U.S.		1.5—33	9.8	7.6
Illinois Basin		2.0—34	7.3	6.0

Cobalt in U.S. coals was studied in some details by Zubovic et al.[173] and Gustkoter.[137] Their results are shown in Table 69.

Average cobalt concentration in 12 samples of North Dakota lignite is reported to be 5 ppm.[122] Hansen et al.[181] have reported cobalt concentration in the ash of high-volatile bituminous coal (8.9% ash) from Utah to be 9 ppm.

Cobalt concentrations in coal samples from Nebraska ranged from 1 to 150 ppm.[3] Illinois Basin coals reported by Wewerka et al.[199] had cobalt in rather narrow range, 9 to 17 ppm. Coal samples from seven countries studied by Loevblad[132] had cobalt within the range 0.6 to 18.4 ppm with the mean value 6.2 ppm.

Average cobalt concentration in U.S. coals is estimated to be 7 ppm (see Reference 2) while the worldwide average is probably around 5 ppm.

Spectrophotometric determination of cobalt in coal is discussed by Chasak.[394]

25. Nickel (Ni)

Gibson and Selvig[185] have written the first review article on nickel in coal. The first reports on nickel in coal date from 1920s.

Mott and Wheeler,[395] in a study of the inherent ash of coal, found nickel in the ash of British coal from the Oarkgate bed, Yorkshire. The residual or inherent ash remaining in a sample of purified durain contained 0.78% nickel. Clarke[396] reported that millerite (nickel sulfide) was found in coal from Belgium, and he mentioned an occurrence of linneaeite, the cobalt-nickel sulfide, in coal from Wales.

Reynolds[397] found from a trace to 2.36% nickel in the ash of certain vitrain samples separated at 1.35 sp gr. Shakhov and Efendi[398] reported that nickel content of ash of coals from the Kuznetsk region generally is low, 0.001 to 0.01%, but some samples ranged from 0.05 to 1.0%.

Jones and Miller[346] reported unusual occurrences of nickel in vitrains that they obtained from coal deposits called cauldrons and horsebacks. The deposits were embedded in the shales overlying the Yard bed in Northumberland and from the Busty and Brokewell beds of Durham, England. Cauldron vitrain deposits were formed in the roots and lower part of tree trunks, whereas horsebacks were formed of fallen branches, buried horizontally in the shale and made lens-shaped by compression. Vitrain samples were cleaned for analysis by float-and-sink separation in a mixture of carbon tetrachloride and gasoline of 1.35 sp gr. Ash from 11 float portions of these samples contained 0.7 to 8.1% nickel. Eight similar English vitrains from Kent coals separated at 1.30 sp gr contained 0.08 to 1.3% nickel in the ash. Grosjean[399] also described an occurrencc of millerite in the rocks associated with coal beds in Belgium. Millerite was found in isolated patches of radiating needlelike crystals along small fracture planes in Australian coal. For additional reports on nickel occurrence in coal see Vorob'ev[360] and Swaine.[96]

Nickel content of coal ashes reported by various investigations are given in Table 70. Concentrations of nickel in some U.S. coals is shown in Table 71.

Average concentration of nickel in North Dakota lignite is reported to be 11.6 ppm.[122] Hansen et al.[181] have reported nickel concentration in the ash of high-volatile bituminous coal (8.9% ash) from Utah to be 25 ppm.

Nickel concentration in coal samples from Nebraska ranged from 5 to 200 ppm.[3] Illinois coal samples studied by Wewerka et al.[199] had only 21 ± 11 ppm of nickel. The average concentration of nickel in U.S. coals and an estimate of worldwide average are the same figure, 15 ppm (see Reference 2).

26. Copper (Cu)

First informations on copper in coal date from the last century.[400] Early review articles on copper in coal are Gibson and Selvig[185] and Abernethy and Gibson.[178] The copper content of coal ash is shown in Table 72.

Analyses of 20 samples of pyrites from English coals by Dunn and Bloxam[402] showed 1 to 170 ppm copper in 19 samples. Savul and Ababi[403] found 1.87 to 14.68 g of copper per ton of Romanian coal.

Fraser[404] has reported finding of a Canadian peat containing up to 10% (dry weight) Cu. In this work, wet chemical techniques and X-ray diffraction were utilized to identify Cu compounds. Copper was organically bound in peat as a chelate complex and

Table 70
NICKEL IN COAL ASH

Source of coal	Conc (%)	Ref.
U.S.		
N. Great Plains	0.0003—0.059	10
Texas, Colorado, North and South Dakota	0.01—1.0	219
South Dakota	0.12	219
West Virginia	0.013—0.079	52
Pennsylvania, anthracite	0.01—0.09	192
Australia	0.001—0.06	401
Nova Scotia	0.0052—0.0645	128
Spitzbergen	0.001—0.01	364
Germany, Newrode	0.11	196
Germany	<0.3	363
England, Newcastle	0.079	196
England, vitrain	0.05—0.3	141
Russia	0.47—1.02	360

From Abernethy, R. F. and Gibson, F. H., Rare Elements in Coal, Rep. BM-IC-8163, U.S. Bureau of Mines, Washington, D.C., 1963.

Table 71
NICKEL IN SOME U.S. COALS[137,173]

Region	Coal	Conc range (ppm)	Arithmetic mean (ppm)	Geometric mean (ppm)
Appalachian	Bituminous	1.1—220	14	11
Interior province	Bituminous	0.87—580	26	17
N. Great Plains	Lignite	0.52—84	2.9	1.8
N. Great Plains	Subbituminous	0.32—67	5.0	3.2
Rocky Mountain	Subbituminous	0.45—69	6.4	4.2
Rocky Mountain	Bituminous	0.35—340	6.5	3.7
All U.S.	Different	0.14—990	13	7.3
Illinois Basin		7.7—68	21	19
Eastern U.S.		6.3—28	15	14
Western U.S.		1.5—18	5.0	4.4

did not occur as an oxide, sulfide, or as elemental Cu. Because of the low S content of peat, the Cu was assumed to be bound to nitrogen or oxygen-containing components. Copper, having a greater affinity for N, tended to form the more stable Cu−N chelate. The element was concentrated as circulating cupriferous groundwaters filtered through the peat.

In the report by Ong and Swanson[405] the degree of absorption of Cu by a peat, a lignite, and a bituminous coal was determined, and a mechanism for the absorption process was proposed. In the course of the investigation the mode of occurrence of Cu in the coals was established. Samples utilized were selected to be representative of the range of organic material known to occur in the three steps of coalification. Analysis for Cu was performed using IR, colorimetric spectroscopy, and wet chemical techniques. Copper was organically associated in the coals and was complexed with HA. The complex involved was probably the carboxylic group. Adsorption of Cu decreased with increased coalification, with peat exhibiting the greatest adsorptive capacity because of its high HA content. Increased compaction of organic materials, accompanied

Table 72
COPPER IN COAL ASH

Source of coal	Conc (%)	Ref.
U.S.		
N. Great Plains	0.002—0.07	10
North Dakota	0.020	374
Pennsylvania	0.001—0.01	192
Anthracite	0.03—0.07	373
Texas, Colorado, North and South Dakota	0.01—0.1	219
West Virginia	0.022—0.10	52
England, Newcastle	0.06	196
England, vitrain	0.08—0.1	141
Germany, brown coal	0.001	46
Germany, Westphalia	0.016—0.054	400
Germany	0.4	78
Portugal, anthracite	0.001—0.01	180

From Abernethy, R. F. and Gibson, F. H., Rare Elements in Coal, Rep. BM-IC-8163, U.S. Bureau of Mines, Washington, D.C., 1963.

Table 73
COPPER IN SOME U.S. COALS[137,173]

Region	Coal	Conc range (ppm)	Arithmetic mean (ppm)	Geometric mean (ppm)
Appalachian	Bituminous	0.13—280	18	14
Interior province	Bituminous	2.7—170	17	14
N. Great Plains	Lignite	2.2—78	6.7	5.5
N. Great Plains	Subbituminous	1.2—80	12	9.1
Rocky Mountain	Subbituminous	0.16—120	14	10
Rocky Mountain	Bituminous	1.5—68	9.4	7.8
All U.S.	Different	0.13—1700	16	12
Illinois Basin		5.0—44	14	13
Eastern U.S.		5.1—30	18	16
Western U.S.		3.1—23	10	8.5

by a decrease in surface area during coalification, resulted in decreased Cu adsorption, and by the formation of the coal phase, adsorption of Cu had essentially ceased.

Copper contents of U.S. coals were investigated by Zubovic et al.[173] and Gluskoter.[137] Their findings are shown in Table 73. Distribution of copper concentration in coal samples analyzed by Gluskoter[137] is shown in Figure 17.

Average concentration of copper in 12 samples of North Dakota lignite is reported to be 23 ppm.[122] Hansen et al.[181] have reported copper concentration in the ash of high-volatile bituminous coal (8.9% ash) from Utah to be 64 ppm.

Copper in coal samples from Nebraska studied by Burchett[3] was within the concentration range 14 to 79 ppm. Raw coals from three Illinois Basin Preparation plants had copper concentrations from 35 to 56 ppm, with mean value 44 ± 9 ppm.[199] Copper concentrations in coal samples from seven countries studied by Loevblad[132] was within the range 5 to 24 ppm, with mean value 14 ppm.

Copper in U.S. coals has an average concentration of 19 ppm (see Reference 2). The highest concentration (27 ppm) is found in anthracite.[183] The worldwide average is 15 ppm. Copper in coal is discussed by several other authors including Mikhailova and Vlasov,[406] Massey,[407] Massey and Barnhisel,[408] Sorensen et al.,[409] and Drever et al.[202]

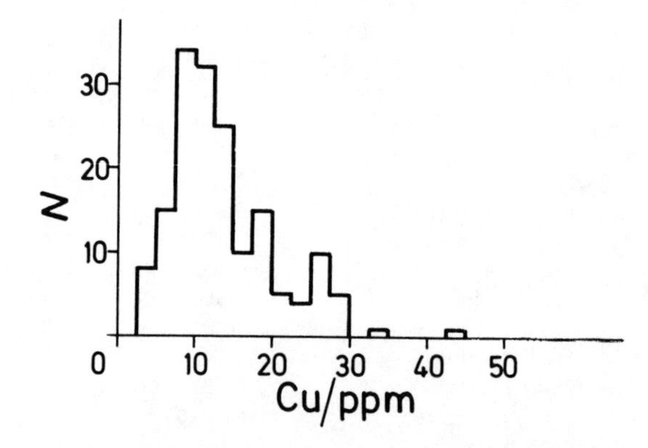

FIGURE 17. Distribution of copper concentration values for U.S. coals. (As reported by Gluskoter.[137])

27. Zinc (Zn)

A review article on zinc occurrence in coal is written by Abernethy and Gibson.[178] As early as 1929 Selvig and Seaman[267] have analyzed a sample of coal from Crittenden County, Ky., and found 1.6% zinc in the ash. The ash of another sample from Hopkins County, Ky., contained 0.16% zinc. Headlee and Hunter[52] in their examination of ash constituents of West Virginia coals reported an average of 0.043% zinc in the ash with a maximum of 0.19%. Examination of the ash of western coals[219] showed that zinc was not present in percentages high enough to be detected (0.01%). Zinc was detected in 21 out of 221 coal ash samples tested from Northern Great Plains.[10] The limit of detection was 0.02% and up to 0.7% zinc in ash was reported.

Hawley[128] reported an average of 0.0218% zinc in the ash of Nova Scotian coals, and the range was 0.0115 to 0.0550%. Newmarch detected zinc in the ash of Kootney coals of British Columbia.[410] An unusual coal ash containing 11.7% zinc was reported by the British Fuel Research Board (1938). Leutwein and Rösler[197] found a maximum of 2.1% in German coal ash.

Zinc sulfide was reported to be associated with certain coal deposits in Missouri.[411,412] The large amounts found show that the coal had been mineralized by ore-bearing fluids. A zinc-bearing peat deposit in New York was described.[413] Sphalerite (ZnS) was identified[414] in nodules of pyrite-marcasite in shale immediately above the coal in the No. 6 bed at Bichnell, Ind.; the pyrite-marcasite was estimated to contain 0.27% sphalerite. Sphalerite was identified also in an English coal from Leicestershire.[415]

Mott and Wheeler[395] found zinc in the inherent ash of clarain and durain from the Parkgate bed, Yorkshire, England. The clarain and durain, from which extraneous mineral matter had been removed, contained 0.31 and 0.35% zinc, respectively, in the ash. Horton and Aubrey[141] reported a maximum of 0.07% zinc in the ash of three English vitrains. The distribution of zinc in coal carbonization products was investigated.[416,417] For additional reports on zinc occurrence in coal see Moss et al.,[418] Savul and Ababi,[403] Thurauf and Assenmacher,[419] and Weaver.[420]

Zinc concentrations in U.S. coals reported by Gluskoter[137] and Zubovic et al.[173] are shown in Table 74. Zinc has a high positive correlation coefficient for Cd; 0.94 for coals of the Illinois Basin,[137] in the mineral sphalerite (ZnS). Cadmium is found in solid solution in the sphalerite in concentrations as high as 65 ppm. The sphalerite in coal is epigenetically deposited along cleats and in clastic clay dikes. Barium, which occurs as the mineral berite ($BaSO_4$) is also closely correlated with zinc and cadmium.

Zinc concentration in North Dakota lignite is reported to be 11 ppm.[122] Hansen et

Table 74
ZINC IN U.S. COALS[137,173]

Region	Coal	Conc range (ppm)	Arithmetic mean (ppm)	Geometric mean (ppm)
Appalachian	Bituminous	1.3—1100	21	12
Interior plains	Bituminous	1.2—5100	50	51
N. Great Plains	Lignite	1.0—88	7.5	4.5
N. Great Plains	Subbituminous	0.88—220	18	11
Rocky Mountain	Subbituminous	0.92—910	29	13
Rocky Mountain	Bituminous	0.85—120	13	8.6
All U.S.	Different	0.03—51000	96	14
Illinois Basin		10—5300	250	87
Eastern U.S.		2.0—120	25	19
Western U.S.		0.30—17	7.0	5.0

al.[181] have reported zinc concentration in the ash of high-volatile bituminous coal (8.9% ash) from Utah to be 61 ppm.

Zinc concentration in coal samples from Nebraska was found by Burchett[3] to be within concentration range 24 to 974 ppm. Illinois Basin coals studied by Wewerka et al.[199] had a zinc concentration of 120 to 200 ppm. Coal samples from seven countries studied by Loevblad[132] had zinc within the range 0.3 to 116 ppm, with mean value of 23 ppm.

28. Gallium (Ga)

Ramage has found small quantities of gallium in some flue dust in England as early as 1927. Morgan and Davies[421] have investigated the problem of recovering gallium. They found varying amounts of these elements in 72 flue dusts and ashes that were collected from gas works. Occurrence of gallium coal ashes was also discussed by Cooke.[422]

Gallium in the dusts ranged from 0.04 to 0.55% with one sample showing 1.58%. Their investigation showed that flue dust is a potential source of gallium. A method of recovery was tested in which gallium trichloride was extracted from the residue with ether (see Abernethy and Gibson[178]).

Gallium also was investigated[423-426] in Japanese coals and methods were developed for its recovery as a by-product of germanium extraction.

Gallium concentration in U.S. coals has been studied by Gluskoter.[137] The results are shown in Table 75. The paper by Volodarskii et al.[427] gives the results of an experimental investigation of the distribution of gallium in brown and hard coals and the products of their combustion. For some types of coals a liner correlation has been established between the amount of gallium in the coal and its ash content. A linear correlation has also been established experimentally between the amount of gallium in the slag and incompletely burnt coal in the combustion of coals in which gallium is concentrated in the organic matter.

Gallium concentration in North Dakota lignite (average for 12 samples) is 4.6 ppm.[122] A very interesting result has been reported by Bonnett and Czechowski.[428] In their experiment, powdered dried coal (92 kg) from Draw Mill Colliery (South Midlands) was extracted with 7% v/v sulfuric acid in methanol for 12 hr. The filtrate was diluted, and much of the organic material was transferred into chloroform, which was washed (aqueous $NaHCO_3$, water) and taken to dryness (42.8 g). This crude extract was submitted to column chromatography on silica gel, the elutriant being benzene-methanol. Porphyrin-containing fractions were combined and submitted to preparative thin layer chromatography (TLC) on silica gel irrigated with benzene-ammoniacal

Table 75
GALLIUM CONCENTRATION IN U.S. COALS

Area	Conc range (ppm)	Arithmetic mean (ppm)	Geometric mean (ppm)
Illinois Basin	0.8—10	3.2	3.0
Eastern U.S.	2.9—11	5.7	5.2
Western U.S.	0.80—6.5	2.5	2.1

After Gluskoter, H. J., Trace Elements in Coal: Occurrence and Distribution, Circ. 499, Illinois State Geological Survey, Urbana, 1977.

methanol. Three red bands, barely separated, appeared at R_F 0.5 to 0.55. These were combined, extracted, and submitted to further purification by TLC. Finally the orange-red material was extracted with dichloromethane. The extract was filtered, concentrated, and treated with petroleum ether to give the metalloporphyrin concentrate (17.8 mg representing 0.19 ppm of coal).

The metalloporphyrin concentrate thus obtained was a purple solid, not obviously crystalline but possesing a metallic sheen, and it did not melt below 300°C. Metal analysis by thermal neutron activation showed that gallium was the principal metal present, with much smaller amounts of sodium, iron, potassium, magnesium, and aluminum. These latter metals possibly represent inorganic impurities, although iron and aluminum may both be present as metalloporphyrins. The mass spectrum indicated a mixture of homologous metalloporphyrins. The largest peak, m/e 517.185 correspond to $C_{30}H_{32}N_4$ ^{69}Ga (the axial ligand having been lost), and was attended by a second peak at m/e 519.188 ($C_{30}H_{32}N_4$ ^{69}Ga). The second most abundant pair of ions (m/e 547.545) correspond formally to the gallium complex of tetraethyltetramethylporphyrin (etioporphyrin). Although heavy ion may well represent an isomeric mixture, the formulations suggested above are favored on geobiogenetic grounds and the general analogy with the petroporphyrins.

Small-scale experiments on other coals gave similar but not identical results. Three other British bituminous coals (Annesley, South Nottinghamshire), contained metalloporphyrins with similar properties, as did a Turkish lignite (Ankara-Beypazari). However, samples of a German brown coal and a Polish brown coal (Konin) contained a porphyrin fraction with different properties. With the amounts of porphyrin detected (0.6 ppm) it is important to appreciate the dangers of contamination, especially in a laboratory where porphyrins have been studied for many years.

The presence of gallium in coal porphyrins is unexpected: the abundance of gallium in the crust of the Earth (15 ppm) is an order of magnitude less than that of vanadium, and although some organisms accumulate gallium, the biological function of this does not seem to be understood. The vanadyl porphyrins which occur in crude oil are thermodynamically and kinetically stable, and it may be argued that once formed they persist through geological time. Whether a similar argument can be applied to gallium porphyrins is not known. Hansen et al.[181] have reported gallium concentration in the ash of high-volatile bituminous coal (8.9% ash) from Utah to be 46 ppm.

Gallium concentration in coal samples from Nebraska is found to be within the range 3 to 30 ppm.[3] Raw coals from three Illinois Basin preparation plants studied by Wewerka et al.[199] had an average value of 12 ± 2 ppm.

Gallium in coal has been studied by many other research groups including Asai and Inagaki,[429] Bertetti,[430] Takacs et al.,[431] Takacs and Horvath,[432] Kranz et al.,[257] Lose et al.[205] and Zharov.[433] Distribution of gallium in brown coal was studied by Gordon and

Saprykin,[434] Idzikowski and Sachanbinski,[362] and Komissarova et al.[343] Manokin and Chernyak[436] have studied gallium behavior during industrial semicoking and gasification of coal. Ga in coals, and its associations with trace elements under different geochemical conditions of humid lithogenesis is described by Saprykin et al.[437] Many authors have discussed analytical methods used for gallium determination in coal (e.g., Naser et al.[48])

29. Germanium (Ge)

Review articles on germanium concentrations in coal are published by Gibson and Selvig[185] and later by Abernethy and Gibson.[178]

The number of papers on germanium occurrence in coal is very large. It was Goldschmidt[4,5,439] who was among the first to find small amounts of germanium in coal, coal ash, flue dust, and coal tar. He estimated the average amount of germanium in coal to be about 0.001%. However, some samples greatly enriched in germanium were found (1.6% GeO_2 in the ash). Goldschmidt suggested that germanium is concentrated in the inherent ash and that a high content of rare elements can be expected only in ash from coals that have a low percentage of ash, but that many low-ash coals may be poor in rare elements.

From analyses of coals containing more than 0.001% germanium, Vistelius[440] presented data indicating that germanium is combined with the organic matter of coal.

Development of germanium transistors by the electronics industry in 1948 increased demand for the metal and created almost a worldwide interest in coal as a possible new source of germanium. Although most of the production of germanium was from by-products of smelting zinc ore, small quantities were known to occur in coal.[441,442]

Stadnichenko et al.[443] have studied the concentration of germanium in the ash of American coals. They have reviewed the available literature on germanium in coals and related the findings to a detailed analysis of over 700 American coal samples. Column samples from each mine were cut into blocks which were then split in half; one half was saved for petrographic analysis, and the remaining portion was ground to −60 mesh and ashed in a cold muffle furnace. The ashed sample was then analyzed for Ge content by spectrographic methods. Detailed results of the analysis on the ashed samples and pertinent geological information and general observations were presented.

Some observations noted from the analyses were that Ge content varied through the coal sample but generally was concentrated at the top and/or bottom of the same; samples of coal from the Lower Kittanning seam taken in Ohio had the largest concentrations of Ge in the ash (up to 0.13%); Ge concentration in the ash decreased as the amount of ash increased; and no correlation was found in the lateral Ge concentration in ash of the lower and Middle Kittanning coal samples. It was generally concluded from these analyses that the higher concentrations of Ge are found in woody coal occurring in isolated sediments (logs, kettle bottoms, etc.), and in the tops or bottoms of the coal beds. However, a few exceptions were noted in some western coal samples. Also more information about the Ge content in the rocks of the surrounding regions are required to make a definite conclusion about the Ge content in coal.

Machin and Witters[444] have determined germanium concentrations in fly ash, collected from several power plants in Illinois, to determine the feasibility of utilizing the ash as a source for Ge. The element was identified using a spectrochemical method. Germanium concentrations ranged from 0.0004 to 0.046% based on the assumption that it was present in the ash as the oxide (GeO_2), and the fly ash studied was not considered to be an economic source of Ge.

Breger and Schopf[445] reported up to 4% germanium in the ash of coalified logs found in the Chattanooga shale in Tennessee and in the Ohio shale. A related occurrence was reported in Japan where lignitic logs associated with lignite beds contained more germanium than the beds. One sample of lignitic log from the Tsukidote area was reported to contain 1970 ppm germanium on the ash-free, dry basis.[446] Bed samples of lignite contained a maximum of 357 ppm.[447]

Headlee and Hunter[194] used a quantitative spectrographic method, and determined germanium in 35 column samples of West Virginia coals representing 16 coal beds. The germanium content of ash in cubes cut from the column samples ranged from less than 0.003%, the lower limit of detection, to 0.17% in ash from the top 1-in. cube of a column from the Sewell bed. Average GeO_2 content of the coal columns computed as parts per million in ash-free coal ranged from 0.4 to 31 or 0.3 to 22 ppm of germanium. Top and bottom layers of most of the column samples contained more germanium than the remainder of the bed. Other investigators[443,448-450] noted a similar preferential distribution of germanium in coal beds.

Germanium in coal samples from Kansas has been analyzed by Schleicher.[451] Spectrograhic analyses of 20 coal seams from 117 locations in Kansas were made to determine the Ga content of these coals in relation to economic recovery of Ge. The concentration of Ge in the coal ashes ranged from 0.0018 to 0.0575% and in whole coal samples ranged from 0.0006 to 0.0116%. The Ge concentration was reported from each sampling site in the study. The author concluded that at least some of the Kansas coals contain sufficient Ge to warrant recovery of this metal from coal ash.

The Bureau of Mines[452,453] investigated the occurrence and distribution of germanium in the stoker ash and fly ash from coal-fired steam boilers. Thirteen boilers were tested that included slag tap and dry bottom pulverized coal furnaces, cyclone-fired furnaces, and those having underfeed and traveling-grate stokers. Results showed that much of the germanium present in the coal is concentrated in the fine fly ash leaving the boiler. Usually no germanium was found in the slag or refuse from the furnaces or in the stoker ash. In those tests where samples were obtained from the dust-collector hoppers, the germanium content of the fly ash ranged from 15 to 60 ppm, whereas the germanium content of the coal fired ranged from 2.4 to 9 ppm. The sample of fly ash from a cyclone furnace contained 290 ppm of germanium, and one from a furnace fired with a chain-grate stoker contained 530 ppm. The concentration of germanium in these two samples was ascribed to the relatively small proportion of ash discharged as fly dust. Although a considerable quantity of germanium is potentially available from the fly ash from some furnaces, the Bureau researchers concluded at the time of the investigation that the concentration generally was too low for fly ash to be considered a commercial source of germanium. They suggested that attention be directed toward methods of coal utilization that tend to concentrate the germanium in the products of combustion. Work in England[454] also indicated that more enrichment of germanium in flue dust may occur in traveling-grate and stoker-fired furnaces than with pulverized coal firing. Recirculation of fly ash in traveling-grate furnaces was studied in Czechoslovakia as a means of producing a product enriched in germanium.[444]

Analyses by the Illinois Geological Survey[455] of 34 samples of fly ash from public utility power plants ranged from 28 to 319 ppm germanium. Spectrographic analyses of ashes from 24 samples of Kansas coals[456] ranged from 36 to 680 ppm germanium in the ash or 7 to 48 ppm in the coal. A second series of 117 samples of ash from Kansas coals[451] showed a maximum of 1070 ppm germanium in the ash (116 ppm in coal), but only 27 of these samples contained more than 200 ppm germanium in the ash.

Hawley[198] has studied the germanium content of some Nova Scotian coals. Samples from each coal seam were ground and then ashed in an electric furnace at 400°C for 24 to 48 hr. The ashed samples were analyzed using rhodium standards (sparks), and bismuth standards (DC arc) spectrographic methods, and by X-ray fluorescence. Also, one sample (Harbour coal seam) was analyzed by the U.S. Geological Survey using

DC arc method. Of the 11 seams, 3 were sampled over a distance of 2.8 to 26 mi in order to detect any lateral and vertical variations of Ge in the coal seams. Analyses for Ge content by the different methods were in reasonably good agreement. The report gives the average results of Ge and the 14 other elements analyzed for in over 180 samples. Other tables included in the report are the average Ge concentrations for the 11 coal seams and some rider seams; the lateral and vertical variations of Ge content in the Backpit, Harbour, Gardner, and Mullins coal seams; the banded ingredients and Ge content in the Harbour coal seam of sampled petrographic intervals; and the percentage of macerals and Ge of the Harbour coal seam. Overall it was found that the higher Ge concentrations occurred at the top or bottom of the seams and that the Ge content of the samples from the Nova Scotian coal seam were somewhat higher than the Ge content found in Pennsylvania coals.

Germanium in coal was investigated in many countries. Gunduz and Onabasioglu,[457] have investigated Turkish coal, Durie and Schafer[249] and Pilkington[458] have investigated coal in Australia; Simek,[459] Simek et al.[460] in Czechoslovakia; Kunstmam and Hammersma[461] in South Africa; Belugou and Dumoutet[462] in France; Wai and Wang[463] in Taiwan; Rouir[464] in Belgium; Schreiter[465] in Germany; Takacs et al.[431] in Hungary; Rishi,[466] Subrahmanyan and Nair,[467] Banerjee et al.,[468] Mukherjee and Dutta,[469] in India; Kranz and Witkowska,[470] Gregorowicz,[471] Mielecki and Mraz[472] in Poland; Mantea et al.[473] in Romania; Bertetti[430] in Italy.

Germanium determinations in samples from most of the coal deposits in Russia[474] showed that the range of 0.1 to 1% in ash occurred in only 19 samples from the Donetz basin, 13 from the Ural and Petckora regions, 1 from Barents Island, and 1 from northern Dvina. In 58 ash samples the germanium content was 0.01 to 0.1%. Germanium was detected in the ash of 11 out of 35 coals that were tested from Kazakhstan, and Zilbermintz[474] discussed recovery of germanium from the ash of Donetz coals.

Investigations of Russian coals[434,475-484] showed that germanium is associated mainly with the organic substance rather than with mineral matter, but it was suggested that, in a peat bog under conditions of high water with probable introduction of additional mineral matter, the germanium is absorbed to a lesser degree by the organic material,[485] and some may occur with the mineral matter.[476] Relatively large quantities of germanium ranging from 0.11 to 7.5% in ash of fusain of a lower Juriassic coal laid down under oxidizing conditions were reported.[486]

Ershov[475] suggested that germanium in coal has a bond with the organic matter that is more stable than a sorption bond, and is present as organic Ge^{++} compounds. Gordon[434] pointed out that losses of germanium during thermal decomposition of coal may be caused by carbon monoxide that removes some germanium in the form of carbonyl compounds. Germanium was found in dusts and tar from coke oven-gas dusts.[487,488]

Extensive research on germanium in coal was conducted in Japan.[423,424,429,489,491]

Coals younger than Tertiary age were found to contain germanium.[423] Most germanium-bearing flue dusts contained about 7 ppm germanium, but some enriched samples from gas producers contained up to 2080 ppm or more and gas liquor contained about 0.07 to 0.7 ppm.[424] According to Beard[492] (1955) in 1955, a production of 20 kilograms of germanium per month was expected from coal being burned in an electric plant. Some of the values reported for germanium concentration in coal are shown in Table 76.

Table 77 shows the values obtained by Gluskoter[137] from germanium concentrations in U.S. coals. They have also studied distribution of germanium in the bench sets. The germanium content of the top bench and/or that of the bottom bench are greater than the germanium content of the other benches in all five sample sets. Earlier efforts[11,135,138] demonstrated that germanium is primarily associated with the organic

Table 76
GERMANIUM IN COAL

Source of coal	Conc (ppm)	Ref.
U.S.		
N. Great Plains	1.6	9
Eastern interior	13	9
Appalachian region	5.8	9
West Virginia	0.3—22	359
Australia	9	458
Belgium	1—10	464
Canada, Nova Scotia	1—8	198
Czechoslovakia		
Cheb basin	1200	449
Sokolov district	190	449
England	~7 av	494—497
England, Dorsetshire lignite	1800	498
Hungary	20	431
India	3—9	499
Italy	30—50	430
Japan, lignite	3—24	491
Japan, Tsukidate area	<357	447
Poland	2—39	471
Romania, lignite	300	473
South Africa	0.4—15.6	461

Table 77
GERMANIUM IN U.S. COALS

Area	Conc range (ppm)	Arithmetic mean (ppm)	Geometric mean (ppm)
Western U.S.	0.10—3.0	0.91	0.50
Eastern U.S.	0.10—6.0	1.6	0.87
Illinois Basin	1.0—43.0	6.9	4.8

From Gluskoter, H. J., Trace Elements in Coal: Occurrence and Distribution, Circ. 499, Illinois State Geological Survey, Urbana, 1977.

fraction of the coals in Illinois and not in mineral matter fraction. This and the observation that the germanium is concentrated at the boundaries of the coal bed, the top and the bottom, suggest that the germanium was introduced into the coal bed after burial and thus its origin is not related to conditions in the swamps in which the coal was formed. Rather, the germanium was transported into the coal bed in solution and was assimilated by the coal when geochemical conditions within the coal bed were favorable for the removal of the germanium from the solutions.

Zubovic et al.[117] present a different interpretation for the concentration of elements at the top and the bottom of the coal beds in the Illinois Basin. They attribute these concentrations to greater availability of mineral matter and mineral-rich solutions toward the beginning and end of the accumulation of the plant debris that eventually becomes coal.

The float-sink, or washability, data can be displayed as washability curves and as histograms (see Gluskoter[137]). The washability curve is a type of cumulative curve from which the expected concentration of an element at any given recovery rate of a coal can be read assuming the separation was based on specific gravity differences. Therefore, the abscissa is "recovery of float coal in percent" and should be applicable to any specific gravity separation without regard to the medium in which it is done or the method used.

Figure 17 shows the washability curve and histogram for germanium in a sample from the Davis Coal Member. Germanium is the element with the highest organic affinity in the coals studied. The negative slope of the curve indicates that germanium is concentrated in the clean coal fractions; this is also apparent from the histogram. The histogram indicates that there is a higher concentration of germanium in the 1.60 to 2.79 sp gr fraction than in the greater-than-2.79 sp gr fraction. Apparently, a greater portion of the germanium is concentrated with the clay minerals than with the sulfide minerals, which compose the majority of the 2.79 sink fraction.

The following generalizations represent an adequate summary of the various aspects of germanium occurrence in coal:[500]

1. Coals with a high-vitrain (woody coal) content are much richer in germanium than coal with a low-vitrain content.
2. Low-ash coals are richer in germanium than coals with a high-ash content.
3. Geologically older coals usually have a lower germanium content than more recent coals.
4. Germanium is believed to be associated with the organic matter and not the mineral matter in coal.
5. Germanium is usually concentrated in the top or bottom few inches of coal beds.

Furthermore, Fisher noted that the future of germanium hinges on the degree of substitution by other semiconductor materials that may be made in the manufacture of electronic devices. Other materials mentioned as substitutes for germanium in this field were high-purity silicon and certain tellurium, selenium, indium, and gallium bimetallics.

In addition to studies described here, germanium in coal has been studied by many other authors. Let us mention some. Germanium compounds and its forms in coal were studied by Alekhina,[501] Alekhina and Adamenko,[502] Alekhina and Lisin,[503] Alekhina et al.,[504] Bekyarova and Rushev,[505] Ryabchenko and Lisin,[506,507] Ryabchenko and Shiryaeva,[508] and Syabryai et al.[509,510] Germanium concentration in coal, its distribution in coal beds is described in papers by Zhou,[511] Zahradnik,[512] Zahradnik et al.,[513] Vnukov and Sirotenko,[514] Thomo et al.,[515] Sobynyakova et al.,[516] Paraeev,[517] Maleshko et al.,[518] Lexow and Maneschi,[519] Hampel,[520] Han,[521] Fletcher,[522] Fokina and Klitina,[523] Fortescue,[524] Fisher,[500] Fisel and Gonteanu,[525] Farnand and Puddington,[526] Bunkina et al.,[527] Balikhina et al.,[528] Aubrey,[494-497] Adamkin,[529-532] Kulinenko,[533] Kostin,[534] Kostin and Meitov,[535] Crawley,[536] and Shpirt and Sendul'skaya.[537] Methods of germanium determination in coal have been discussed by many authors including Salikova and Sevryukov,[538] Zagorodnyuk et al.,[539] Selenkina et al.,[540] Patzek,[541-543] Perkova and Pecherkin,[544] Ruschev and Bekyarova,[545] Menkovskii and Aleksandrova,[546] Nurminskii,[547,548] Houzim and Volf,[549] Kekin and Marincheva,[550] Khizhnyak et al.,[551] Kul'skaya and Vdovenko,[210] Ganguly and Dutta,[499] Andrianov and Koryukova,[552] and Adamenko et al.[553]

Investigations of germanium in lignites have been reported by Abbolito,[554] Hallam and Payne,[498] Breger and Schopf,[445] and others.

Origin of germanium in coal has been discussed by Ryabchenko,[555] Kulinenko,[556] Ryabchenko et al.,[557] and Shirokov.[558,559] Migration of germanium with natural waters and its accumulation in coals is described by Vekhov et al.[560] Forms of germanium migration and deposition of germanium in coal have been discussed by Sedenko and Kuznetsova.[561] Sharova et al.[562] have discussed the nature of bonding of germanium with acids in coals.

Kuznetsova and Vekhov[563] have studied sorption of germanium by brown coal. Lazebnik et al.[564] have studied correlation between germanium and sulfur contents in different coals from the U.S.S.R.

The fate of germanium and its compounds during coal utilization processes has been described by Adamenko,[565] Adamenko and Yavorskii,[566,567] Akulova et al.,[568] Bentisianov et al.,[750] Aubrey and Payne,[497] Medvedev and Akimova,[569] and Medvedev et al.[570-573]

Germanium concentrations in ash, fly ash, and the possibility of germanium recovery are discussed by Singh and Mathur,[574] Steward,[575] Pogrebitskii,[576] Headlee,[577] Iyer and Sundaram,[578] Kranz and Witkowska,[470] Thompson and Musgrave,[579] Williams,[580] Chirnside and Cluley,[581] Egorov and Kalinin,[582] Eidel'man et al.,[583] El'khones et al.,[584] Bylyna et al.,[585] and Aleksandrova et al.[586-588]

30. Arsenic (As)

Early work on arsenic determination in coal is summarized by Gibson and Selvig.[185] One of the first reports on arsenic in coal is work by Simmersbach.[590]

Goldschmidt and Peters[189] found 0.05 to 0.1% arsenic in the ash of certain coals and pointed out that dust in industrial cities may carry arsenic. They found that atmospheric dust in Hamburg, Germany, contained 0.007 to 0.033% arsenic. Somewhat similar results were found in Leeds, England, where the public analyst reported that six dusts collected from public buildings contained 0.003 to 0.023% arsenic.

Simmersbach[590] found 0.0037% arsenic in a sample of beehive coke from the U.S., and from 0.002 to 0.011% in three out of eight German coke samples tested. Arsenic contents of coal and coal ash reported by several investigators is shown in compilation by Abernethy and Gibson.[178]

Duck and Himus[591] have determined arsenic concentrations in 25 coals and their respective ashes, and in 6 commercial cokes. The form of occurrence as As was established in addition to its behavior during combustion. Coals were combusted at temperatures ranging from 290 to 1050°C (approximate temperature attained in a coke oven). Float-sink tests were performed utilizing mixtures of toluene and carbon tetrachloride. Elemental As was determined by a modification of the Gutzeit process. Arsenic occurred as arsenopyrite in the coals, and to some extent as arsenites and arsenates in the ash. Over 50% of the As was retained in the ash, and coke contained a higher As concentration than the original coke. A portion of the As was said to be in organic association in the coals. Arsenic in U.S. coals as determined by Zubovic et al.[173] and Gluskoter[137] is shown in Table 78. In the work by Gluskoter[137] arsenic is found to be enriched in the samples of coals from the eastern USA (EF = 8.2). In general, arsenic is associated with the sulfide-rich fraction of the coal and most likely is in the solid solution in the ferrous disulfides in coal: pyrite and marcasite.

Gluskoter[137] has reported the washability curve for As to have very steep positive slope. According to him, arsenic is more strongly associated with the mineral matter fraction of the coal than is chromium. It is assumed to be present in solid solution in the iron sulfide minerals.

Arsenic concentrations in North Dakota lignite were reported by Somerville and Elder[122] to be 10.1 ppm (average for 12 samples). Arsenic concentrations in eight samples of coal from Nebraska were in the range 3 to 225 ppm.[3] Arsenic in raw coals from Illinois was in the concentration range 9 to 30 ppm, with mean value 18 ± 8 ppm.[199] Loevblad[132] has summarized values reported worldwide and found the mean value to be 3.2 ppm.

The distribution of arsenic in coal is also discussed by Drever et al.,[202] Kunstmann and Bodenstein,[592] and others. Kogan and Evdokimov[593] have discussed the kinetics of the separate and combined sorption of arsenic and germanium on coals.

Table 78
ARSENIC IN U.S. COALS[137,173]

Region	Coal	Conc range (ppm)	Arithmetic mean (ppm)	Geometric mean (ppm)
Appalachian	Bituminous	0.12—354	20	9
Interior province	Bituminous	0.70—240	16	9.2
N. Great Plains	Lignite	0.70—110	7.6	4.7
N. Great Plains	Subbituminous	0.20—420	5.9	2.3
Rocky Mountain	Subbituminous	0.10—125	6.2	2.2
Rocky Mountain	Bituminous	0.19—60	2.2	1.3
All U.S.	Different	0.10—1650	14	5.1
Illinois Basin		1.0—120	14	7.4
Eastern U.S.		1.8—100	25	15
Western U.S.		0.34—9.8	2.3	1.5

Table 79
SELENIUM IN U.S. COALS[137,173]

Area	Coal	Conc range (ppm)	Arithmetic mean (ppm)	Geometric mean (ppm)
Appalachian	Bituminous	0.12—150	3.7	2.9
Interior province	Bituminous	0.02—75	3.4	2.5
N. Great Plains	Lignite	0.10—3.4	0.69	0.59
N. Great Plains	Subbituminous	0.10—16	0.99	0.73
Rocky Mountain	Subbituminous	0.10—9.6	1.7	1.4
Rocky Mountain	Bituminous	0.10—13	1.5	1.2
All U.S.	Different	0.10—150	1.7	2.56
Illinois Basin		0.4—7.7	2.2	2.0
Eastern U.S.		1.1—8.1	4.0	3.4
Western U.S.		0.4—2.7	1.4	1.3

Determination of arsenic in coal and environmental samples has been discussed by many authors including: Edgcombe and Gold,[594] Crook and Wald,[589] Crawford et al.,[595] Ault,[596] and Benenati.[597] Several methods were applied for arsenic determination in coal. For example, Kekin and Marincheva[598] and Hall and Lovell[599] described spectroscopic determination of arsenic in coal and coal ashes. X-ray fluorescence spectroscopic determination of trace amounts of arsenic in coal is described by Jackwerth and Kloppenburg.[600] Abernethy and Gibson[613] have used colorimetric method for determination of arsenic in coal. AA spectroscopy for arsenic determination in coal has been applied by Guscavage[601] using a graphite tube atomizer, and also by Spielholtz et al.[602] Neutron activation analysis has been used by many research groups; here we shall mention only work by Santoliquido[603] and Orvini et al.[604]

31. Selenium (Se)

First reports on selenium in coal date back to the end of last century.[605] Selenium in U.S. coals was discussed in reports by Zubovic et al.[173] and Gluskoter.[137] Their values for selenium concentrations are shown in Table 79. Figure 18 shows the distribution of selenium in coals analyzed as reported by Gluskoter.[137]

Selenium is the only element enriched in all three regions of the U.S. (Illinois Basin, EF = 40; eastern U.S., EF = 68; western U.S., EF = 26). Selenium is the most strongly enriched of all the elements (see Table 79). Selenium is present in both organic and inorganic combination in coals. It is suggested that at least a portion of the selenium in the coal may be inherited directly from the Se concentrated by plants in the original coal swamp.

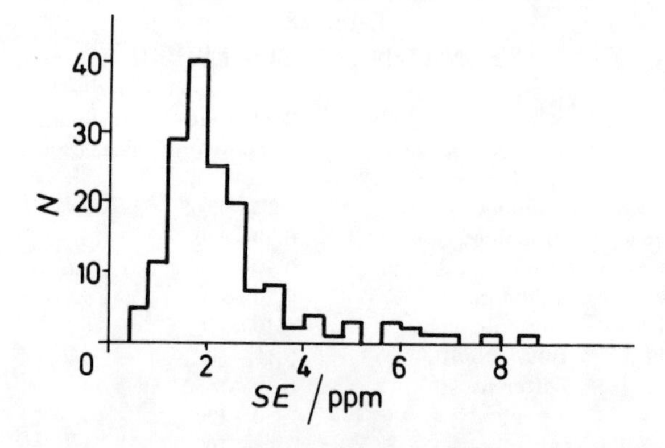

FIGURE 18. Distribution of selenium concentration values for U.S. coals. (As reported by Gluskoter.[137])

In the report by Andren et al.[606] the distribution and form of occurrence of selenium was determined for coal utilized in, and fly ash and flue gas produced at, a typical coal-fired steam plant located in Memphis, Tenn. Coal samples were compiled from grab samples taken at 15-min intervals during flue gas sampling. Flue gases were sampled via a modified ASTM method. Precipitator fly ash samples were composited for selenium analysis. Gas chromatography (GC) was used in conjunction with a microwave emission spectrometric detection system in experiments designed to determine the chemical characterization of selenium by establishing its oxidation states in the combustion by-products. This was based on chemical properties of the following selenium compounds expected in emissions from combustion processes: SeO, SeO_2, SeO_3^{-2}, and SeO_4^{-2}. Extensive fractionation of selenium occurs during coal combustion. Minimum concentrations of the element are present in the slag, while approximately 70% is present in the fly ash, the remainder being vaporized. All selenium in fly ash, slag, and the vapor phase exists as elemental Se, and it is postulated that excess SO_2 produced during combustion acts as a reducing agent in the formation of elemental Se from complex oxides. Sulfuric acid is formed as a reaction by-product during the reduction.

Selenium in North Dakota lignite was determined by Sommerville and Elder[122] to be 0.85 ppm, average value for 12 coal samples. Selenium in eight coal samples from Nebraska was found to be within concentration range 0.7 to 2.9 ppm.[3] Slightly higher values are reported by Wewerka et al.[199] for raw coals from the Illinois Basin: 3 to 7 ppm (mean 5 ± 2 ppm).

Selenium concentration in coal samples from seven countries studied by Loevblad[132] were within range 0.5 to 3.7 ppm, with mean value 1.5 ppm.

Other reports which discuss the occurrence of selenium in coal include Pillay et al.,[607] Belopolskaya and Serikov,[608] Barnes and Lapham,[609] and Gutenmann et al.[610]

32. *Bromine (Br)*

Values for bromine concentrations as reported by Gluskoter[137] are shown in Table 80. Bromine concentration in North Dakota lignite is reported to be 1.7 ppm.[122] Wewerka et al.[199] have reported bromine concentration in raw coals from three Illinois Basin preparation plants to be within the range 0.3 to 4.5 ppm, with mean value 2.7 ± 1.6 ppm.

Coal samples from seven countries studied by Loevblad[132] had bromine within the range 2 to 99 ppm. The highest concentration of 99 ppm was a coal sample from England. The mean value for all samples investigated was 17 ppm. Bromine concentra-

Table 80
BROMINE IN U.S. COALS

Area	Conc range (ppm)	Arithmetic mean (ppm)	Geometric mean (ppm)
Illinois Basin	0.6—52	13	10
Eastern U.S.	0.71—26	12	8.9
Western U.S.	0.50—25	4.7	2.1

After Gluskoter, H. J., Trace Elements in Coal: Occurrence and Distribution, Circ. 499, Illinois State Geological Survey, Urbana, 1977.

Table 81
RUBIDIUM IN U.S. COALS

Area	Conc range (ppm)	Arithmetic mean (ppm)	Geometric mean (ppm)
Western U.S.	0.30—29	4.6	2.4
Eastern U.S.	9.0—63	22	19
Illinois Basin	2.0—46	19	17

After Gluskoter, H. J., Trace Elements in Coal: Occurrence and Distribution, Circ. 499, Illinois State Geological Survey, Urbana, 1977.

tions in U.S. subbituminous coals and lignites are reported to be 2.3 ppm and 3.2 ppm, respectively.[183]

33. Rubidium (Rb)

Little information is available on the rubidium content of coal.[178] Headlee and Hunter[359] found that the ash of column samples of West Virginia coals contained from a minimum of less than 0.027% rubidium, the lower limit of detection, to a minimum of 0.050% with an average of 0.027%. Maximum content found in ash of cube samples from the coal columns was 0.11%. Butler[364] detected rubidium in some of the 35 samples analyzed from two coal beds at Spitzbergen. Rubidium and strontium were found in vitrain ashes from coals in Nova Scotia.[611]

Values for rubidium concentrations in U.S. coals, as reported by Gluskoter[137] are shown in Table 81. Average rubidium concentration in North Dakota lignites (12 samples) is reported to be 4.1 ppm. Rubidium concentration in raw coals from three Illinois Basin preparation plants was within the range 22 to 100 ppm with the mean value of 68 ppm.

Swanson et al.[183] have reported rubidium concentration in U.S. subbituminous coals and lignites to be 5.3 ppm and 0.98 ppm, respectively. The average concentration of rubidium in U.S. coals (2.9 ppm) is much lower than the estimate of worldwide average of 100 ppm (see Reference 2).

34. Strontium (Sr)

Strontium occurs in coals in rather high concentrations. For example, Headlee and Hunter[359] found the strontium content of ash from 31 column samples of West Virginia coals ranged from 0.09 to 0.89%. Individual cube samples cut from the columns showed strontium contents ranging from 0.008 to 10.1% of ash.

Semiquantitative spectrographic analyses by Duel and Annell[193] of western coals showed a maximum concentration of 0.1 to 1.0% strontium in the ash. Chemical

Table 82
STRONTIUM IN U.S. COALS

Area	Conc range (ppm)	Arithmetic mean (ppm)	Geometric mean (ppm)
Western U.S.	93—500	260	220
Eastern U.S.	28—550	130	100
Illinois Basin	10—130	35	30

After Gluskoter, H. J., Trace Elements in Coal: Occurrence and Distribution, Circ. 499, Illinois State Geological Survey, Urbana, 1977.

analysis of one sample of lignite ash from Harding County, S. D., showed 0.45% strontium.

Hawley[128] reported an average of 0.056% strontium and a range of 0.0225 to 0.075% in the ash of Sydney coal from Nova Scotia. Qualitative tests by Newmarch[410] showed the presence of strontium in the ash of Kootney coals from British Columbia.

Butler[364] found 0.1 to 2% strontium in the ash of Spitzbergen coal; De Brito[180] found 0.1 to 2% strontium in the ash of Portuguese anthracite; and Thilo[196] found 0.14% strontium in ash of Newrode coal from Germany. Leutwein and Rösler[199] reported a maximum of 0.1% in German coal. Other early reports of Sr in coal include work by Rafter[220] and Tupper.[611]

The review of reports on strontium occurrence in coals is presented by Abernethy and Gibson.[178] Table 82 shows data on strontium concentration in U.S. coals as reported by Gluskoter.[137] Average strontium concentration in 12 samples of North Dakota lignite is reported to be 0.10%.[122]

Hansen et al.[181] have reported strontium concentration in the ash of high-volatile bituminous coal (8.9% ash) from Utah to be 140 ppm. Strontium in coal samples from Nebraska was studied by Burchett.[3] His eight samples had Sr concentration ranging from 15 to 70 ppm. Strontium in coal samples from seven countries studied by Loevblad[132] was within the range 15 to 286 ppm, with mean value 100 ppm.

Strontium in anthracite, bituminous and subbituminous coals from U.S. is found in concentrations of 100 ppm.[183] Its concentration in lignites is much higher: 300 ppm. It should be noted that the estimated worldwide average concentration is even higher: 500 ppm (see Reference 2).

35. Yttrium (Y)

Abernethy and Gibson[178] have summarized the knowledge on yttrium occurrence in coals. Only a few measurements are reported so far.

From less than 0.001 to 0.052% yttrium was reported in ash of coals from the Northern Great Plains.[173] Duel and Annell [219] detected a maximum of more than 0.1% in the ash of certain western coals, but the range was usually 0.001 to 0.01%. Arnautov and Shipilov[614] reported 0.002 to 0.02% yttrium in coal ashes they examined.

Average yttrium concentration in North Dakota lignite is reported to be 23 ppm.[122] Masursky[342] has suggested that yttrium was probably deposited in coal by water passing through coal beds. Hansen et al.[181] have reported yttrium concentration in the ash of high-volatile bituminous coal (8.9% ash) from Utah to be 66 ppm.

Yttrium in coal samples from Nebraska was found by Burchett[3] to be within the concentration range 3 to 100 ppm. Wewerka et al.[199] have reported the range 8.2 to 15 ppm for yttrium concentration in coals from the Illinois Basin.

The average yttrium concentration in coal worldwide is assumed to be 10 ppm (see Reference 2). This is in agreement with the value accepted for the average concentration of yttrium in U.S. coals.

Table 83
ZIRCONIUM IN U.S. COALS

Area	Conc range (ppm)	Arithmetic mean (ppm)	Geometric mean (ppm)
Western U.S.	12—170	33	26
Eastern U.S.	8—88	45	41
Illinois Basin	12—130	47	41

From Gluskoter, H. J., Trace Elements in Coal: Occurrence and Distribution, Circ. 499, Illinois State Geological Survey, Urbana, 1977.

36. Zirconium (Zr)

Goldschmidt[4] reported that some coal ashes may contain as much as 0.5% zirconium. The zirconium content of ashes examined by Duel and Annell[193] was highly variable, ranging from less than 0.001 to more than 0.1%. Headlee and Hunter[359] reported that ash from column samples of West Virginia coals contained from 0.014 to 0.034% zirconium with a maximum of 0.066% zirconium in ash from cube samples cut from the columns.

Spitzbergen coals tested by Butler[364] contained from 0.004 to 0.11% zirconium in the ash, and De Brito[180] reported 0.01 to 0.1% in anthracite ash from Portugal. Horton and Aubrey[141] found 0.01 to 0.05% in ash of three English vitrains. Degenhardt[615] mentioned a Yorkshire coal that contained 106 ppm zirconium. The summary of these works is presented by Abernethy and Gibson.[178]

Table 83 shows values for zirconium concentrations in U.S. coals as reported by Gluskoter.[137] The average value of zirconium concentration in North Dakota lignite is 68.4 ppm.[122] Hansen et al.[181] have reported Zr concentration in high-volatile bituminous coal (8.9% ash) from Utah to be 220 ppm. Zirconium in coal samples from Nebraska was found to be within the range 10 to 100 ppm.[3] Raw coals from three Illinois Basin preparation plants had mean Zr concentration of 56 ± 6 ppm.

37. Niobium (Nb)

Somerville and Elder[122] have reported measurement of niobium in North Dakota lignite from Mercer County and Dunn county. The average concentration for 12 samples was 3.86 ppm. Hansen et al.[181] have reported Nb concentration in the ash of high-volatile bituminous coal (8.9% ash) from Utah to be 22 ppm. Niobium concentration in coal samples from Nebraska was reported to be within the range 2 to 15 ppm.[3]

The average concentration of niobium in U.S. coals is calculated from the analysis of 799 samples to be 4.5 ppm (see Reference 2). Lignite has somewhat smaller Nb concentration (2.7 ppm) than subbituminous coal (5.4 ppm) (see Swanson et al.[183]).

38. Molybdenum (Mo)

In their report published in 1944 Gibsen and Selvig[185] mentioned only two cases of molybdenum occurrence in coal. One of the cited works was by Ter Meulen,[616] who has discussed the occurrence of Mo in coal and some plants. Traces of a metal sulfide were detected in the soluble portion of Na sulfide during analysis of a coal ash. After a sufficient quantity was isolated, the metal was identified as Mo, and an average concentration of Mo in coal was found to be approximately 21 parts in a hundred million.

Reports by Jorissen[605,617] showed that small quantities of molybdenum were present in the ash and soot of coal from Liege, Belgium. In the more recent summary paper by Abernethy and Gibson[178] there were reports of a few more investigations of molybdenum occurrence in coal. Their summary is shown in Table 84.

Table 84
MOLYBDENUM IN COAL ASH

Source of Coal	Conc in ash (%)	Ref.
U.S.		
N. Great Plains	<0.005—0.0065	10
Texas, Colorado, North and South Dakota	<0.1—1.0	193
South Dakota, Harding County	0.15	
West Virginia	0.003—0.027	359
Pennsylvania, anthracite	0.001—0.009	192
Nova Scotia	0.0018—0.0105	128
Portugal, anthracite	0.001—0.01	180
Germany, Newrode	0.007	196
England, Newcastle	0.033	196
England, vitrain	0.008—0.020	141

From Abernethy, R. F. and Gibson, F. H., Rare Elements in Coal, Rep. BM-IC-8163, U.S. Bureau of Mines, Washington, D.C., 1963.

Table 85
MOLYBDENUM IN U.S. COALS[122,137,173]

Region	Coal	Conc range (ppm)	Arithmetic mean (ppm)	Geometric mean (ppm)
Appalachian	Bituminous	0.18—29	2	1.8
Interior province	Bituminous	0.37—128	5.3	3.2
N. Great Plains	Lignite	0.35—280	3.4	1.5
N. Great Plains	Subbituminous	0.13—41	2.0	1.2
Rocky Mountain	Subbituminous	0.19—19	2.6	1.8
Rocky Mountain	Bituminous	0.24—57	1.6	1.1
All U.S.	Different	0.01—280	3.0	1.8
Western U.S.	—	0.10—30	2.1	0.59
Eastern U.S.	—	0.10—22	4.6	1.8
Illinois Basin	—	0.3 —29	8.1	6.2
North Dakota	Lignite	4—63	22.2	

Molybdenum concentration in North Dakota lignite is reported by Somerville and Elder[122] to be 22.2 ppm (average for 12 samples) (see Table 85). Masursky[342] has suggested that molybdenum was probably deposited in coal by water passing through coal beds.

Molybdenum concentration in coal samples from Nebraska was reported by Burchett[3] to be within the range 3 to 15 ppm. Wewerka et al.[199] have reported Mo concentration range to be 6.8 to 25.0 ppm for Illinois Basin coals.

Coal samples from seven countries studied by Loevblad[132] have Mo concentrations between 0.26 and 2.68 ppm with mean value of 1.2 ppm. Distribution of molybdenum in coal was also discussed by Almassy and Szalay,[354] Fiskel et al.,[618] Korolev et al.,[619] Idzikowski and Sachenbinski,[362] and Drever et al.[202] Processes of absorption of Mo on coal were studied by Matsuda and Abrao.[620] Determination of molybdenum in coal was recently discussed by Dyatel and Vasserman.[621]

39. Ruthenium (Ru)

Somerville and Elder[122] have studied North Dakota lignites using spark source MS. Detection limit of the system used was 0.1 ppm. No ruthenium was observed in the studied samples.

Table 86
SILVER IN U.S. COALS

Area	No. of samples	Conc range (ppm)	Arithmetic mean (ppm)	Geometric mean (ppm)
Western U.S.	22	0.01—0.07	0.03	0.02
Eastern U.S. (Appalachian)	13	0.01—0.06	0.02	0.02
Illinois Basin	37	0.02—0.08	0.03	0.03

After Gluskoter, H. J., Trace Elements in Coal: Occurrence and Distribution, Circ. 499, Illinois State Geological Survey, Urbana, 1977.

40. Rhodium (Rh)

Rhodium in Noth Dakota lignites is reported to be present below detection limit (0.1 ppm) of the method used.[122]

41. Palladium (Pd)

No palladium was observed to be present in concentrations above the detection limit (0.1 ppm) of the method used by Somerville and Elder[122] to study North Dakota lignites.

42. Silver (Ag)

Silver was detected in only a few coal ashes from western coals examined by Duel and Annell;[193] the maximum amount found was 1 to 10 ppm.

Headlee and Hunter[359] found more silver in northern than in southern coals of West Virginia; they reported an average of 9 ppm silver in ash with a range from 5 to 28 ppm. Maximum concentration reported in the ash of coal cubes from the columns was 84 ppm.

De Brito[180] detected silver in the ash of Portuguese anthracite, and Newmarch[612] in Kootney coals of British Columbia. Briggs[622] also reported that an appreciable amount of silver was found in pyrites from the roof of a South Wales coal bed. Popović[623] detected silver in the ash of some coals of the Timok Basin. Leutwein and Rösler[197] reported a maximum of 0.006% silver in German coal ashes.

The review articles on the subject are published by Gibson and Selvig[185] and Abernethy and Gibson.[178] The recent findings by Gluskoter[137] about silver concentration in U.S. coals is shown in Table 86.

Silver was detected in one coal sample from Nebraska studied by Burchett[3] with the concentration of 0.1 ppm. Silver in subbituminous coals from the U.S. is reported to be present in concentration of 0.17 ppm, while in lignite its concentration is slightly higher: 0.26 ppm.[183] The average silver concentration in U.S. coals is 0.20 ppm, while estimated worldwide avearge is 0.50 ppm (see Reference 2).

43. Cadmium (Cd)

The investigation of Cd in coal has not kept pace with the studies of many other trace elements in coal. There are few references on Cd in coal and coal ash published prior to 1969.[624,626] Recently additional analyses have been reported.[16,98,597,604,627] In the only study designed to investigate coals systematically, Swanson[98] analyzed 71 coal samples and found that 61 had a Cd content below detectable limits, which for the analytical method used was approximately 0.5 ppm in a 1.0-g sample of coal ash. The highest amount of Cd in coal reported was 0.7 ppm.[627] Gluskoter and Lindahl[167] have analyzed several Illinois coal seams for cadmium content in correlation with the amount of zinc present to support the author's hypothesis that cadmium occurs pri-

Table 87
CADMIUM IN U.S. COALS[122,137,173]

Region	Coal	Conc range (ppm)	Arithmetic mean (ppm)	Geometric mean (ppm)
Appalachian	Bituminous	0.01—3.1	0.12	0.08
Interior province	Bituminous	0.01—170	5.2	0.43
N. Great Plains	Lignite	0.06—2.7	0.33	0.18
N. Great Plains	Subbituminous	0.04—2.7	0.24	0.16
Rocky Mountain	Subbituminous	0.03—3.7	0.43	0.29
Rocky Mountain	Bituminous	0.02—0.99	0.18	0.13
All U.S.	Different	0.01—180	1.2	0.14
Western U.S.	—	0.10—0.60	0.18	0.15
Eastern U.S.	—	0.10—0.60	0.24	0.19
Illinois Basin	—	0.1 —65	2.2	0.59
North Dakota	Lignite		0.21	

marily as a replacement for zinc in the mineral sphalerite. Portions of 23 representative samples of either face channel or drill cores of Illinois coal seams were ashed using high- and low temperature procedures, then treated with the acid-digestion bomb technique for cadmium and zinc determination by AA spectrophotometry. These results were compared to cadmium determinations done on selected ashed samples by anodic stripping voltammetry and neutron activation analysis. Sphalerite samples obtained from a high specific gravity fraction of a low-temperature ash and an unashed coal sample were digested by nitric acid and analyzed for cadmium and zinc concentrations by AA. Results of analysis were: cadmium content ranged from 0.3 to 28 ppm and zinc content ranged from 18 to 300 ppm in ashed samples; zinc:cadmium ratios in sphalerite samples suggest that most, if not all, of the cadmium in the coal is within sphalerite; presence of sphalerite in discrete particles of low-temperature ashes and coal fractions with high specific gravity suggests that sphalerite occurs as an epigenetic material; a significant reduction in zinc and cadmium contents in some coals can be effected by "washing", due to the high specific gravity of sphalerite.

Schultz et al.[172] have reported cadmium concentration in Pittsburg seam coal to be 0.14 ± 0.05 ppm. Cadmium concentration in ash was in the range 0.36 to 1.22 ppm depending on the combustor properties. The cadmium concentration in various U.S. coals is shown in Table 87.

Petrographic analysis of Illinois Basin coals indicated a separate, epigenetic, zinc-sulfide mineralization phase had concentrated sphalerite containing some Cd in three coals mined in northwestern Illinois.[628] Eighty-one coals from Illinois and Kentucky were analyzed; ratios of Zn:Cd were determined and compared with ratios found in sphalerite samples from the Upper Mississippi Valley district and the Illinois-Kentucky district. Hand-picked sphalerite samples were dissolved in nitric acid, and determinations were made by AA for Zn, Cd, and Fe. Areal comparison of Zn:Cd ratios could not preclude genetic similarities between sphalerite samples from different coal areas or from other lithotypes. The coals contained a Zn:Cd ratio of 48:1 to 358:1, while samples from other strata contained ratios of from 29:1 to 1050:1. Sphalerite in northwestern Illinois coals was found to occur as complex cleat fillings and was associated with quartz, kaolinite, pyrite, and calcite. Mineralization stages consisting of sulfide, silicate, and carbonate stages of paragenetic sequence are discussed and illustrated.

Cadmium in coal samples from Nebraska was found by Burchett[3] to be within the range 0.2 to 6.6 ppm. Raw coals from three Illinois Basin preparation plants had Cd in the range 0.09 to 2.1 ppm, with the mean value 0.5 ppm.

Loevblad[132] has found smaller Cd concentrations in coal samples from seven coun-

Table 88
INDIUM IN SOME U.S. COALS

Area	Concentration range (ppm)	Arithmetic mean (ppm)	Geometric mean (ppm)
Western U.S.	0.01—0.25	0.10	0.07
Eastern U.S.	0.13—0.37	0.23	0.22
Illinois Basin	0.01—0.63	0.16	0.13

After Gluskoter, H. J., Trace Elements in Coal: Occurrence and Distribution, Circ. 499, Illinois State Geological Survey, Urbana, 1977.

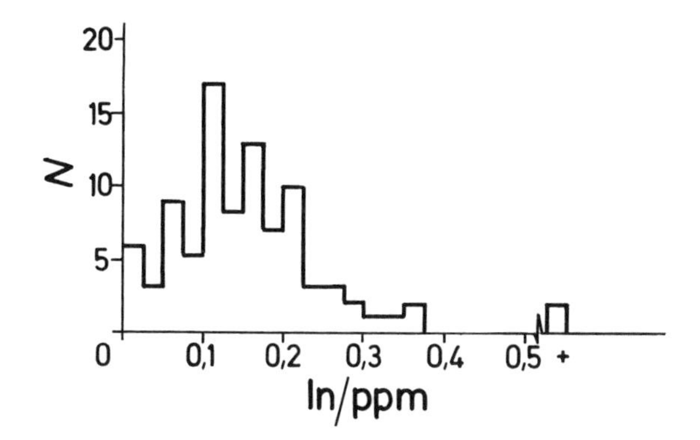

FIGURE 19. Distribution of indium concentration values for U.S. coals. (As reported by Gluskoter.[137])

tries (Norway, England, U.S., W. Germany, Poland, U.S.S.R., Australia). The measured values were within the range 0.1 to 0.3 ppm.

Cadmium concentrations in U.S. coals are reported by Swanson et al.[183] to be anthracite, 0.3 ppm; bituminous coal, 1.6 ppm; subbituminous coal, 0.2 ppm; lignite, 1.0 ppm. The U.S. average value is calculated to be 1.3 ppm (see Reference 2).

44. Indium (In)

Distribution of indium in U.S. coals analyzed by Gluskoter[137] is shown in Figure 19 and Table 88. Loevblad[132] has studied coal samples from seven countries. Cd concentration values found by him were within the range 0.005 to 0.074 ppm, with some kind of world average of 0.022 ppm.

45. Tin (Sn)

Duel and Annell[193] reported that tin showed an erratic distribution pattern in the coals they examined. It was detected in only 7 of 151 samples from Harding County, S. D., but was found in all 48 samples collected in Milam County, Tex., and 5 of these samples contained from 0.1 to 1.0%.

Headlee and Hunter[359] found that tin content in ash from 7 of the 31 column samples examined from West Virginia was less than the limit of detection of 0.008% tin, and that the southern coals were richer in tin than the northern coals. Bed average ranged from less than 0.008 to 0.055% tin in the ash.

Abernethy and Gibson[178] have summarized the knowledge on tin concentrations in coal ash; this is shown in Table 89. Tin concentration in coal was reported to be 12.6 ± 1.7 ppm[126] while the concentration in ash was found to be 111.2 ± 0.21 ppm. Average

Table 89
TIN IN COAL ASH

Source of coal	Conc (%)	Ref.
U.S.		
N. Great Plains	0.002—0.01	10
Pennsylvania, anthracite	0.01—0.09	192
England, vitrain	0.005—0.02	141
England, Newcastle	0.039	196
Germany, Newrode	0.0016	196
Germany	0.1	613
Nova Scotia	0.0009	198
Portugal, anthracite	0.0001—0.1	180
U.S.S.R., Kuznetsk basin	0.03	629

After Abernethy, R. F. and Gibson, F. H., Rare Elements in Coal, Rep. BM-IC-8163, U.S. Bureau of Mines, Washington, D.C., 1963.

Table 90
ANTIMONY IN SOME U.S. COALS[122,137,173]

Region	Coal	Conc range (ppm)	Arithmetic mean (ppm)	Geometric mean (ppm)
Appalachian	Bituminous	0.04—35	1	0.68
Interior province	Bituminous	0.04—16	1.6	0.78
N. Great Plains	Lignite	0.10—4.5	0.52	0.35
N. Great Plains	Subbituminous	0.06—9.1	0.53	0.35
Rocky Mountain	Subbituminous	0.04—43	1.2	0.59
Rocky Mountain	Bituminous	0.10—7.3	0.58	0.39
All U.S.	Different	0.04—43	1.0	0.58
Illinois Basin	—	0.1—8.9	1.3	0.81
Eastern U.S.	—	0.25—7.7	1.6	1.1
Western U.S.	—	0.18—3.5	0.58	0.45
North Dakota	Lignite	0.27—0.9	0.31	

tin concentration in North Dakota lignite was reported to be 5.08 ppm.[122] The calculated average tin concentration in U.S. coals is 1.6 ppm (see Reference 2). Tin in coals of the Kuznetsk Basin in the U.S.S.R. was studied by Borovik and Ratinskii.[629] This value is obtained from the measurement of tin concentration in some 800 coal samples.

46. Antimony (Sb)

There are only a few reports on the antimony occurrence in coal.[178] In a study of ash composition of 31 column samples of coal from 15 beds in West Virginia, Headlee and Hunter[359] reported that antimony was below the limit of detection of 0.004%. Certain parts of the beds contained some antimony, usually at the top or bottom; the greatest amount found was 0.023%.

Horton and Aubrey[141] found 0.01 to less than 0.05% antimony in the ash of three English vitrain samples. According to Otte[78] the maximum antimony content of German coal ash is 0.3%.

Antimony concentrations in different U.S. coals are shown in Table 90. Antimony concentrations in coal samples from Nebraska studied by Burchett[3] were in the range 1.5 to 15 ppm.

Coal samples from seven countries studied by Loevblad[132] had antimony concentrations within the range 0.2 to 2.47 ppm, with the mean value of 0.83 ppm.

Table 91
ANTIMONY IN COAL

Coal	Conc (ppm)
Anthracite	0.9
Bituminous	1.4
Subbituminous	0.7
Lignite	0.7
U.S. av	1.1
Worldwide av	3.0

From U.S. National Committee for Geochemistry, Trace Element Geochemistry of Coal Resource Development Related to Environmental Quality and Health, National Academy Press, Washington, D.C., 1980.

Table 92
IODINE IN SOME U.S. COALS

Area	Conc range (ppm)	Arithmetic mean (ppm)	Geometric mean (ppm)
Illinois Basin	0.24—14	1.7	1.2
Eastern U.S.	0.33—4.9	1.7	1.4
Western U.S.	0.20—1.0	0.52	0.46

After Gluskoter, H. J., Trace Elements in Coal: Occurrence and Distribution, Circ. 499, Illinois State Geological Survey, Urbana, 1977.

Table 91 shows average antimony concentrations in U.S. coals and an estimated worldwide average.[2] Determination of antimony in coals by AA spectrometry is discussed in report by Guscavage,[601] Pollack and West,[630] and some others.

47. Tellurium (Te)

Somerville and Elder[122] have reported tellurium concentration of 0.27 ppm in lignite from Mercer County, N.D. Tellurium concentration in U.S. subbituminous coals and lignites is reported to be 0.1 ppm.[183] This value is assumed to be an overall average for tellurium in U.S. coals.

48. Iodine (I)

There are only few reports on iodine occurrence in coals; they are summarized by Abernethy and Gibson[178] (see also Gibson and Selvig[185]).

Wilke-Dorfurt and Romersperger[631] reported that the iodine content of 12 mid-European coals ranged from about 1 to 11 ppm. Wache[632] found 1 to 6 ppm of iodine in moisture- and ash-free German brown coal and about 1 ppm in bituminous coal. In an examination of a wide range of solid fuels from East Germany, Stolper[633] concluded that iodine is present in these fuels but is almost entirely lost during combustion and gasification; he calculated that the 1950 coal output of East Germany contained 280 to 560 ton of iodine. Although the estimated total amount is large, it averages only 0.001 to 0.002% of the coal.

Gluskoter[137] has reported iodine concentration in U.S. coals, his results are shown in Table 92. Average iodine concentration in 12 lignite samples from North Dakota is reported to be 0.39 ppm.[122] Iodine concentration in U.S. subbituminous coals and lignites is reported to be 1.0 ppm and 1.3 ppm, respectively.[183]

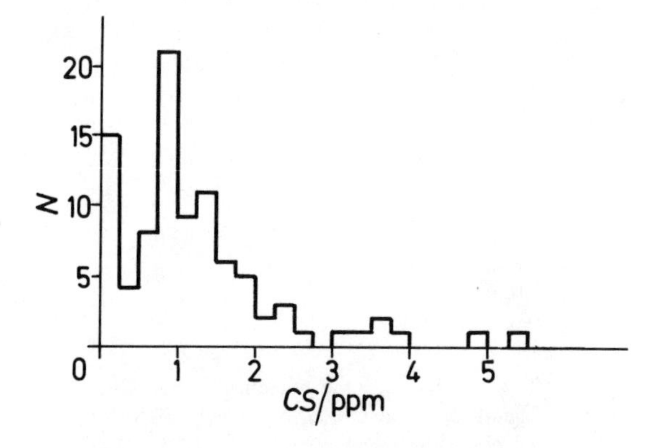

FIGURE 20. Distribution of cesium concentration values for U.S. coals. (As reported by Gluskoter.[137])

Table 93
CESIUM IN SOME U.S. COALS

Area	Conc range (ppm)	Arithmetic mean (ppm)	Geometric mean (ppm)
Western U.S.	0.02—3.8	0.42	0.16
Eastern U.S.	0.40—6.2	2.0	1.6
Illinois Basin	0.5—3.6	1.4	1.2

After Gluskoter, H. J., Trace Elements in Coal: Occurrence and Distribution, Circ. 499, Illinois State Geological Survey, Urbana, 1977.

49. Cesium (Cs)

Data on cesium concentration in U.S. coals as reported by Gluskoter[137] are shown in Figure 20 and Table 93. Cesium concentration in lignite from North Dakota is reported to be 0.26 ppm (average for 12 samples, as reported by Somerville and Elder[122]).

Raw coals from three Illinois Basin preparation plants are reported to have cesium concentration within range 2.9 to 5.0 ppm, with mean value 3.8 ± 0.8 ppm.[199] Average cesium concentration in 800 coal samples from different parts of the U.S. is found to be 0.4 ppm (Reference 2).

50. Barium (Ba)

Review articles on barium occurrence in coals are published by Gibson and Selvig[185] and Abernethy and Gibson.[178] Reynolds[397] found that small quantities of barium commonly occur in English coals. Munch, et al.[634] stated that barium was found in the cleavage planes where mine waters containing barium chloride penetrated into the coal. Reynolds also mentioned several unusual and localized barium deposits associated with British coals. Barite was reported in the roof rocks above a coal bed in Leicestershire and in a cavity of the sandstone rock above a coal bed in Yorkshire.[635] Briggs[622] described a true fissure vein of barite found in a colliery near Durham, England. Barite actually was mined with the coal in 1914; about 200 ton a week were obtained. Witherite, the carbonate of barium, also was present.

Barium contents reported in coal ash by several investigators are given in Table 94. A more recent report on barium content of U.S. coals is presented by Gluskoter.[137] His results are shown in Table 95.

Table 94
BARIUM CONCENTRATION IN COAL ASH

Source of coal	Conc (%)	Ref.
West Virginia	0.05—0.44	61
North Dakota	0.15	16
Alaska, Nenana field	0.4—0.8	144
Engtand	0.0—4.3	136
Nova Scotia	0.0018—0.22	60
Germany, Newrode	0.22	160
Germany, brown coal	0.0001	44
Germany	>0.1	103
Portugal, anthracite	0.01—0.1	18
Spitzbergen	0.1—0.2	19

After Abernethy, R. F. and Gibson, F. H., Rare Elements in Coal, Rep. BM-IC-8163, U.S. Bureau of Mines, Washington, D.C., 1963.

Table 95
BARIUM CONCENTRATION IN SOME U.S. COALS

Area	Conc range (ppm)	Arithmetic mean (ppm)	Geometric mean (ppm)
Western U.S.	160—1600	500	430
Eastern U.S.	72—420	200	170
Illinois Basin	5.0—750	100	75

After Gluskoter, H. J., Trace Elements in Coal: Occurrence and Distribution, Circ. 499, Illinois State Geological Survey, Urbana, 1977.

Somerville and Elder[122] have reported average concentration of barium in 12 lignite samples from North Dakota to be 230 ppm. Hansen et al.[181] have reported barium concentration in the ash of high-volatile bituminous coal (8.9% ash) from Utah to be 940 ppm. Barium concentrations of eight coal samples from Nebraska were reported by Burchett[3] to be in the range 30 to 150 ppm.

Coal samples from seven countries studied by Loevblad[132] had Ba concentrations from 76 to 337 ppm, with mean value of 180 ppm. Barium concentration in U.S. coals as reported by Swanson et al.[183] are anthracite and bituminous coals, 100 ppm; subbituminous coal and lignites, 300 ppm. The all-coal average for U.S. is 150 ppm (Reference 2). The role of barium as catalyst and sulfur scoranger in the stream gasification of coal chars is discussed by Otto et al.[339]

51. Lanthanum (La)

Reports on lanthanum occurrence in coals are summarized by Abernethy and Gibson.[178] In the analysis of 31 column samples of West Virginia coals by Headlee and Hunter[359] lanthanium concentrations ranged from less than 0.030%, the lower limit of detection, to 0.043% in the ash. Lanthanum was the only rare earth that could be detected, but it was noted that several others may be present in similar concentration. The maximum concentration in a part of one bed was 0.082% lanthanum in the ash.

Duel and Annell[193] reported that lanthanum was not present in percentages high enough to be detected in most of the western coals tested. The maximum concentration

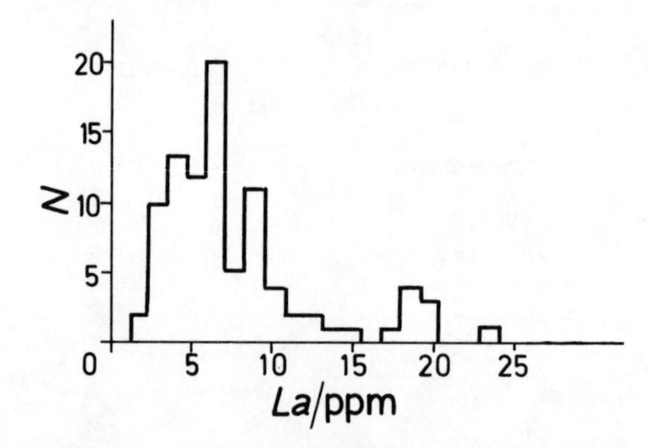

FIGURE 21. Distribution of lanthanum concentration values for U.S. coals. (As reported by Gluskoter.[137])

detected by a semiquantitative method was in the range of 0.01 to 0.1% lanthanum in the ash. De Brito[180] reported a similar maximum concentration in ash of Portuguese anthracites. Butler[364] detected lanthanum in the ash of Svalbard coals. Zubovic et al.[9] reported a range from less than 0.003% to 0.036% lanthanum in ash of coals of the Northern Great Plains.

A recent determination of lanthanum in U.S. coals is reported by Gluskoter.[137] His results are shown in Figure 21 and Table 96. Lanthanum in North Dakota lignite is present with concentration of 5.8 ppm.[122]

Masursky[342] has suggested that La was probably deposited in coal by water passing through coal beds. Hansen et al.[181] have reported La concentration in the ash of high-volatile bituminous coal (8.9% ash) from Utah to be 76 ppm.

Lanthanum concentration in coal samples from Nebraska[3] were reported to be within range 7 to 100 ppm. Illinois Basin coals studied by Wewerka et al.[199] had La concentrations from 15 to 31 ppm. All-coal average value for the U.S. is reported to be 6.1 ppm (see Reference 2); this is close to an estimated worldwide average of 10 ppm.[184]

52. Rare Earths (Lanthanides)

The lanthanides include Ce, Pr, Nd, Pm, Sm, Eu, Gd, Tb, Dy, Ho, Er, Tm, Yb and Lu.

The most extensive report on rare earths occurrence in coal is the paper by Gluskoter;[137] Table 97 shows the results. Information on cerium, samarium, europium, ferbium, dysrpsium, ytterbium, and lutetium are shown.

Somerville and Elder[122] have studied the elemental composition of lignite from North Dakota using spark source MS. Their results are shown in Table 98. Masursky[342] has suggested that Nd was probably deposited in coal by water passing through coal beds. Hansen et al.[181] have reported Yb concentration in high-volatile bituminous coal (8.9% ash) from Utah to be 7 ppm. Rare earth elements in some coal basins of Bulgaria were recently reported by Eskenazy.[636]

Terbium was also studied by Cahill et al.[130] in 70 coal samples, of which 31 are from the Illinois Basin which includes parts of Indiana, Illinois, and western Kentucky. There are 14 samples from eastern states, 22 from western states, and 3 from Iowa. The vertical distribution of trace elements was studied in five coal seams from the Illinois Basin. Elements such as Tb often show relatively flat distributions.

Figure 22 shows distributions of cerium, samarium, europium, dysporsium, ytterbium, and lutetium in U.S. coals.

Table 96
LANTHANUM IN SOME U.S. COALS

Area	Conc range (ppm)	Arithmetic mean (ppm)	Geometric mean (ppm)
Illinois Basin	2.7—20	6.8	6.4
Eastern U.S.	6.1—23	15	14
Western U.S.	1.8—13	5.2	4.5

After Gluskoter, H. J., Trace Elements in Coal: Occurrence and Distribution, Circ. 499, Illinois State Geological Survey, Urbana, 1977.

Wewerka et al.[199] have studied raw coals from three Illinois Basin preparation plants. They have measured concentrations of the following rare earths: Ce, Sm, Eu, Tb, Dy, Yb, and Lu. The concentrations found by them are shown in Table 99.

Concentrations of rare earths in coal is shown in Table 100 which is obtained from Reference 2. Concentration values for U.S. subbituminous coal and lignite are after work of Swanson et al.,[183] while the estimates for worldwide averages are after work of Bertine and Goldberg.[184]

Hafnium (Hf)

Data on hafnium concentration in U.S. coals are reported by Gluskoter[137] (see Table 101). Hafnium in raw coals from three Illinois Basin preparation plants are reported by Wewerka et al.[199] to be within the range 1.3 to 2.9 ppm, with the mean value of 1.8 ppm. The U.S. all-coal average hafnium concentration is estimated to be 0.60 ppm (Reference 2).

54. Tantalum (Ta)

Data on tantalum concentration in some U.S. coals are reported by Gluskoter[137] (see Table 102). Wewerka et al.[199] reported tantalum concentration in Illinois Basin coals to be in the range 0.3 to 1.0 ppm, with the mean value of 0.7 ppm.

55. Tungsten (W)

Little information is available on the amount of tungsten in coal ash.[178] Headlee and Hunter[359] reported that tungsten was below the limit of detection (80 ppm) in ash from the southern and several of the northern West Virginia coals they tested. The maximum bed average detected was 150 ppm tungsten in the ash. Analyses of cube samples from a column of the Lower Kittanning bed gave the maximum of 0.044% tungsten in the ash.

Nunn et al.[192] found that the tungsten content in ash from Pennsylvania anthracites ranged from 10 to 90%. Gluskoter [137] has studied tungsten in U.S. coals and his findings were as follows: Illinois Basin, range 0.04 to 4.2 ppm, with arithmetic mean 0.82 ppm and geometric mean 0.63 ppm for 56 investigated samples; eastern U.S., range 0.22 to 1.2 ppm, with arithmetic mean 0.69 ppm and geometric mean 0.62 ppm for 14 samples investigated. Western U.S. range 0.13 to 3.3 ppm, with arithmetic mean 0.75 ppm and geometric mean 0.58 ppm for 22 samples investigated.

Tungsten concentration in North Dakota lignites (Dunn County) is reported by Somerville and Elder[122] to be 0.58 ppm (average of 12 samples). Tungsten concentration in Illinois coal was reproted by Wewerka et al.[199] to be within range 0.5 to 1.2 ppm.

56. Rhenium (Re)

Rhenium concentration in North Dakota lignite is reported to be below detection

Table 97
CONCENTRATION OF RARE EARTH ELEMENTS IN SOME U.S. COALS

	Illinois Basin			Eastern U.S.			Western U.S.		
Element	Range (ppm)	Arithmetic mean (ppm)	Geometric mean (ppm)	Range (ppm)	Arithmetic mean (ppm)	Geometric mean (ppm)	Range (ppm)	Arithmetic mean (ppm)	Geometric mean (ppm)
Cerium	4.4—46	14	12	11—42	25	23	2.8—30	11	9.1
Samarium	0.4—3.8	1.2	1.1	0.87—4.3	2.6	2.4	0.22—1.4	0.61	0.56
Europium	0.1—0.87	0.26	0.25	0.16—0.92	0.52	0.47	0.07—0.60	0.20	0.16
Terbium	0.04—0.65	0.22	0.18	0.06—0.63	0.34	0.28	0.06—0.58	0.21	0.17
Dysprosium	0.5—3.3	1.1	1.0	0.74—3.5	2.3	2.0	0.22—1.4	0.63	0.57
Ytterbium	0.27—0.15	0.56	0.53	0.18—1.4	0.83	0.73	0.13—0.78	0.38	0.34
Lutetium	0.02—0.44	0.09	0.08	0.04—0.40	0.22	0.18	0.01—0.43	0.07	0.05

After Gluskoter, H. J., Trace Elements in Coal: Occurrence and Distribution, Circ. 499, Illinois State Geological Survey, Urbana, 1977.

Table 98
RARE EARTHS IN LIGNITE

Element	Conc range (ppm)	Av of 12 samples (ppm)	Conc in ash (ppm)
Ce	11—34.6	14.1	37—190
Pr	0.5—1.5	0.85	2—8
Nd	0.8—2.7	0.96	3—18
Pm	—	—	—
Sm	0.45—1.07	0.47	2—7
Eu	0.3—0.4	0.26	0.5—4
Gd	<0.1—0.8	0.23	0.9—5
Tb	<0.1—0.67	0.15	0.6—3
Dy	<0.1—0.67	<0.1	2—8
Ho	<0.1—0.4	<0.1	0.6—5
Er	<0.1	<0.1	0.5—4
Tm	<0.1	<0.1	0.2—0.5
Yb	<0.1	<0.1	1—4
Lu	<0.1	<0.1	0.1—0.5

After Somerville, M. H. and Elder, J. L., in Environmental Aspects of Fuel Conversion Technology III, Ayer, G. A. and Massoglia, M. F., Eds., Environmental Protection Agency, Washington, D.C., 1977.

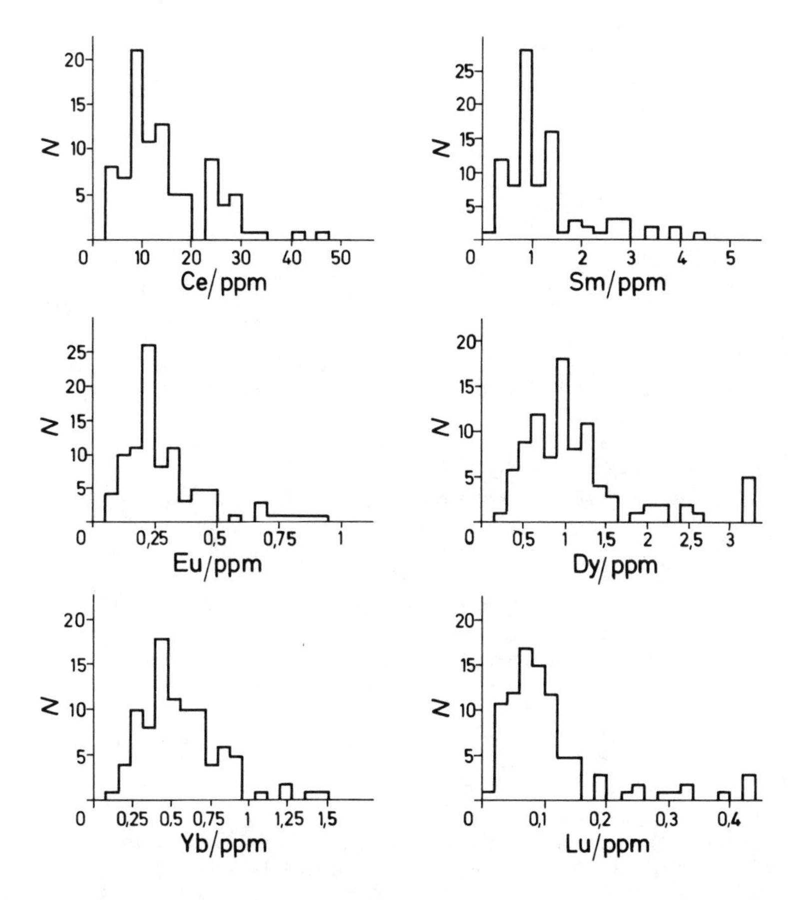

FIGURE 22. Distributions of cerium, samarium, europium, dysporsium, ytterbium, and lutetium in U.S. coals. (As reported by Gluskoter.[137])

Table 99
RARE EARTHS IN RAW COALS FROM ILLINOIS BASIN PREPARATION PLANTS

Element	Conc (ppm)	Mean (ppm)
Ce	28—61	46
Sm	2.3—5.6	3.8
Eu	0.5—1.4	0.8
Tb	0.7—1.0	0.8
Dy	1.8—4.1	2.8
Yb	1.0—2.8	1.7
Lu	0.14—0.39	0.25

After Wewerka, E. M., Williams, J. M., and Wanek, P. L.[199]

Table 100
CONCENTRATION OF RARE EARTH ELEMENTS IN COAL (PPM)

Element	U.S. subbituminous	U.S. lignite	U.S. av	Worldwide av
Cerium (Ce)	5.5	12.3	7.7	11.5
Praseodymium (Pr)	6.1	2.7	2.7	2.2
Neodymium (Nd)	50	11	37	4.7
Samarium (Sm)	0.50	0.27	0.42	1.6
Europium (Eu)	0.61	0.13	0.45	0.7
Gadalinium (Gd)	0.13	0.21	0.17	1.6
Terbium (Tb)	0.1	0.1	0.1	0.3
Dysprosium (Dy)	2.7	1.4	2.2	—
Holmium (Ho)	0.13	0.06	0.11	0.3
Erbium (Er)	0.46	0.16	0.34	0.6
Thulium (Tm)	0.07	0.07	0.07	—
Ytterbium (Yb)	0.5	1.5	1	0.5
Lutetium (Lu)	0.09	0.05	0.08	0.07

From U.S. National Committee for Geochemistry, Trace Element Geochemistry of Coal Resource Development Related to Environmental Quality and Health, National Academy Press, Washington, D.C., 1980.

Table 101
HAFNIUM IN SOME U.S. COALS

Area	Conc range (ppm)	Arithmetic mean (ppm)	Geometric mean (ppm)
Illinois Basin	0.13—1.5	0.54	0.49
Eastern U.S.	0.58—2.2	1.2	1.1
Western U.S.	0.26—1.3	0.78	0.70

After Gloskoter, H. J., Trace Elements in Coal: Occurrence and Distribution, Circ. 499, Illinois State Geological Survey, Urbana, 1977.

Table 102
TANTALUM CONCENTRATION IN SOME U.S. COALS

Area	Conc range (ppm)	Arithmetic mean (ppm)	Geometric mean (ppm)
Illinois Basin	0.07—0.3	0.15	0.14
Eastern U.S.	0.12—1.1	0.33	0.26
Western U.S.	0.04—0.33	0.15	0.12

After Gluskoter, H. J., Trace Elements in Coal: Occurrence and Distribution, Circ. 499, Illinois State Geological Survey, Urbana, 1977.

limit (0.1 ppm) of the method used by Somerville and Elder.[122] Rhenium occurrence in nature is also discussed in papers by Kalinin et al.[637] and Martin and Garcia-Rossell.[638,639]

57. Osmium (Os)

The concentration is below detection limit (0.1 ppm) in North Dakota lignite.[122]

58. Iridium (Ir)

Iridium concentration in North Dakota lignite is below 0.1 ppm.[122]

59. Platinum (Pt)

Concentration of platinium is below 0.1 ppm in North Dakota lignite.[122]

60. Gold (Au)

Only a few reports on gold in coal can be found in the literature. They are rather old and summarized in papers by Gibson and Selvig[185] and Abernethy and Gibson.[178]

61. Mercury (Hg)

Early works on mercury determination in coal are summarized by Abernethy and Gibson.[178] Only a few reports could be accounted for in the literature, one of the first being by Kirby.[640]

Recently the interest in mercury determinations in coals has increased because of pollution control (e.g., see Joensun,[641] Poelstra et al.,[642] Crockett and Kinnison,[643] Klein,[625] and Ensminger[271]).

The Ensminger[271] studies of the mercury content of commercial coals in 1971 and 1972 revealed concentrations of less than 1 ppm. Most coals have mercury contents of less than 0.2 ppm, but contents of 1 ppm have been found in a few specific locations. Ensminger[271] has proposed a good average for U.S. coals might be around 0.15 ppm (see also Hall et al.[644]). Several research groups have studied mercury mass balance at a coal-fired power plant.[645-648] Ruch et al.[648] have reported mercury concentration determination in 66 raw coals sampled from 21 seams located in Illinois and 6 other states, in addition to specific gravity fractions of 3 selected coals prepared in the laboratory. Neutron activation analysis was used for all Hg determinations. Cinnabar is often found in natural association with pyrite, but no correlation of Hg content with total S, pyritic S, or ash was ascertained from the raw coal analyses of Illinois coals. No more than a random distribution of Hg concentration was shown for these coals. Raw coal samples from the six other states correlated well within those from Illinois with respect to Hg content. All coals subjected to specific gravity separations demonstrated a decrease in Hg content of at least 50% in the lightest fraction, while showing a two- to fivefold increase in Hg concentration in the heaviest specific gravity fraction (>1.60). The Hg content of the fractions of one coal increased with increasing specific gravity. Mercury concentrations in the gravity fractions of this same coal showed good correlation with pyritic S content. However, the lightest fraction was found to contain half the Hg concentration of the raw coal, suggesting organic association of some of the Hg. Determination of mercury in coal has been discussed by several authors (see Vasilevskaya et al.,[649,650] Rook et al.,[651,652] Huffman et al.,[653] and Heinrichs.[654]

Because of difficulties in Hg determination in coal, mercury concentrations were determined in a series of U.S. coals by several independent laboratories using a coal standard of proven homogeneity with an accurate Hg concentration determined by the National Bureau of Standards.[651] Eleven coals representative of U.S. production, both deep-mined and stripped, and also both washed and raw were included in the study. All analyses were performed by neutron activation. Results were indicative of a wide range of Hg concentrations in U.S. coals, with a low value of 0.05 ppm for a Colorado coal to a high value of 0.5 ppm for an Ohio coal. Average Hg content varied from 0.1 to 0.2 ppm. The need for better standardization of Hg analyses was indicated. O'Gorman et al.[655] have reported analysis of ten lignite, bituminous, and anthracite coal samples for Hg content by three different analytical procedures to compare the relative advantages and accuracies of the techniques and to determine the mode of occurrence of Hg in some of the coals. Gold amalgamation-flameless atomic absorption, neutron activation, and the combustion-solution technique were employed for Hg analysis. A large percentage of Hg in the lignite and bituminous coals studied was organically associated and volatilized during low-temperature ashing. Mercury in the anthracite was not organically bound and thus was relatively unaffected by ashing temperature. Analysis of coals from increasing depths with the same seam was indicative of the concentration of Hg in the bottom half of the seam. Generally, Hg distribution in the coals studied was very nonuniform due in part to the heterogeneous nature of the coal.

There are several reports on the fate of mercury during coal combustion. We shall mention some reports from the literature. Diehl et al.[656] have tested three different types of coal that were run on a bench scale 100 g/hr combustor, and a 500 lb/hr combustor to determine the fate of trace Hg in coal during combustion. Samples of the feed coal, fly ash, flue gas, and scrubber solutions from the two pulverized coal combustion units were analyzed for Hg content by double gold amalgamation-flameless AA procedures. Mercury content in samples from coal combustion units were then compared with the Hg content of fly ash samples from three different types of coal-burning power plants. Results of all samples analyzed accounted for 9 to about 70% of the Hg in the coal that was burned. In all cases observed in this study some Hg remained with the fly ash. It was generally concluded that contact between fly ash and flue gas after cooling was the reason for the retention of Hg in the fly ash, and that further contact at a lower temperature might lead to greater retention of Hg by the fly ash.

Billings and coworkers[657] have studied mercury emission during coal combustion. In their report, mass balance for mercury is presented in a large steam generator fired with pulverized coal by utilizing samples of coal, bottom ash water, fly ash, and flue gas. Coal samples were obtained over a 3-day period from collections taken for routine analyses. Ash samples were taken from the appropriate hoppers. Water samples were obtained from the adjacent river. Suspended fly ash was sampled isokinetically at several locations, and effluent gas samples containing mercury vapor were obtained using appartus suggested by the Environmental Protection Agency. Three highly sensitive analytical techniques designed for trace metals were utilized for sample analysis: anodic stripping voltammetry (ASV), radio-frequency helium plasma emission spectro-

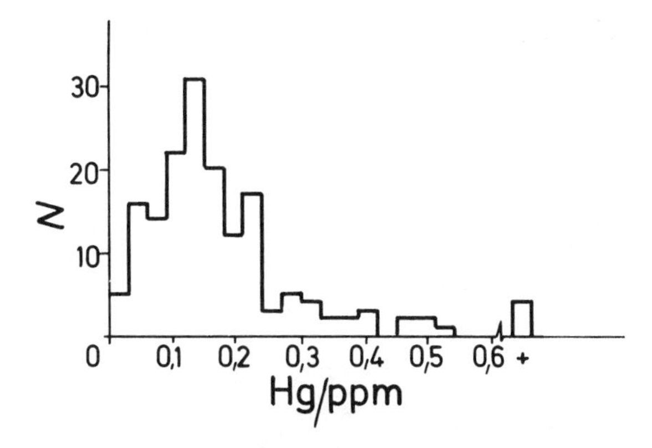

FIGURE 23. Distribution of mercury concentration values for U.S. coals. (As reported by Gluskoter.[137])

Table 103
MERCURY IN SOME U.S. COALS[122,137,173]

Region	Coal	Conc range (ppm)	Arithmetic mean (ppm)	Geometric mean (ppm)
Appalachian	Bituminous	0.01—3.2	0.20	0.12
Interior province	Bituminous	0.01—1.5	0.15	0.11
N. Great Plains	Lignite	0.01—12	0.18	0.10
N. Great Plains	Subbituminous	0.01—3.8	0.11	0.07
Rocky Mountain	Subbituminous	0.01—8.0	0.11	0.06
Rocky Mountain	Bituminous	0.01—0.90	0.08	0.05
All U.S.	Different	0.01—12	0.16	0.09
Western U.S.	—	0.02—0.63	0.09	0.07
Eastern U.S.	—	0.05—0.47	0.20	0.17
Illinois Basin	—	0.03—1.6	0.2	0.16
North Dakota	Lignite	0.11—0.2	0.20	

photometry (PE), and neutron activation analysis (NAA). An average daily mercury input to the furnace was calculated, and based on analyses of all sample materials, most of the mercury appeared to be discharged as vapor in the flue gas. The average mercury balance (input vs. output) was not in total agreement because of the use of varied analytical techniques, the small number of samples taken, and the wide range of data used in compiling averages.

Similar research is performed by Schultz et al.[172] The coal they investigated contained 0.18 ± 0.04 μg Hg per gram coal. Four types of ash samples were analyzed and reported: bottom, initial mechanical, final mechanical (multicones), and electrostatic ash. Because of the inability to empty an initial mechanical, a final mechanical, and an electrostatic ash hopper simultaneously prior to the start of a test, normally only three ash samples were collected during any one test. About 12% of mercury was found to be in fly ash. Studies of U.S. coals for mercury concentration are done in detail by Gluskoter,[137] (see Fig. 23), and Zubovic et al.[173] (see Table 103).

Mercury in coal samples from Nebraska was measured by Burchett;[3] he found concentrations ranging from 0.06 to 0. 15ppm. Loevblad[132] in his study of coal samples from seven different countries detected mercury in concentrations below 0.015 ppm up to 0.177 ppm. The mean value of all measurements was 0.08 ppm.

62. Thallium (Tl)

Gluskoter[137] has reported thallium in Illinois Basin coals to be in the range 0.12 to

1.3 ppm, with arithmetic mean 0.66 ppm and geometric mean 0.59 ppm for 25 investigated samples.

Thallium concentration in U.S. subbituminous coals is reported to be 0.1 ppm.[183] The all-coal average is also estimated to be 0.1 ppm (Reference 2).

63. Lead (Pb)

First reports on occurrence of galena (PbS) in coal were published at the beginning of this century. Hinds[411] and Jenney[412] reported occurrences of galena (PbS) associated with certain Missouri coals in which the coal obviously had been mineralized by ore-bearing waters.

Moss et al.[418] have examined samples of coal and ash deposits from a boiler by chemical and spectroscopic methods to determine the presence of Pb in coal. Analysis of ash deposits from a boiler indicated the presence of potash and soda in chloride forms as well as trace amounts of Pb and Zn. Traces of lead were found in the coals, and a dirt parting from one of the beds contained 0.024% lead.

Dunn and Bloxam[402] found lead in pyrites from several English coals. They reported that 19 out of 20 pyrite samples examined contained 0.0001 to 0.0461% lead and demonstrated that atmospheric dust from a coke works carried enough lead compounds to contaminate the neighboring pastures. Review articles on lead occurrence in coal are published by Gibson and Selvig[185] and Abernethy and Gibson.[178]

Chow and Earl[658] have discussed the origin of Pb in Pennsylvania anthracite from four major fields and in an unwashed meta-anthracite from Rhode Island by examining the isotopic relationships of these elements. Coal samples were ashed at 425°C and the Pb extracted with acid, isolated by ion exchange, purified by dithizone extractions, converted to the sulfide form, and subjected to MS analysis. Lead concentrations were determined by the isotopic dilution method. It was concluded that Pb was incorporated into the ancient, bedded plant material in carboniferous times prior to coalification, and was probably extracted from circulating groundwater while the plant material grew or shortly after the accumulation of plant detritus.

Lead isotopes in North American coals were discussed also by Chow and Earl.[658] It was possible to distinguish coal Pb pollutants from those of gasoline by examining their isotopic compositions, the Pb isotopes (^{204}Pb, ^{206}Pb, ^{207}Pb, ^{208}Pb) in coal being much more radiogenic than those found in Pb from the continental crust (i.e., Pb used as gasoline additives). MS was performed on 21 North American coals after they were pulverized, ashed, the Pb extracted by nitric and perchloric acids, isolated by a standard ion-exchange technique, purified by several dithizone extractions, and converted to the sulfide form for isotopic analysis. The isotopic composition of the Pb in the coal samples showed no correlation with the age of the plant accumulation or coalification; however, the radiogeneity of the Pb in the coal is inversely proportional to the Pb concentration. Thus, the isotopic composition of Pb may make it possible to determine the source of emission (automobile or power plant) where Pb pollution is concerned. The Pb content of 143 coal samples was determined, the average being 11 ppm Pb (96 ppm on ashed basis).

Lead contents of coal, coal ash and fly ash were recently discussed by Block and Dams.[659] Lead concentrations were determined for ten coals and their corresponding ashes, and for six fly ash samples collected from a combustion facility utilizing low-ash coal. Coal combustion was found to be a significant atmospheric pollutant source for lead. Coal and ash samples were decomposed by heating in a Teflon® bomb in an HNO_3/HF matrix. Absorption measurements were performed using a double-beam AA spectrophotometer. Six of the coals tested were utilized for home heating, two in power station, and two for coke production. Lead concentrations in home heating and coking coals were relatively constant and about five to ten times lower than the levels

Table 104
LEAD IN U.S. COALS[122,137,173]

Region	Coal	Conc range (ppm)	Arithmetic mean (ppm)	Geometric mean (ppm)
Appalachian	Bituminous	0.37—86	9.2	6.6
Interior province	Bituminous	0.78—590	44	18
N. Great Plains	Lignite	1.4—17	4.0	3.5
N. Great Plains	Subbituminous	0.72—58	5.5	4.1
Rocky Mountain	Subbituminous	0.95—76	8.0	6.0
Rocky Mountain	Bituminous	0.76—137	7.3	5.2
All U.S.	Different	0.06—1300	15	7.0
Illinois Basin	—	0.8—220	32	15
Eastern U.S.	—	1.0—18	5.9	4.7
Western U.S.	—	0.70—9.0	3.4	2.6
North Dakota	Lignite	1.5—8	5.44	

of lead found in power-station coals. Lead concentrations in coal ash are not directly proportional to the ash content, the highest levels being attended in low-ash (home heating) coals. No significant amount of lead was volatilized upon ashing. A study of lead distribution vs. fly ash particle size demonstrated maximum concentration in the submicron particle range. Lead is enriched in fly ash as compared to the original coal, and being preferentially deposited on submicron particles, is not effectively retained in the ash. Coal combustion is projected to contribute 6% or more to total atmospheric lead emissions.

Methods of lead determination in coal are described rather well in the literature. Block[660] presents a detailed description of the use of the AA method, while Block and Dams[661] have described the use of NAA. Lead concentration in a Pittsburg seam coal was reported by Schultz et al.[172] to be 7.7 ± 0.5 ppm, while its concentration in ash ranged from 25 ppm to 71 ppm, depending on combustor characteristics. Lead in coal is also discussed by Savul and Ababi.[751]

Table 104 shows lead concentrations in U.S. coals reported by Zubovic et al.[173] Gluskoter,[137] and Somerville and Elder.[122] Masursky[342] has suggested that lead was probably deposited in coal by water passing through coal beds. Hansen et al.[181] have reported lead concentration in the ash of high-volatile bituminous coal (8.9% ash) from Utah to be 42 ppm.

Lead in coal samples from Nebraska measured by Burchett[3] was in the concentration range 36 to 201 ppm. Smaller values are reported by Wewerka[199] for coals from the Illinois Basin; range 13 to 21 ppm, mean value 17 ppm. Loevblad[132] has reported values from 10 ppm to 60 ppm for coal samples from seven different countries.

64. Bismuth (Bi)

Fuchs[46] reported that the ash of a German coal contained 0.1 to 1 ppm bismuth. Headlee and Hunter[359] reported from less than 36 to 63 ppm bismuth in ash of West Virginia coals. More discussion on bismuth in coal can be found in an article by Abernethy and Gibson.[178] The estimate for worldwide average concentration of bismuth in coal is 5.5 ppm (see Reference 2).

65. Radon (Rn)

Very few reports on radon can be found in the literature. Wetherill[662] has reported the field measurements of radon in the air of a coal mine in southwestern Alberta. Few other reports of similar nature can be found.

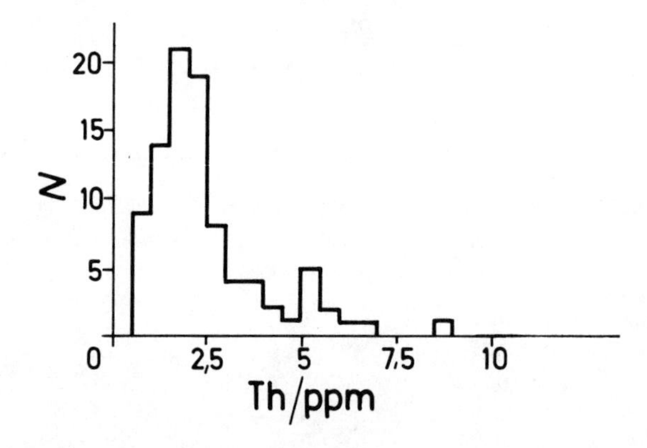

FIGURE 24. Distribution of thorium concentration values for U.S. coals. (As reported by Gluskoter.[137])

Table 105
THORIUM CONCENTRATION IN SOME U.S. COALS

Area	Conc range (ppm)	Arithmetic mean (ppm)	Geometric mean (ppm)
Western U.S.	0.62—5.7	2.3	1.8
Eastern U.S.	1.8—9.0	4.5	4.0
Illinois Basin	0.71—5.1	2.1	1.9

After Gluskoter, H. J., Trace Elements in Coal: Occurrence and Distribution, Circ. 499, Illinois State Geological Survey, Urbana, 1977.

66. Radium (Re)

There are only a few reports on radium content of coals. For example, the amount of Ra present in samples of Alabama coals was determined by Lloyd and Cunningham,[663] as part of a larger study being made by the Alabama Geological Survey to analytically characterize the coals of the state. Representative samples of the coal seams of Alabama were analyzed for Ra content by the acid digestion solution method. Results of these analyses were: the average of Ra per gram of coal and per gram of ash was 0.166×10^{-12} and 2.15×10^{-12}, respectively. No correlation was found between the concentration of Ra in the coals and the S content, or volatile or fixed C content. It was noted that the average value of Ra in the coal ash was significantly higher than concentrations in sedimentary rocks.

Other early works on radium occurrence in coals are summarized by Gibson and Selvig.[185] Bayliss and Whaite[664] have described their study of radium alpha activity of coal, coal ash, and particulate emission of a Sidney power station.

67. Thorium (Th)

Thorium distribution in some U.S. coals is shown in Figure 24 and Table 105 as reported by Gluskoter.[139] Thorium concentration in lignite from North Dakota was studied by Somerville and Elder.[122] The average value for 12 samples from Dunn County was found to be 3.64 ppm. Eight coal samples from Nebraska studied by Burchett[3] had thorium concentration from 5.6 to 25.0 ppm. Raw coals from three Illinois basin preparation plants were reported by Wewerka et al.[199] to have thorium concentration within the range 3.1 to 8.0 ppm, with mean of 5.8 ppm.

The all-coal average thorium concentration for U.S. is estimated to be 1.9 ppm (Reference 2). Thorium in coal is also discussed in reports written by Martin and Garcia-Rossel,[665] Erdtmann,[666] and Evcimen and Cetincelik.[667]

68. Uranium (U)

Uranium association with coal has a long history (see Abernethy and Gibson[178]). In 1875 Berthold[668] reported the occurrence of 0.2 to 2% uranium in coal from the Leyden mine near Denver, Colo. The samples were collected from a mineralized section of the coal bed. This mine was soon abandoned, but Wilson visited the site in 1922 and observed specimens of carnotite on the dump of the old workings. He states that the carnotite occurs as yellow incrustations and inclusions in fractured and partly silicified coal. The uranium content of the samples he collected ranged from 0.076% U_3O_3 for unsilicified coal from a bed near the old workings to 1.3% U_3O_8 for picked specimens of carnotite-stained coal. Carnotite also contains vanadium, but the amount was not determined in these samples. Wilson concluded that commercial development was not feasible.[185]

After World War II, a very intensive uranium search was initiated. The measurements of coal radioactivity were performed in many countries; however only a few are documented. For example Josa and collaborators[669] have measured uranium concentrations in lignites from Spain (Huesca, Lerida, Ternel, Galicia, Murcia) and reported concentration values 20 to 1200 ppm. Recent bibliographies on uranium in coal are written by Akers[145] and Alderman et al.[670]

The older work on uranium in U.S. coals is summarized in a paper by Abernethy and Gibson,[178] and the reported values on uranium concentration are in the range 10 to 7300 ppm (see Table 106). Here are some of these reports.

Staatz and Bauer[671] have determined levels of radioactivity in five lignite beds located in southeastern Nevada. Measurements were taken using a Geiger-Mueller Counter and ranged from 0.003 to 0.059% equivalent U. No uranium minerals were identified, and the lignites were generally underlain by clays with selenite (gypsum) commonly occurring on the bedding planes.

Gott[672] has determined uranium distribution in lignites, shales, and limestones from throughout the U.S., and a possible mechanism for U accumulation in lignites was suggested. Highest U concentrations were prevalent in lignites from the Dakotas, Wyoming, and Montana (0.01%), and from a high-ash Nevada lignite which contained up to 0.05% U. It was postulated that uranium was possibly concentrated in lignite by the action of percolating surface waters after having been leached from volcanic ash. Uranium-bearing coal in the Red Desert area in Wyoming has been studied by Masursky; his findings are documented in several reports.[342,673-676] In the first report[673] core and channel samples taken from the Red Desert area in Wyoming were used to investigate the origin of U in the coal of the region. Specific uraniferous zones examined included the Sourdough, Monument, Battle, and Luman zones. Areas which were topographically higher and in which coal was overlain by conglomerate showed the highest U concentration. Studies of core samples revealed that U concentration in the coal beds correlates well with the degree of permeability of adjacent rocks. Where coal beds are overlain or underlain by sandstone, the greatest concentrations of U occur at the top and/or bottom of the bed. In the shales analyzed, contact with sandstone increased the U content in the ash by as much as a factor of 4. This information tends to support an epigenetic origin of U in coal. If such is the case, U could be emplaced in coal epigenetically not only from adjacent uraniferous permeable rocks, but also from source rocks miles away in contact with a sandstone aquifer acting as a conduit to transfer U to coal beds.

In another report Masursky[675] presents tests on rocks and coal from the Red Desert

Table 106
URANIUM IN SOME U.S. COALS

Region	Uranium conc in coal (%)	Ref.
California	<0.02	723
Idaho		
Bonneville County	0.02	678
Bonneville County	<0.13	677
Cassia County	0.0 to 0.1	695
Cassia County	<0.097	699
Illinois	0.001 to 0.008	724
Indiana	0.001	725
Montana	0.001 to 0.034	700
Montana	<0.013	700
Nevada		
Esmeralda County	0.003	723
Churchill County	0.059	726
New Mexico		
Sandoval County	0.001 to 0.62	698
North Dakota	<0.045	682
North Dakota	<0.14	727
Ohio	0.001	725
Pennsylvania	0.002 to 0.014	728
Pennsylvania	<0.019	724
Pennsylvania	0.001	729
South Dakota	0.08 to 0.73	688
South Dakota	0.005 to 0.02	691
South Dakota	0.01	687
South Dakota	0.005	701
Utah	0.002	730
N. West Virginia	0.001 to 0.003	724

From Abernethy, R. F. and Gibson, F. H., Rare Elements in Coal, Rep. BM-IC-8163, U.S. Bureau of Mines, Washington, D.C., 1963.

area in Wyoming conducted to determine the source and mode of occurrence of U in the area. A sample of Precambrian Sweetwater granite (0.002% U) was pulverized and leached in a steam bath for 30-min periods with distilled H_2O, Na_2CO_3, and HNO_3 to investigate the mobility of the U. The H_2O removed 0.12% of the U contained in the granite, Na_2CO_3 removed 1.31%, and HNO_3 (pH 1.55) leached 7.70%. Thin sections of the granite were covered with nuclear emulsion and exposed from 2 to 6 weeks. Alpha tracks indicated that quartz, potassium feldspar, and interstatial material were alpha particle emitters. A −100 mesh coal sample, mixed in a 990-ppm uranyl nitrate solution for 27 days, extracted 95% of the U from solution. The coal was then reimmersed in a 550-ppm uranyl nitrate solution. At the end of 120 days the coal had extracted 35% of the U in solution. X-ray studies made after each test failed to show the presence of any U minerals in the coal. On the basis of these tests the author suggests that U in the Sweetwater granite may be mobilized by weathering and erosion and emplaced in the coal by the groundwater flow.

Additional information on uranium and some other trace elements in coal in the Red Desert, Wyoming, are presented by Masursky[342] in a paper published in 1956. Geologic mapping and core drilling were used to obtain data concerning the geochemistry and uranium content of coals in the Red Desert area of Wyoming. Chemical analyses for U were made on 1700 core samples and 500 surface samples, and semiquantitative spectrographic analyses were performed on about 100 samples. Lab tests were conducted to determine if epigenetic emplacement of U could be a possible source of the U in coal. Water and shroekingerite were mixed to form a solution of similar mineral content (60 ppb U) to some groundwaters in the region. This solution was passed over coal from the Red Desert area, and the coal adsorbed 95% of the U from solution, indicating that the U in coal could be emplaced by water passing through coal beds. The trace metals La, Mo, Nd, Pb, Sc, and Y may also have been deposited in this way.

In his 1962 report Masursky[676] determined concentrations of U and 61 associated trace elements in coal sampled from 9 major seams in south central Wyoming. When coupled with additional studies of the geography, mineralogy, and permeability of the coal beds and their associated rock formations, a hypothesis for the origin, occurrence, and distribution of U in coal was formulated. Both core and auger samples of coal were collected from seams ranging in thickness from a few inches to 42 ft. Element and mineral composition were determined by spectroscopy and X-ray diffraction. Uranium was identified as an organometallic complex in the coals and did not occur in a mineral matrix. Positive correlation existed between U and Fe, Mo, Fl, Ga, Pb, Sc, and V concentrations. The highest U level (0.054%) was found in the upper part of the stratigraphically highest coal beds, and in coal seams adjacent to coarse-grained permeable sandstone beds. Within a single coal seam U concentration was lowest in the central part of the relatively impermeable pure coal splits and highest in the impure coal adjacent to the partings. The coals studied were effective in their ability to absorb U from solution. The origin of U was epigenetic.

Vine and Moore[677] have reported a search for uraniferous coal, lignite, and associated carbonaceous rocks in the Fall Creek Area of Idaho. Potential uraniferous deposits were located by radiometric measurements with a Geiger-Mueller counter, and 52 samples were collected and analyzed for U, U in ash, and U in the whole sample. They were also examined for uranium minerals by mineralogic and petrographic studies. Analysis for U in the whole samples and sample ashes revealed U ranging from 0.001 to 0.131% in the whole sample and from zero to 0.145% in the sample ashes. The average U content of carbonaceous rocks studied in the Fall Creek area was 0.045% in the top foot of coal and 0.082% in the sample ashes. The amount of U was found to decrease from top to bottom in the coal beds of the Fall Creek area. No uranium minerals were identified in the samples. In his report of 1962 Vine[678] has reviewed the available literature and data concerning U deposits in coaly rocks. A historical summary of the discovery and development of U deposits in coaly rock, and the general features of coaly carbonaceous rocks (including individual reports) was made, which included lignite coal, subbituminous coal, coaly shale, lignitic shale, and carbonaceous shale from deposits in Colorado, Idaho, Montana, New Mexico, North Dakota, and South Dakota. The general geological features of uraniferous coal deposits and the geochemistry of U emplacement in the deposits were discussed. Some uranium minerals (carnotite, autunite, tolbernite, metazeunerite, and coffinite) were found occurring in uraniferous coals; however they occurred only in those samples containing greater than 0.1% total U.

The concentration of U in lignites and lignitic shales from several regions in eastern Montana was described in another report by Vine.[679] About 16.5 million tons of lignite in the Ekalaka Hills region of Montana was established to contain an average of 0.005% U, which was too low to be considered as a primary source of U at the time of the study. The U concentration in lignite beds of the Lone Pine Hills deposit ranges from 0.003 to 0.014%; however, insufficient data exists to estimate the total tonnage of uranium bearing lignite in the region. Two possibilities exist for commercial utilization of uranium-bearing lignite in this region: first, recovery of uranium from the ash of lignite burned as fuel and second, direct recovery of uranium from lignite without regard for the fuel value of the lignite.

Petrologic studies were performed on uraniferous coals from North Dakota, South Dakota, Idaho, and Wyoming by Schopf et al.[680] Certain maceral components appear to be associated with areas in the coal beds of high U content. Highly uraniferous layers of Idaho and Wyoming samples contained unusually high percentages of amorphous waxy matter. Samples from these two states did not, however, exhibit the normal, top-preferential patterns of U occurrence in the coal bed. In the Dakota coals, highly decayed and finely particulate plant debris was generally associated with those layers of the bed of highest U concentration. Frequently, layers of coal high in U were overlain by a layer high in detrital matter.

Schopf and Gray[681] have found that organic petrographic constituents of coal could not be correlated with U content of three lignite beds from South Dakota and Idaho. Drill cores were obtained, and layers of differing U content were analyzed. Petrographic studies utilized thin-sections, and U content was measured by radioactivity and wet chemical methods. Though U content may be related to the amount of degraded humic matter in the layer, no significant correlation could be made, especially a relationship that could be used to predict U occurrences.

A total of 63 samples of carbonaceous rocks from 38 locations in California and adjacent areas in western Nevada and southern Oregon were analyzed for U content. Coal samples tested ranged in rank from high-volatile bituminous to partially coalified wood, but were predominately either lignite or subbituminous coals. Samples overlain by rocks of volcanic origin and rhyolitic composition were given special attention. Although most of the samples tested showed U concentrations of no more than a few thousands of a percent, one lignite sample from southern California contained 0.020% U.

Moore et al.[682] have also studied uranium-bearing lignite in southwestern North Dakota. The lignite beds of the Bullion Butte area contain an average of less than 0.001% U, the Sentinel Butte beds average 0.007% U, the Chalky Butte bed average 0.008% U, and the HT Butte lignite bed average 0.015% U. Reserve figures for uraniferous lignite in the southwestern North Dakota area were presented, and the epigenetic origin of U in the lignite beds was discussed.

In the report of Breger and Deul[683] information concerning the presence of U in water, petroleum compressed gas, limestone, and carbonaceous materials was reviewed in order to detect possible geochemical explanations for the occurrence of U in carbonaceous materials. Significant attention was given to hypothetical mechanisms responsible for the presence of U in coal. In the explanation cited for coal, the UO_2^{++} ion, which has an affinity for coal, forms a uranyl humate when it comes in contact with a coal bed via aqueous transportation. Although U is not genetically associated with coaly substances (i.e., present in the original plant material) it may be picked up by coal during the course of migration. The fact that coalified wood materials in close proximity may or may not contain U demonstrates the effect of localized channeling of uranium-bearing waters.

Breger and coworkers[684,685] have made analytical studies of partially weathered Red Desert subbituminous coal from Wyoming to determine the mineralogical composition and geochemical characteristics of the sample in relation to the presence of U in the coal. The coal, obtained from the Luman No. 1 bed, was subjected to semiquantitative spectrographic analysis (28 elements) and X-ray diffraction. Spectrographic analysis was reported on three size fractions (−20 × 50 mesh, −140 × 200 mesh, and −325 mesh) and the original whole coal. Mineral components were removed from the coal material by specific gravity separations and reported as percentage estimates because of indistinct separations. Clay mineral and quartz identification were by X-ray diffraction. The approximate mineral content of the Luman No. 1 coal was gypsum, 6%; kaolinite, 1.0%; quartz, 0.3%; calcite and limonite, trace. Size fractions and sink-float fractions were ashed and analyzed for U content which was found to decrease as ash increases. The concentration of U in the mineral portion of the sample was only 0.006%, as compared to 0.002% U in the whole coal sample. Acid extraction tests (6 *N* HCl) removed almost 90% of the U in the coal. The author suggests that the presence of gypsum and limonite and the absence of pyrite indicates the sample was subjected to weathering. Breger[686] has also studied the role of organic matter in the accumulation of uranium.

High-grade uraniferous lignites in Harding County, S. D. were studied by King and Young.[688] The mode of occurrence and distribution of U in lignites was determined. Several other accessory minerals were also identified. Elemental analyses were carried out using semiquantitative spectroscopy. Uranium occurred primarily as a urano-organic complex associated with the coal substance, although U minerals including meta-autunite, metatorbernite, and metazeunerite were also identified. Among the accessory minerals identified were analcite (probably of syngenetic origin), pyrite, limonite, jarosite, and gypsum. Arsenic and Mo occurred in lignite ash in concentrations analogous to those of U. Uranium was deposited in lignite beds by the predominantly lateral movement of solutions, and this process was probably facilitated by pre-Oligocene weathering of the coals.

Zeller and Schoph[687] have described analysis of core drilling for uranium-bearing lignite. Lignite samples from 20 cores in North and South Dakota were analyzed to determine the U content of lignite beds in the area. Semiquantitative spectrographic analyses for 35 elements were performed on statigraphic sections of the lignite cores to determine the vertical distribution of U within the beds and to discover if any correlation existed between the occurrence of a particular trace element and the occurrence of U in the lignite. Uranium and Mo were both found to be more highly concentrated at the top of lignite beds in other sections of the beds and were also found to parallel each other in occurrence and concentration. The author estimates that 9 million tons of lignite averaging 0.01% U are present in the areas sampled. The concentration of U is highly variable in the study area and may average as much as 0.03% U in some small deposits. The ash content of the lignite tested contained as much as 20% U. A general description of the geology of the areas sampled is included in the report.

A summary of uranium-bearing coal, lignite, and carbonaceous shale investigations in the Rocky Mountain Region in the early 1950s is presented in a report by Denson.[689] The author suggested that U presence in coal is due to percolating groundwaters descending into the coal beds from overlying radioactive volcanic rocks of volcanic derivation. The concept is supported by the following information: (1) in every area where uraniferous coal or lignite occurs, these carbonaceous beds were or presently are overlain by radioactive rocks of volcanic origin; (2) chemical analyses of springs emanating from rocks of volcanic origin show abnormally high levels of U; (3) variations in the permeability of the rock overlying mineralized zones in coal or lignite beds reflect the intensity of mineralization; (4) in a series of flat-lying coal or lignite beds, the topographically higher beds in the strata are more radioactive than the lower beds; (5) the U content of coal beds in the Dakotas, Montana, and Wyoming is independent of the age of the formation but closely correlated with the topographic location of the bed and the permeability of the overlying rocks.

Denson and Gill[690] have studied uranium-bearing lignite and its relation to volcanic tuffs. Approximately 500 channel samples and 1000 core samples of uraniferous formations in Montana, North Dakota, and South Dakota were analyzed for U content, trace element content, and mineral composition. Chemical and semiquantitative spectrographic analyses were used to identify trace elements and petrologic studies were used to identify minerals in the lignite. Metazeunerite, autunite, and torbernite were identified in lignite and associated rocks in the Cave Hills area. Sandstone and carbonaceous shale in the Slim Buttes area were found to contain uranophane and aetatyuyamunite, respectively. Carnotite was found to occur locally in lignitic rocks and associated sandstones. The studies indicated that U was leached from overlying tuffs and epigenetically emplaced in the underlying lignite beds.

In another report Denson and coworkers[691] have studied uraniferous lignite in South Dakota, Montana, and North Dakota (see also Denson[692,693]). Uranium minerals, which are not usually present in carbonaceous materials, were discovered in the lignite and carbonaceous sandstone of the Cove Hills and Slim Buttes areas of South Dakota. Impregnated megascopic autunite, zeunerite, torbernite, and metatyuyamunite were identified in the fractures and joints in the thin, impure lignite beds and enclosing sandstone. Petrographic and mineralogic studies of other lignite samples from North Dakota and South Dakota revealed the presence of gypsum, analcite, jarosite, limonite, and quartz, but no U minerals; however, no relationship between the presence of these minerals and the U content of the lignite was discerned. Semiquantitative spectrographic analysis of lignite core samples indicated that U and Mo show a marked decrease in concentration from the top to the bottom of the lignite bed. The author considered lignite as a much better host material than higher rank coals for the emplacement of U because of its soft, porous nature. Certain specific lignitic constituents may also increase the affinity of lignite for U.

Pipiringos[694] described the mode of deposition of uranium in the coal beds of the central part of the Great Divide Basin in Sweetwater County, Wyoming. The observations noted in this paper were taken from a previous geological investigation of the central Great Divide Basin by the author and a review of pertinent literature. Generally, the uranium content in the coal seams of the area decreased as the permeability of the overlying rocks decreased. The relationship supported the theory that uranium was transported to the coal beds by groundwater passing through the overlying uraniferous sandstone. Gray[695] has studied low-rank coals and carbonaceous shale deposits from Idaho petrographically to determine the mode of occurrence and distribution of U in these materials. Generally carbonaceous shales associated with altered tuffaceous sediments were more radioactive than coals associated with unaltered tuffaceous material. No quantitative relationship was evident between organic composition of the coals and U concentration, although the majority of the U occurred in the humic matter (see also Porter[696] for the discussion of uranium occurrence in lignites).

In his work, White[697] identified and characterized 12 megascopic U minerals collected from 11 North and South Dakota lignite deposits. The minerals were hand picked from lignite samples and characterized by optical, chemical, radiometric, and X-ray diffraction techniques. Elemental analyses of each mineral was performed using X-ray fluorescence and emission spectrometry. Eleven of the minerals belonged to the torbernite-meta-torbernite series of hydrous U arsenates and phosphates, including Na-autunite, metauranocircite, meta-autunite, selenite, sabugalite-selenite, H-autunite, and avernathyte. Sodium-autunite was the most common and abundant species. The minerals were generally found coating cleat and fracture surfaces, and favored blocky lignite where the coatings showed no affinity for any lignite maceral. Other common nondetrital minerals in these lignites were gypsum, jarosite, barite, and analcite. Recent deposition of the U minerals is evidenced mainly by radioactive Ra-bearing barite which sets a maximum age of 15,000 years for the minerals. Uraniferous deposits are generally confined to near-outcrop portions of the seams, or where the overburden was less than 30 ft. The ultimate source of the U was overlying volcanic beds, but the minerals studied resulted from secondary movement of the U within the lignite and its redeposition in cleats and fractures.

Bachman et al.[698] have discussed uranium concentrations determined in coal, carbonaceous shale, and carbonaceous sandstone located in Sandoval County, New Mexico, and the possible source and mode of deposition of U in the coal. Highest U concentrations (up to 0.62%) occurred in the stratigraphically highest coal seams with essentially no U occurring in stratigraphically lower coal beds. All major U deposits were found in the coal, and no uranium minerals were identified. The source of U deposits in the coal of Sandoval County was the Bendelier rhyolite tuff of Pleistocene Age and the deposits were probably epigenetic in origin and derived from groundwater solutions. Mapel and Hail[699] have presented a detailed geological description of the Goose Creek district located in southern Idaho and in parts of Utah and Nevada. Included as part of the geological description is the measurement of the U concentrations in the lignite deposits found in this area and a possible explanation of the source and mode of deposition. Results of the U measurements of the lignite deposits and surrounding rock types are presented, and an examination of the data indicated the vertical distribution of U to be irregular in the thin deposits; however, in deposits of 3 ft or more the highest concentration appeared at the top and decreased towards the bottom of the deposit. The authors found evidence to support the hypothesis that the U was leached from the volcanic ash in the Salt Lake formation and deposited in the lignite beds by groundwater.

Gill[700] and Gill et al.[701] have studied uranium-bearing lignite in South Dakota and Montana. They have reported some lignite deposits containing as much as 0.1% uranium. Some physical and chemical properties of vitrains associated with uranium have been studied by Ergun et al.[702]

Studies of the chemical composition, density, reflectance, and X-ray scattering of nine coalified uraniferous wood samples indicate that these properties are affected by the U radiation emitted from within the coal material. The samples were taken from mines in Utah and Colorado and were analyzed for U content. Geochemical studies indicate that the U entered the coal in aqueous solution as a complex alkali or alkaline earth UO_2CO_3 and that the UO_2^{++} ion was absorbed by the coal and reduced to UO_2. Autoradiographic studies revealed that the UO_2 is dispersed throughout the coal. Certain effects of U on the physical and chemical properties of the coalified logs are reported. The alpha radiation emitted by the U causes the coal to lose hydrogen, which effects the organic structure by causing the formation of alicyclic units. This conception is supported by the density, reflectance, and X-ray scattering studies performed on the sample.

Astheimer et al.[703] have studied uranium enrichment from seawater by absorption on brown coal, while uranium isotopes in coal are discussed by Zverev et al.[746] There are a number of more recent measurements of uranium in coal (see for example Hail and Gill,[704] Green,[705] Mitchell,[706] Steinberg,[707] Cameron and Birmingham,[708] Chow and Earl,[709] Jeczalik,[710] Little and Dirham,[711] Noble,[712] Danchev et al.,[713-715] Ree and Emmermann,[716] Gentry et al.,[717] Logomerec,[718] Evcimen and Cetincelik,[667] Koglin,[719] Shin et al.,[720] and Nadal et al.[721]).

Cameron and Birmingham[708] have measured radioactivity in western Canadian coals. The U, Mo, and V content of several Tertiary Age coals (lignite bituminous, and subbituminous) of Western Canada was determined both by field and laboratory measurements to see if any approached sufficient concentration to be commercially mineable. Uranium content was determined using scintilometry and fluorimetry. Vanadium and Mo concentrations were measured by X-ray fluorescence. One Saskatchewan lignite was considered to be of ore-grade quality and contained 0.08% U on a whole coal basis. Molybdenum and V concentrations in all coals analyzed were below ore-grade levels.

Gentry et al.[717] have obtained some evidence relating to the time of uranium introduction and coalification. A study of U-rich inclusions in Mesozoic coalified wood of the Colorado Plateau suggested that U was introduced much more recently than previously believed. The compositions of the inclusions and their resultant halos were studied in thin-sectioned coalified wood by electron microprobe X-ray fluorescence and ion microprobe MS techniques. The eliptical shape of the halos suggested that they formed prior to complete coalification. Measured U/Pb ratios in the inclusions date them, and therefore the coalified wood, at several thousand years of age which greatly contradicts the geologically established Mesozoic age. The original source of the U appeared to be not far from the coal examined, and the study suggested a search for the ore deposit in and around the Colorado Plateau.

About 5000 samples of coal have been analyzed for uranium by the U.S. Geological Survey, U.S. Bureau of Mines, U.S. Atomic Energy Commission, U.S. Energy Research and Development Administration, and numerous state and private agencies. In the report by Facer,[722] the locations of more than 2000 coal samples analyzed for uranium content by the U.S. Geological Survey are shown and the corresponding listings of data are arranged by state and county in tabular form. The locations of 151 of the samples, which contained 5 ppm U or greater, are shown in a separate map, and the numerical data are also listed. Most of the uranium analyses reported were done by neutron activation and are considered very reliable. Although the thicknesses of the coal units sampled are given, there is no information concerning the size of coal deposits or any plans for the utilization of the coal. There are about five locations, out of the hundreds of samples, where it would be desirable to obtain more information. Table 107 shows some of the values reported by Facer.[722]

Data by Zubovic et al.[173] and Gluskoter[137] are presented in Table 108. They contain information on mean uranium concentrations in coals from different regions of the U.S.

Correlations of uranium concentrations in coal with other coal properties are discussed in detail by Breger.[686] Some of his findings are presented in Figures 25, 26, 27, and 28. In Figure 25 relationship between volatile matter and uranium in coalified logs is shown. Logs with smaller amounts of volatile matter contained higher uranium concentrations. This is in agreement with other reports.

Figure 26 shows the relationship between calorific values and uranium in coalified logs. Some coals rich in uranium are of very low calorific values; however there are some exemptions. In some locations in Europe, high uranium concentrations are found even in high calorific coal.

Figure 27 shows the relationship between organic sulfur and uranium in coalified logs. For Jurassic logs, organic sulfur concentration is constant, while the uranium concentration is variable. For Triassic logs, higher uranium concentrations are related to higher organic sulfur concentrations. The role of organic substances in the geochemistry of uranium is also discussed by Borovec,[731] Danchev and Strelyanov,[714] Kakimi et al.,[732] Szalay,[124] Uspenskii and Pen'kov,[733] Agiorgitis and Schermann,[734] Horr et al.[735] and Schmidt-Collerus.[745]

Figure 28 shows the relationship between hydrogen and uranium in coalified logs. For all logs there is decrease in hydrogen content with increase of uranium content.

In 1975, the U.S. Bureau of Mines' Metallurgy Research Center at Salt Lake City analyzed fly ash samples from 11 major coal-burning power plants located in Arizona, Alabama, Colorado, Kentucky, New Mexico, North Dakota, Tennessee, and Wyoming. The uranium content varied from 13 to 25 ppm with a mean value of 18 ppm.

Another analytical survey (Furr et al.[47]) by four universities (in 1975 to 1976) of elements in fly ash from 23 coal-burning power plants, located in 21 states, showed a range of uranium concentration from 0.8 to 19 ppm with mean and median values of 7.3 ppm (Table 109).

In 1976, the Grand Junction Office of the then Energy Research and Development Administration (ERDA) obtained, and had analyzed for uranium, samples of ash from

Table 107
COAL SAMPLES WITH URANIUM CONCENTRATION HIGHER THAN 100 PPM

Location	Uranium (ppm)
Illinois	
Perry	103.30
Montgomery	94.10
Wyoming	
Sweetwater	75.38
Illinois	
Perry	43.39
Iowa	
Wapello	42.93
Wapello	34.57
Wapello	29.59
Pennsylvania	
Northumberland	25.24
Illinois	
Montgomery	20.51
Wyoming	
Sweetwater	19.47
Missouri	
Macon	19.28
Iowa	
Wapello	18.70
Appanoose	17.93
Nebraska	
Otoe	17.06
Pawnee	16.74
Mississippi	
Scott	16.70

From Facer, J. F., Uranium in Coal, Rep. GJBX-56(79), U.S. Department of Energy, Grand Junction Office, Colo., May 1979.

Table 108
URANIUM CONCENTRATIONS IN SOME U.S. COALS[137,173]

Region	Coal	Arithmetic mean (ppm)	Geometric mean (ppm)	Conc range (ppm)
Appalachian	Bituminous	1.6	1.2	0.10—19
Interior province	Bituminous	3.2	1.7	0.20—59
N. Great Plains	Lignite	1.6	1.2	0.21—13
N. Great Plains	Subbituminous	1.6	0.98	0.9—16
Rocky Mountain	Subbituminous	2.8	1.9	0.06—76
Rocky Mountain	Bituminous	2.0	1.4	0.13—42
All U.S.	Different	2.6	1.3	0.06—2700
Illinois Basin	—	1.5	1.3	0.31—4.6
Eastern U.S.	—	1.5	1.3	0.40—2.9
Western U.S.	—	1.2	0.99	0.30—2.5

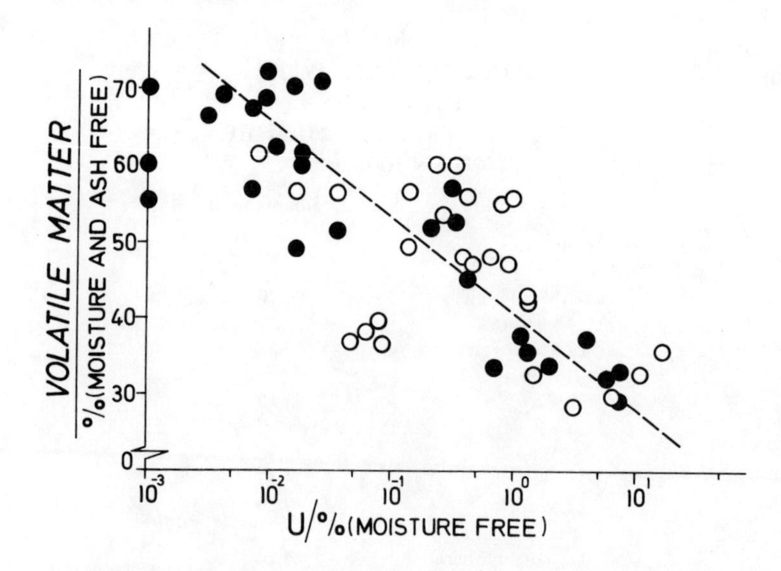

FIGURE 25. Relationship between volatile matter and uranium coalfield logs. Open circles represent Jurassic logs while solid ones are Triassic logs. (After Breger.[686])

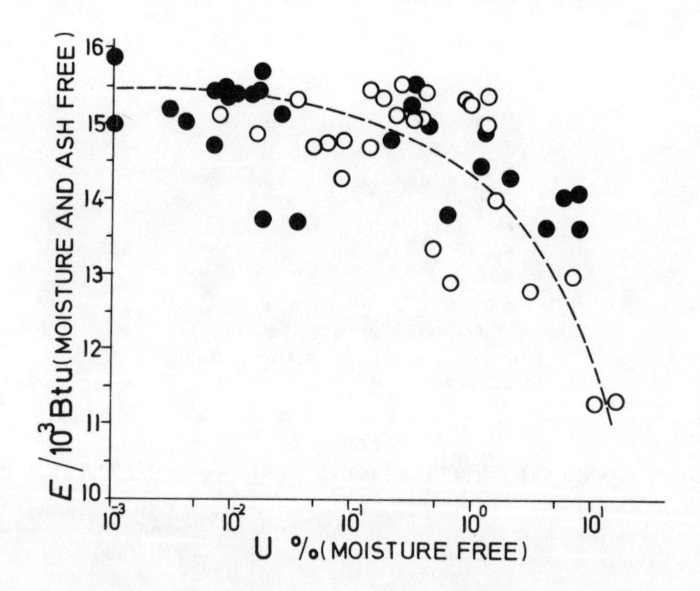

FIGURE 26. Relationship between calorific values and uranium in coalified logs. Open circles represent Jurassic logs while solid ones are Triassic logs. (After Breger.[686])

11 power plants burning lignite or subbituminous coal from Montana, North Dakota, Texas, and Wyoming. These samples contained 5 to 17 ppm U with a mean value of 12 ppm. The analyses are shown in Table 110.[722] Analyses by others of lignite and ash from some of the same and neighboring power plants are in general agreement with these results.

Let us also mention the work by Calvo,[125] who has reported a short experiment on the metallogenetic aspect of the uranium-organic matter association in nature. The author carried out several experiments on uranium fixation by organic matter with

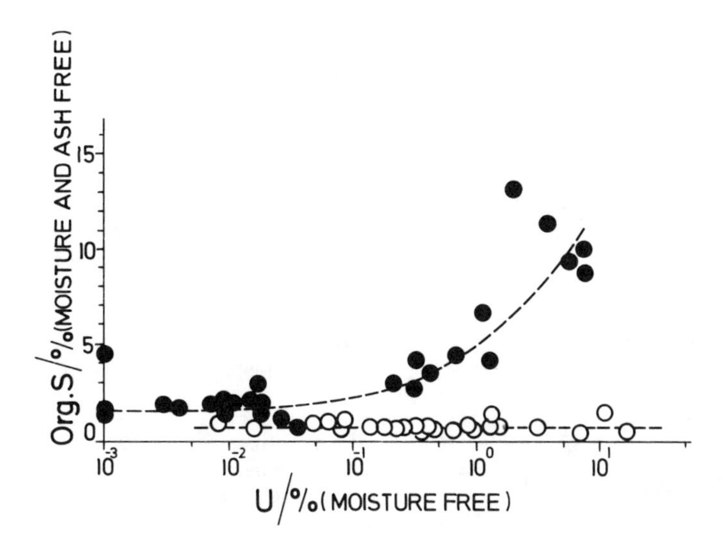

FIGURE 27. Relationship between organic sulfur and uranium in coalified logs. Open circles represent Jurassic logs, while solid ones are Triassic logs. (After Breger.[686])

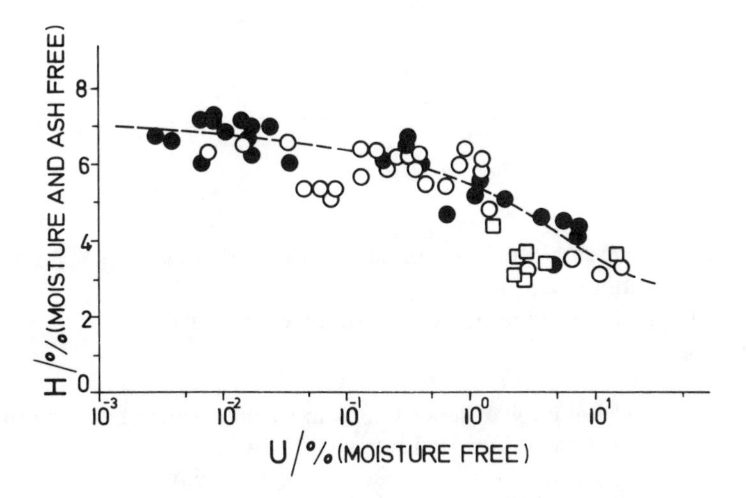

FIGURE 28. Relationship between hydrogen and uranium in coalified logs. (After Breger.[686])

different degrees of maturity. Although further studies are considered necessary, the results already obtained lead to the following conclusion: when uraniferous solutions come into contact with humic organic matter, the efficiency and nature of the uranium-organic matter association thus established depends on the degree of carbonization attained by the humic materials. At the beginning of the development process, the organic matter has high chemical activity, and the uranium tends to occur in stable organic phases. As the process continues, a stage is reached from which uranium tends to form its own minerals independently of organic matter. From the coalification stage onward, the tendency toward uranium-organic matter association becomes increasingly less in any form. Petrographically, the humic constituents to which uranium is closely and positively related belong to the huminite maceral group. Lastly, it is considered that when the organic matter is present as an accessory constituent in other

Table 109
URANIUM IN FLY ASH

Location of power station	Uranium in fly ash (ppm)	Type of coal	% ash
Alabama	9.7	Bituminous	14.1
Colorado	7.3	Subbituminous	5.0—5.5
Delaware	7.4	Bituminous	10—16
Georgia	12	Bituminous	6—26
Iowa	5.2	Bituminous	6.9
Kentucky	9.0	Bituminous	—
Maryland	7.4	Bituminous	12.2
Massachusetts	4.8	—	11
Michigan 1	3.9	Bituminous	4
Michigan 2	5.7	Bituminous	12
Minnesota	8.2	Bituminous and subbituminous	10
Montana	0.8	Lignite	9
New Hampshire	7.5	Bituminous	7.5
New Mexico 1	4.6	Subbituminous	22
New Mexico 2	8.2	Subbituminous	—
New York	7.2	Bituminous	16—18
North Carolina	9.0	Bituminous	11.4
Ohio	7.2	Bituminous	17.0
South Carolina	19	Bituminous	10.7
South Dakota	6.4	Subbituminous	7.0
Utah	4.1	Bituminous	10—14
West Virginia	5.0	Bituminous	14
Wisconsin	9.0	Bituminous	10

From Furr, A. K. et al., *Environ. Sci. Technol.*, 11, 1194, 1977.

sedimentary rocks and with quite low maturity indices, there is greater favorability for the formation of uranium deposits.

Three hypotheses advanced to explain the occurrence of uranium in some coals were described by Denson[692] as follows:

1. Syngenetic — Uranium was deposited from surface waters by living plants or in dead organic matter in swamps prior to coalification.
2. Diagenetic — Uranium was introduced into the coal during coalification by waters bringing the uranium from areas marginal to the coal deposits or from the consolidating enclosing sediments.
3. Epigenetic — Uranium was introduced in the coal after coalification and after consolidation of the enclosing sediments by groundwater deriving uranium from hydrothermal sources or from unconformably overlying volcanic rocks.

The accumulation of uranium in coal may vary markedly from place to place, and the occurrence of uranium in each deposit should be interpreted in relation to the geologic history of the region. Field evidence favors the epigenetic hypothesis of the origin of uranium in U.S. western coals. Secondary concentration of uranium in coal may occur when solution of small quantities of uranium by groundwater from overlying volcanic rocks is followed by downward percolation of these waters through previous strata until the uranium is taken up and retained by the highest of the underlying lignite beds. Application of this theory led to the discovery of uranium-bearing coal in Wyoming, Montana, Idaho, and New Mexico. In general, the uranium-bearing coal in each of these areas forms the topmost coal bed of a sequence overlain unconform-

Table 110
ANALYSES OF ASH POWER PLANTS WHICH BURN LIGNITE OR SUBBITUMINOUS COAL

Power plant	Source of lignite or coal	Uranium (ppm)	
		Ash	Fly ash
Montana-Dakota Utilities Co., Sidney, Mont., Lewis and Clark Station	Montana-Richland	13	13
The Montana Power Co., Billings, Mont., Corette Plant	Montana-Rosebud	14	11
Black Hills Power and Light Co., Rapid City, S.D.	Wyoming-Campbell	5	
Texas Utilities Generating Co., Fairfield, Tex., Big Brown Station	Texas-Freestone	8	14
Texas Utilities Generating Co., Mount Pleasant, Tex., Monticello Station	Texas-Morris	11	14
United Power Assoc., Stanton, N.D.	North Dakota-Mercer	13	17
Minnkota Power Co-operative, Grand Forks, N. D., Franklin P. Wood Station	North Dakota-Burke	12	10
Montana-Dakota Utilities, Mandon, N.D., R.M. Heskett Station	North Dakota-Mercer	14	12
Otter Tail Power Co., Fergus Falls, Minn., Hoot Lake Plant	North Dakota-Mercer	15	12
Otter Tail Power Co. Big Stone City, S.D.	North Dakota-Bowman	12	14
Basin Electric Power Coop., Stanton, N.D.	North Dakota-McLean	12	14

From Facer, J. F., Uranium in Coal, Rep. GJBX-56(79), U.S. Department of Energy, Grand Junction Office, Colo., May 1979.

ably by layers of silicic volcanic materials or other strata from which uranium may have been leached by groundwaters. The uranium content of succeeding lower coal beds decreases to the vanishing point. The enrichment of uranium in coals which lie near the erosion zone of granite mountains was noted also in Hungary.

Moore[723] has published a detailed study on uranium extraction from solution by coal. His results are summarized in Figure 29; the possibility of uranium extraction from the solution of uranil sulfate in nitric and sulfuric acids is presented. Figure 29 shows that nearly 100% extraction of uranium can be achieved by peat, lignite, and subbituminous coal.

In his experiments, a solution of uranyl sulfate containing 1.000 g of uranium was prepared by dissolving 1.793 g of powdered uranium oxide in a mixture of concentrated nitric and sulfuric acids. This solution was evaporated to dryness and fumed to remove the nitrate ion, and the residue was then dissolved in 1 ℓ of 0.01 *N* sulfuric

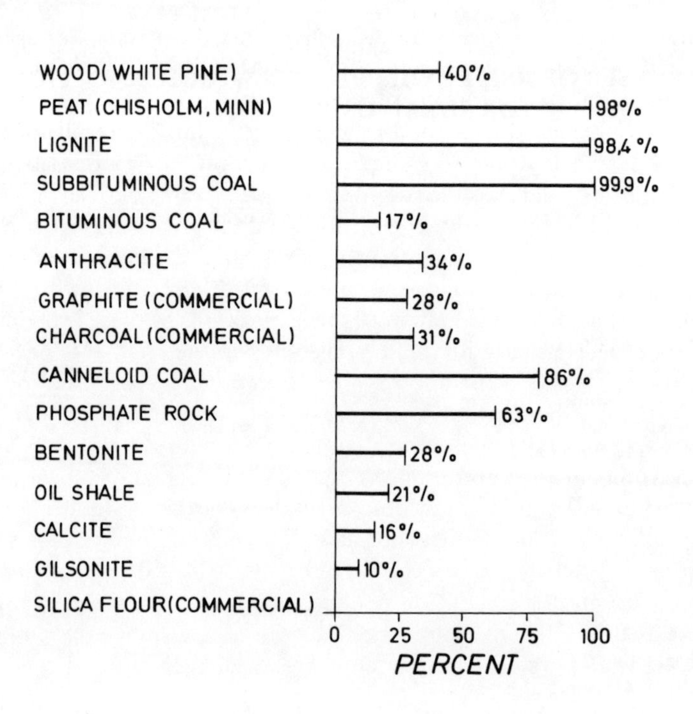

FIGURE 29. Percent uranium extracted by coal and other materials from uranyl sulfate solutions containing 200 ppm uranium. (Data from Moore.[723])

acid to provide a pH of 2. .The solution of UO_2SO_4 thus prepared contained 1000 ppm uranium. For purposes of the experiments the solution was further diluted with water until it contained about 200 ppm uranium at a pH of 2.45. The low pH value was selected to prevent the possible formation of insoluble hydrates.

Coal was ground and screened until it was composed of grains between 40 and 80 mesh (0.42 to 0.177 mm). This granular coal was placed in an apparatus which provided a continuous circulation of the solution. A solution (350 mℓ) containing 196 ppm of uranium was placed in this apparatus with 28 g of coal and the solution circulated for 12 days.

Some of the patents for uranium recovery from solutions including seawater are based on these facts. Results obtained by Moore[723] show that uranium presence in coal cannot be a consequence solely of its presence in the original plant material.

The first and most obvious fact shown by these experiments is that the low-rank coals were more effective in extracting uranium than any of the other materials used. A maximum of 99.9% uranium was removed from solution by subbituminous coal; phosphate rock follows subbituminous coal, lignite, and peat as an extracting agent, for it removed 63% of the uranium from solution. These results are in harmony with the assocation of uranium in nature with coal, coalified logs, and carbonaceous shale and with phosphate rock and fossil bones.

These results are of a preliminary nature, based in most cases on a single sample for each rank of coal, so additional studies may alter the pattern that seems indicated. If these results are accepted as approximating those that would be obtained regardless of the number of samples used, the chief factors influencing the extraction of uranium by coal may be considered. These are surface absorption, ion exchange, chemical reduction, change in pH, and the formation of metallo-organic compounds.

The fact that the uranium is held irreversibly by the coal suggests that surface adsorption phenomena are not important in determining the affinity for uranium. Also, Breger and Deul have shown by base-exchange studies that the uranium in coal is not held to any appreciable extent by ion exchange.

Coal is generally regarded as a good reducing agent, but the results of these experiments are inconclusive as to the role chemical reduction may play in the extraction of uranium. Bituminous coal, anthracite, and charcoal are relatively poor extracting agents for uranium, but there is no chemical reason known to the writer for regarding these as less effective reducing agents in general than the low ranks of coal. Until further studies are made, it is suggested that chemical reduction is not an important factor in the precipitation of uranium under the conditions of these experiments.

Since the uranium is apparently held irreversibly in the coal, it is possible that the uranium is precipitated as a metallo-organic compound as suggested by Breger and Deul on the basis of experimental work on natural uranium-bearing lignite. If this is the mechanism, the organic compound that combines with the uranium may reach its maximum development in subbituminous coal. Further metamorphism of subbituminous coal to bituminous coal could destroy the organic compound important in extracting uranium. Breger and Whitehead have shown by thermographic studies that a relatively strong exothermic peak occurs at about 650°C with subbituminous A and high-volatile C bituminous coals. This peak is not present in subbituminous C coal or in lignite. It is possible that the same conditions that give rise to these thermographic characteristics may also reflect changes that make the higher ranks of coal less effective extracting agents for uranium.

Uranium concentration in coal ash has been subject to numerous studies. This subject will be discussed in some detail in the chapter about uranium recovery. Methods of uranium determination in coal have been discussed by many authors; e.g., Ujhelyi,[736] Szonntagh et al.,[737] Perricos and Belkas,[738] Korkisch et al.,[739] Fujii et al.,[740] Guttag and Grimaldi,[741] Cronin and Leyden,[742] Weaver,[743] and Kovalov.[744]

REFERENCES

1. **Torrey, S., Ed.,** *Trace Contaminants from Coal,* Noyes Data Corporation, Park Ridge, N.Y., 1978.
2. U.S. National Committee for Geochemistry, Panel on the trace elements geochemistry of coal resource development related to health, Trace Element Geochemistry of Coal Resource Development Related to Environmental Quality and Health, National Academy Press, Washington, D.C., 1980.
3. **Burchett, R. R.,** Coal Resources of Nebraska, Rep. NP 23879, University of Nebraska, Lincoln, 1977.
4. **Goldschmidt, V. M.,** Rare elements in coal ashes, *Ind. Eng. Chem.,* 27, 1100, 1935.
5. **Goldschmidt, V. M.,** in *Geochemistry,* Muir, A., Ed., Clarendon, Oxford, 730, 1954.
6. **Gluskoter, H. J.,** Mineral matter and trace elements in coal, in *Trace Elements in Fuel,* Babu, S. P., Ed., American Chemical Society, Washington, D.C., 1975.
7. **Miller, W. G.,** Relationships Between Minerals and Selected Trace Elements in Some Pennsylvania Age Coals of Northern Illinois, M.S. thesis, University of Illinois, Urbana-Champaign, 145, 1974.
8. **Nicholls, G. D.,** The geochemistry of coal-bearing strata, in *Coal and Coal Bearing Strata,* Murchison, D. and Westoll, T. S., Eds., New York, 1968, 269.
9. **Zubovic, P., Stadnichenko, T., and Sheffey, N. B.,** The Association of Minor Elements with Organic and Inorganic Phases of Coal, Prof. Paper 400-B, U.S. Geological Survey, Reston, Va., 1960, B84.
10. **Zubovic, P., Stadnichenko, T., and Sheffey, N. B.,** The Association of Minor Element Association in Coal and Other Carbonaceous Sediments, Prof. Paper, 424-D, Article 411, U.S. Geological Survey, Reston, Va., 1961, D345.
11. **Gluskoter, H. J.,** Inorganic geochemistry of Illinois agglomerating coals, in Proc. Coal Agglomeration and Conversion Symp., Morgantown, W.Va., 1975, 9.

12. **Lyon, W. S., Lindberg, S. E., Emery, J. F., Carter, J. A., Ferguson, N. M., Van Hook, R. I., and Raridon, R. J.,** Analytical determination and statistical relationships of fourty-one elements in coal from three-coal fired steam plants, in *Nuclear Activation Techniques in the Life Sciences,* International Atomic Energy Agency, Vienna, 1978.
13. **Taylor, S. R.,** Abundance of chemical elements in the continental crust, A new table, *Geochem. Cosmochem. Acta,* 28, 1273, 1964.
14. **Turekian, K. K. and Wedepohl, K. H.,** *Distribution of Elements in Some Major Units of the Earth's Crust,* Bull. 72, Geological Society of America, Boulder, 1964, 175.
15. **Abel, K. H. and Rancitelli, L. A.,** Major, minor, and trace element composition of coal and fly ash, as determined by instrumental neutron activation analysis, *Adv. Chem. Ser.,* 141, 118, 1975.
16. **Abernethy, R. F., Peterson, M. J., and Gibson, F. H.,** Spectrochemical analysis of coal ash for trace elements, U.S. Bur. Mines Rep. Invest., 7281, 30, July 1969.
17. **Abernethy, R. F., Peterson, M. J., and Gibson, F. H.,** Major ash constituents in U.S. coals, U.S. Bur. Mines Rep. Invest., p. 9, 1969.
18. **Alberts, J. J., Burger, J., Kalhorn, S., Seils, C., and Tisue, T.,** The relative availability of selected trace elements from coal fly ash and Lake Michigan sediment, *Proc. Int. Conf. Nucl. Methods Environ. Energ. Res.,* 3, 379, 1977.
19. **Anfimov, L. V.,** Distribution of some trace elements in the ash of coals from the Chelyabinsk basin, *Tr. Inst. Geol. Geokhim. Akad. Nauk SSSR (Ural. Filial),* 90, 64, 1971.
20. **Averitt, P., Breger, I. A., Swanson, V. E., Zubovic, P., and Gluskoter, H.,** Minor Elements in Coal — a Selected Bibliography, Prof. Paper 800-D, U.S. Geological Survey, Reston, Va., 1972, 169.
21. **Averitt, P., Hatch, J. R., Swanson, V. E., Breger, I. A., Coleman, S. L., Medlin, J. H., Zubovic, P., and Gluskoter, H. J.,** compilers, Minor and Trace Elements in Coal — a Selected Bibliography of Reports in English, Open File Rep. 76-481, U.S. Geological Survey, Reston, Va., p. 16, 1976.
22. **Azambuja, D. S. and Bristoti, A.,** Concentration of some minor elements in the coal of the Leao Mine Metal, ABM, 35, 79, 1979.
23. **Baria, D. N.,** A Survey of Trace Elements in North Dakota Lignite and Effluent Streams from Combustion and Gasification Facilities, Engineering Experiment Station, University of North Dakota, Grand Forks, May 1975, 64.
24. **Beckner, L. J.,** Trace element composition and disposal of gasifier ash, *Proc. Synth. Pipeline Gas. Symp.,* 7, 359, 1975.
25. **Bella, L. and Szava Benocs, K.,** Trace element content of the brown coal field of Nograd. Banyasz. Kut. Intez., *Kozlem,* 12, 65, 1968.
26. **Block, C. and Dams, R.,** Inorganic composition of Belgian coals and coal ashes, *Environ. Sci. Technol.,* 9, 146, 1975.
27. **Block, C., Dams, R., and Hoste, J.,** Chemical composition of coal and fly ash, in *Meas. Detect. Control Environ. Pollut. Proc. Int. Symp.,* International Atomic Energy Agency, Vienna, 1976, 101.
28. **Bordon, V. E.,** Distribution of impurity elements in Belorussian coals, *Vestsi Akad. Navuk Belarus. SSR, Ser. Khim. Navuk,* 5, 105, 1973.
29. **Botoman, G. and Stith, D. A.,** Analysis of Ohio coals, *Ohio Div. Surv., Inf. Circ.,* 47, 1, 1978.
30. **Bowden, D. N. and Roberts, H. S.,** Analysis and fusion characteristics of New Zealand coal ashes, *N.Z. J. Sci.,* 18, 119, 1975.
31. **Boyadzhiev, G., Nikolov, Zdr., Nenov, N., and Stefanov, G.,** Elemental impurities in coal dust from the Dobrudja basin, *Izv. Geol. Inst., Bulg., Akad. Nauk., Ser. Geokhim., Mineral. Petrogr.,* 20, 31, 1971.
32. **Brown, H. R. and Swaine, D. J.,** Inorganic constituents of Australian coals, *Coal Res. Commonw. Sci. Ind. Res. Org.,* p.15, 1964.
33. **Brown, J. and Guest, A.,** Continuing Analysis of Trace Elements in Coal, The Hydro-Electric Power Commission of Ontario Research Division, Ontario, Canada, June 1973.
34. **Brown, R. L., Caldwell, R. L., and Fereday, F.,** Mineral constituents of coal, *Fuel,* 31, 261, 1952.
35. **Cameron, C. C. and Wright, N. A.,** Some peat bogs in Washington County, Maine: their formation and trace-element content, *Interdisciplinary Studies of Peat and Coal Origins,* Microform Publ. Geological Society of America, Boulder, Colo., 1977, 50.
36. **Chatterjee, P. K. and Pooley, F. D.,** Examination of some trace elements in south Wales coals, *Aust. Inst. Min. Metall. Proc.,* 263, 19, 1977.
37. **Corbett, R. G., Nuhfer, E. C., and Phillips, H. W.,** Trace elements in bituminous coal mine drainage and associated sulfate minerals, *Proc. W. Va. Acad. Sci.,* 39, 311, 1967.
38. **Damagalowa, M.,** Trace elements in ashes of carboniferous lepidophytes within the upper Silesian coal basin, *Przegl. Geol.,* 17, 245, 1969.
39. **Davison, R. L., Natusch, D. F. S., Wallace, J. R., and Evans, C. A.,** Trace elements in flyash-dependence of concentration on particle size, *Environ. Sci. Technol.,* 8, 1107, 1974.

40. Dixon, K., Skipsey, E., and Watts, J. T., The distribution and composition of inorganic matter in British coals, III. The composition of carbonate minerals in the coal seams of the East Midlands coalfields, *J. Inst. Fuel,* 43, 229, 1970.
41. Dobrovol'skaya, I. A., Forms of occurrence of trace elements in sediments of the brown coal formation in the northern part of the Zhitkovichi deposit, *Vopr. Geol. Beloruss.,* p. 20, 1974.
42. Dreesen, D. R., Gladney, E. S., Owens, J. W., Perkins, B. L., Wienke, C. L., and Wangen, L. E., Comparison of levels of trace elements extracted from fly ash and levels found in effluent waters from a coal-fired power plant, *Environ. Sci. Technol.,* 11, 1017, 1977.
43. Dreesen, D. R., Wangen, L. E., Gladney, E. S., and Owens, J. W., Solubility of trace elements in coal fly ash, *Dep. Energ. Symp. Ser.,* 45, 240, 1978.
44. Elejalde, C. and Martin, A., Relation between amino acid content and trace elements in Spanish coals, *Bol. Real Soc. Espan. Hist. Nat. Secc. Geol.,* 66, 339, 1968.
45. Fang, P., Tasai, P.-C., and Yung, K.-H., Distribution of trace elements in Mesozoic deposits in the Ordos Basin, *Ti Ch'iu Hua Hsueh,* 4, 286, 1973.
46. Fuchs, W., Rare elements in German Brown-Coal Ashes, *Ind. Eng. Chem.,* 27, 1099, 1935.
47. Furr, A. K., Parkinson, T. F., Heffron, C. L., Reid, J. T., Haschek, W. M., Gutenmann, W. H., Pakkala, I. S., and Lisk, D. J., National survey of elements and radioactivity in fly ashes, *Environ. Sci. Technol.,* 11, 1194, 1977.
48. Gindy, A. R., El-Askary, M. A., and Khalil, S. O., Differential thermal and thermogravimetric studies and some trace-element contents of some coals and carbonaceous shales from west-central Sinai, Egypat, *Chem. Geol.,* 22, 267, 1978.
49. Ginzburg, A. I., Petrographic composition of Paleogene brown coals of the European portion of the USSR, and their trace element distribution, *Tr. Vses. Nauch. Issled. Geol. Inst.,* 132, 264, 1968.
50. Gulyawa, L. A. and Itkina, E. S., *Microelementy Uglei, Goryuchikh Stantsev i Ikh Bituminoznykh Komponentov,* (Trace elements of coals, combustible shales, and their bituminous components), 1974, 92.
51. Hatch, J. R. and Swanson, V. E., Trace elements in Rocky Mountain coals, in Proc. Geology of Rocky Mountain Coal, Murray, D. K., Ed., Golden, Colo., 1976.
52. Headlee, A. J. W. and Hunter, R. G., The Inorganic Elements in the Coals, Characteristics of Minable Coals of West Virginia, West Virginia, Geological Survey, 1955, 36.
53. Hidalgo, R. V., Inorganic Geochemistry of Coal, Pittsburgh Seam, Ph.D. dissertation, West Virginia University, Morgantown, 1974, p. 125.
54. Juhasz, A., Investigation of coal seams in Kelet Borsd for trace elements, *Banyasz. Lapok,* 101, 208, 1968.
55. Karmazin, P. S., Content of some elements in the sedimentary rocks of the Donets coal basin, *Vopr. Prikl. Geokhim. Petrofiz.,* p. 61, 1978.
56. Kautz, K., Kirsch, H., and Laufhuette, D. W., Trace element content in coals and fine dust arising from them, *VGB Kraftwerkstechnik,* 55, 672, 1975.
57. Kitaev, I. V., Trace elements in Jurassic-Cretaceous of coals and carbonaceous rocks the Bureya and Tyrminsk Synclines, *Vop. Geol., Geokhim. Metallogen. Sev. Zapad Sekt. Tikhookean. Pojasa, Mater. Nauch. Sess.* p. 207, 1970.
58. Kitaev, I. V., Distribution of some rare elements in coals of the Bureinsk and Tyrminsk basins, *Vop. Litol. Geokhim. Vulkanogenno-Osad. Obrazov. Yuga Dal'nego Vostoka, Akad. Nauk SSSR,* p. 193, 1971.
59. Kohls, D. W., Reconnaissance of Trace Elements in Toxic Coal and Lignite, Mineral Resource Circular, No. 43, Bureau of Economic Geology, University of Texas, Austin, 1962.
60. Kulinenko, O. R., Trace elements in the coal-bearing formations of the Ukraine, *Vop. Geol. Mineral. Rud. Mestorozd. Ukr.,* 4, 241, 1971.
61. Kulinenko, O. R., Quantitative evaluation of trace element mobilities during coal accumulation (in the Donets and Lvov-Volyn basins), *Geol. Uh.,* 32, 86, 1972.
62. Kuznetsov, V. A. and Shimanovich, S. L., Distribution of trace elements in oligocene-miocene sandy formations of the brown coal field near Zhitkovichi in the Gomel district, *Dokl. Akad. Nauk B. SSSR,* 19, 360, 1975.
63. Laktinova, N. V., Egorov, A. P., Eremin, I. V., and Popinako, N. V., Determination of trace elements in coals, *Otkrytiya, Izobret. Prom. Obraztsy. Tovarnye Znaki,* 158, 20, 1979.
64. Laktinova, N. V., Egorov, A. P., and Popinako, N. V., Spectral determination of trace elements in coal, *Khim. Tverd. Topl. (Moscow),* 6, 112, 1978.
65. Lee, S. H. D., Johnson, I., and Fischer, J., Study of the volatility of minor and trace elements in Illinois coal, *Energ. Res. Abstr.,* p. 3, 1978.
66. Lyon, W. S., Ed., Trace Element Measurements at the Coal-Fired Steam Plant, CRC Press, Boca Raton, Fla., 1977.
67. Majewski, S. and Poniewierska, H., Trace elements in brown coal of the Konin deposit, *Geologia (Warsaw),* 3, 45, 1977.

68. **Mechacek, E.**, Microelements in beds of the Handlova-Novaky coal basin, *Geol. Zb. Bratislava,* 23, 311, 1972.
69. **Mechacek, E.**, Microelements in coal seams of the Modry Kamen basin, Dolina Mine, Southern Slovakia, *Acta. Geol. Georg. Univ. Comenianae, Geol.,* 28, 135, 1976.
70. **Medlin, J. H., Coleman, S. L., Wood, G. H., Jr., and Rait, N.**, Differences in minor and trace element geochemistry of anthracite in the Appalachian basin, *Geol. Soc. Am. Abstr. Progr.,* 7, 1198, 1975b.
71. **Medvedev, K. P.**, Role of macro and trace elements in coal formation processes, *Khim. Tverd. Topl.,* 4, 3, 1971.
72. **Miller, R. N.**, Geochemical Study of the Inorganic Constituents in some Low-Rank Coals, Ph.D. thesis, Pennsylvania State University, University Park, 1977.
73. **Minchev, D. and Eskenazi, Gr.**, Trace elements in the coals of Bulgaria, Zonel distribution of germanium in erratic vitrain fragments, *Spis. Bulg. Geol. Druzh.,* 30, 105, 1969.
74. **Minchev, D. and Eskenzai, Gr.**, Trace elements in the coals of Bulgaria, Trace elements in the coals of the Mariza-East coal basin, *God. Sofii. Univ. Geol. Geogr. Fak.,* 64, 263, 1972.
75. **Mukherjee, B. and Ghosh, A.**, Distribution and behaviour of trace elements in some Permian coals of India, *Indian Mineral.,* 17, 23, 1977.
76. **O'Gorman, J. V.**, Mineral matter and trace elements in North American coals, *Diss. Abstr. Int. B,* p. 193, 1972.
77. **O'Neill, R. L. and Suhr, N. H.**, Determination of trace elements in lignite ashes, *Appl. Spectrosc.,* 14, 45, 1960.
78. **Otte, M. U.**, Trace elements in some German mineral coals, *Chem. Erde.,* 16, 237, 1953.
79. **Parr, S. W.**, The Chemical Composition of Illinois Coals, Bull. 16, Illinois State Geological Survey Yearbook for 1909, Urbana, 1909, 205.
80. **Parr, S. W.**, *The Analysis of Fuel, Gas, Water and Lubricants,* McGraw-Hill, New York, 1932, 49.
81. **Plaska, Ya. P., Tikhonova, V. S., and Shabo, Z. V.**, Regular characteristics of trace element distribution in various coals from the Lvov-Volynia basin, *Geol. Geokhim. Goryuch. Iskop,* 24, 64, 1971.
82. **Pollock, E. N.**, Trace impurities in coal, *Am. Chem. Soc., Div. Fuel Chem. Prepr.,* 18, 92, 1973.
83. **Casagrande, D. J. and Erchull, L. D.**, Organic geochemistry of Okefenokea peats: metal consituents, *Interdisciplinary Studies of Peat and Coal Origins,* Microform Publ. 7, The Geological Society of America, Boulder, Colo., 1977, 72.
84. **Radmacher, W. and Hessling, H.**, Trace elements in coal and their spectroanalytical determination, *Z. Anal. Chem.,* 167, 172, 1959.
85. **Rao, C. P.**, Distribution of certain minor elements in Alaskan coals, Rep. 15, Mineral Industry Research Laboratory, Alaska University, Fairbanks, 1968, p. 47.
86. **Rees, O. W.**, Composition of the Ash of Illinois Coals, Circular 365, Illinois State Geological Survey, Urbana, 1964, p. 20.
87. **Ruch, R. R., Cahill, R. A., Frost, J. K., Camp, L. R., and H. J. Gluskoter**, Trace elements in coals of the United States determined by activation analysis and other techniques, *Am. Nucl. Soc. Trans.,* 21, 107, 19765.
88. **Ruch, R. R., Cahill, R. A., Frost, J. K., Camp, L. R., and Gluskoter, H. J.**, Survey of trace elements in coal and coal-related materials by neutron activation analysis, *J. Radioanal. Chem.,* 38, 415, 1977.
89. **Saprykin, F. Ya., Kler, V. R., and Kulachkova, A. F.**, Geochemical characteristics of rare element concentration in various types of coal and bituminous shale-bearing formations, *Zh., Geol. K.* 1971; Abstr. 2K57, Uglenosn. Formatsii Ikh. Genezis, p. 88, 1970.
90. **Saprykin, F. Ya., Kier, V. P., and Kulachkova, A. F.**, Geochemical characteristics of the concentration of rare element in the diverse types of coal-bearing formations, *Uglenos. Form. Ikh. Genezis, Dokl. Vses. Geol. Ugol. Soveshch.,* 4, 126, 1973.
91. **Sim, P. G.**, Concentration of some trace elements in New Zealand coals, *N.Z. Dep. Sci. Ind. Res. Bull.,* 218, 132, 1977.
92. **Sage, W. L. and McIlroy, J. B.**, Relationship of coal ash viscosity to chemical composition, *Combustion,* 31, 41, 1959.
93. **Smirnova, N.**, Rare elements in coals and their processing products, *Z. Angew. Geol.,* 23, 42, 1977.
94. **Sun, C. C., Vasquez-Tosas, H., and Augenstein, D.**, Pennsylvania anthracite refuse, a literature survey on chemical elements in coal and coal refuse, Special Research Report SR-83, Pennsylvania Department of Environmental Resources, Harrisburg, 1971.
95. **Swaine, D. J.**, Trace elements in coals, *Ocherki Sovrem. Geokhim. Anal. Khim.,* p. 482, 1972.
96. **Swaine, D. J.**, Trace elements in coal, in *Recent Contributions to Geochemistry and Analytical Chemistry,* Tugarinov, A. E., Ed., John Wiley & Sons, NY, 1976.
97. **Swaine, D. J.**, Trace elements in fly ash, *N.Z. Dep. Sci. Ind. Res. Bull.,* 218, 127, 1977.

98. **Swanson, V. E.,** Composition and Trace Element Content of Coal and Power Plant Ash, Southwest Energy Study, Part 2, Appendix J, Open File Rep. U.S. Geological Survey, Reston, Va., 1972, p. 61.
99. **Szalay, A. and Szilagyi, M.,** Accumulation of microelements in peat humic acids and coal, *Adv. Org. Geochem., Proc. Int. Meet.,* 4th, 567, 1969.
100. **Thoem, T. L.,** Coal fired power plant trace element study, *Energ. Environ.,* 5, 223, 1978.
101. **Tikhonova, V. S., Fudshuchak, M. Yu., and Kazakov, S. B.,** Rare and trace elements in coals of the Lvov-Volyn Basin, *Geol. Geokhim. Goryuch. Iskop,* 34, 104, 1973.
102. **Timasheva, E. E.,** Characteristics of trace elements in coal-bearing formations of the Dnieper-Donets syncline, *Geol. K,* 1968; Abstr. 5K52, *Nauch. Tekh. Inform. Min. Geol., SSSR, Ser. Geol. Mestorozd. Polez. Iskop Reg. Geol.,* 9, 86, 1967.
103. **Trunov, B. D.,** Characteristics of the distribution of trace element in coal-bearing rocks of the Gukovo-Zverovo region of the eastern Donets basin, *Vopr. Geol., Mineral. Geokhim. Uglenosn. Otlozh. SSSR,* p. 131, 1975.
104. **Urdininaa, J. S. A. and Pintaude, D. A.,** Minor elements in coal mines of the carboniferous basins of Butia-Leao and Candiota, *RS, Mineracao, Met.,* 331, 10, 1972.
105. U.S. Geological Survey and Montana Bureau of Mines and Geology, Preliminary Report of Coal Drill-Hole Data and Chemical Analysis of Coal Beds in Sheridan and Campbell Counties, Wyoming, and Big Horn County, Montana, Open-file Rep. 73-351, U.S. Geological Survey, Reston, Va., 1973, p. 57.
106. **Uzunov, I. and Karadzhova, B.,** Distribution of rare and trace elements in a productive horizon of the Burgas coal basin, *Petrografiya,* 17, 21, 1968.
107. **Varga, Mrs. E., Bella, Mrs. M., and Szava Benocs, Mrs. K.,** Comparative survey of the trends of trace elements concentration in Hungarian coal field, *Publ. Hung. Min. Res. Inst.,* 15, 221, 1972.
108. **Varga, Mrs. I., Bella, Mrs. L., and Szava Benocs, K.,** Comparative investigation of enrichment of trace elements in Hungarian hard coals, *Banyasz. Kohasz. Lapok, Banyasz.,* 105, 395, 1972.
109. **Wangen, L. E. and Wienke, C. L.,** A Review of Trace Element Studies Related to Coal Combustion in the Four Corners Area of New Mexico, NTIS LA-6401-MS, Los Alamos Scientific Laboratory, National Technical Information Service, Springfield, Va., 1976, p. 53.
110. **Wesson, T. C. and Armstrong, F. E.,** Elemental composition of coal mine dust, U.S. Bur. Mines Rep. Invest., 7992 p. 25, 1974.
111. **Williams, J. M., Henslry, W. K., Wewerka, E. M., Wanek, P. L., and Olsen, J. D.,** Trace Element Distribution in Several Coal Conversion Processes: An Exchange Program Between the Los Alamos Scientific Laboratory and National Coal Board of England, Energy Res. Abstr. 1978, Abstr. No. 32594, Report, 1978, p. 26.
112. **Winnicki, J.,** Occurrence and manner of binding of some rare elements in Polish coals, *Pr. Nauk. Inst. Chem. Niorg. Met. Pierwiastkow Rzadkich Politech. Wroclaw,* 18, 45, 1973.
113. **Youh, C. C.,** Distribution of minor elements in the Miocene coals of the Keelung-Taipei Region, Taiwan. II. Mineral matter and trace elements, *Proc. Natl. Sci. Counc., Repub. China,* 1, 27, 1977.
114. **Youh, C. C.,** A study on the mineral matter and trace elements in Miocene coals from the Keelung-Taipei region, Taiwan, *K'uang Yeh,* 22, 107, 1978.
115. **Yudovich, Ya. E., Korycheva, A. A., Obruchnikov, A. S., and Stepanov, Yu. V.,** Average content of trace elements in coals, *Geokhimiya,* 8, 1023, 1972.
116. **Zubovic, P., Sheffey, N. B., and Stadnichenko, T.,** Distribution of Minor Elements in Some Coals in the Western and Southwestern Regions of the Interior Coal Province, Bull. 1117-D, U.S. Geological Survey, Reston, Va., 1967, p. 33.
117. **Zubovic, P., Stadnichenko, T., and Sheffey, N. B.,** Distribution of minor elements in coal beds of the Eastern Interior region; Bull. 1117-B, U.S. Geological Survey, Reston, Va., 1964, p. 41.
118. **Zubovic, P., Stadnichenko, T., and Sheffey, N. B.,** Distribution of Minor Elements in Coals of the Appalachian Region, Bull. 1117-C, U.S. Geological Survey, Reston, Va., 1966, p. 37.
119. **Zul'fugarly, D. I., Orlenko, S. F., Mokienko, V. F., and Filipov, V. P.,** Characteristics of the distribution of trace elements in petroleums of coal fields in the Volgograd land along the Volga, *Zh. Khim.,* 1977; Abstr. 3P151, *Uch. Zap. Azerb. Un-t Ser. Khim. N.,* 1, 25, 1976.
120. **Zverev, L. N.,** Correlation between chemical and spectral analyses of coal-bearing deposits during an evaluation of the concentration of rare and minor elements, *Kontr. Tekhnol. Protsessov Obogashch, Polez. Iskop.,* 3, 7, 1971.
121. **Zverev, L. N.,** Relation and selection of correlation indexes between chemical and different spectral analysis of coal-bearing deposits during an evaluation of element concentration, *Kontr. Tekhnol. Protsessov Obogashch. Polez. Iskop,* 3, 16, 1971.

122. **Somerville, M. H. and Elder, J. L.**, Comparison on trace element analysis of North Dakota lignite laboratory ash with Lurgi gasifier ash and their use in environmental analysis, in Environmental Aspects of Fuel Conversion Technology III, Ayer, G. A. and Massoglia, M. F., Eds., Proc. Symp. Env. Aspects Fuel Conv. Tech., Rep. EPA-600/7-78-063, Environmental Protection Agency, Washington, D.C., 1977.
123. **Ibarra, J. V., Osacar, J., and Gavilian, J. M.**, Retention of metallic cations by lignites and humic acids, *Fuel*, 58, 827, 1979.
124. **Szalay, A.**, Accumulation of uranium and other trace metals in coal and organic shales and the role of humic acids in these geochemical enrichments, *Ark. Mineral. Geol.*, 5, 23, 1969.
125. **Calvo, M. M.**, Consideracions sobre el papel que desempenar les sustancios organicas naturales de caracter humico en la concentracion del uranio, in *Proc. Symp. Formation of Uranium Ore Deposits*, Athens, May 6 to 10, International Atomic Energy Agency, Vienna, 1974, 125.
126. **Hamrin, C. E.**, Catalytic Activity of Coal Mineral Matter, Annu. Rep. FE-2233-3, National Technical Information Service, Springfield, Va., 1977.
127. **Haught, O. L.**, On the occurrence and distribution of minor elements in coal, *Mo. Sch. Mines Metall. Tech. Ser.*, 85, 17, 1954.
128. **Hawley, J. E.**, Spectrographic study of some Nova Scotia coals, *Can. Min. Metall. Bull.*, 48, 712, 1955.
129. **Ctvrnicek, T. E., Rusek, S. J., and Sandy, W. C.**, Evaluation of Low-Sulfur Western Coal Characteristics, Utilization and Combustion, EPA-650/2-75-046, Environmental Protection Technology Series, NTIS PB-234 911, p. 555, National Technical Information Service, Springfield, Va., 1975.
130. **Cahill, R. A., Kuhn, J. K., Dreher, G. B., Ruch, R. R., Gluskoter, J. H., and Miller, W. G.**, Occurrence and distribution of trace elements in coal, *Am. Chem. Soc. Div. Fuel Chem. Prepr.*, 21, 1976.
131. **Casagrande, D. J. and Erchull, L. D.**, Metals in Okefenokea peat-form environments: relation to constituents found in coal, *Geochim. Cosmochim. Acta*, 40, 387, 1976.
132. **Loevblad, G., Ed.**, Trace Element Concentrations in Some Coal Samples and Possible Emissions from Coal Combustion in Sweden, Rep. IVL-B-358, 1977.
133. **Walker, F. K.**, Bibliography and index of U.S. Geological Survey Publications Relating to Coal, 1971—1975, Circ. 742, U. S. Geological Survey, Reston, Va., 1976, p. 36.
134. **Ruch, R. R., Gluskoter, H. J., and Shimp, N. F.**, Distribution of Trace Elements in Coal, EPA-650-2-74-118, Environmental Protection Technological Series, Washington, D.C., 1974, 49.
135. **Ruch, R. R., Gluskoter, J. H., and Shimp, N. F.**, Occurrence and Distribution of Potentially Volatile Trace Elements in Coal: a Final Report, Environ. Geol. Notes, 72, Illinois Geological Survey, Urbana, 1974, p. 96.
136. **Ruch, R. R., Kuhn, J. K., Dreher, G. B., Thomas, J., Frost, J. K., and Cahill, R. A.**, Potentially volatile trace elements in coal, in 168th Natl. Meet., American Chemical Society, Philadelphia, 1975.
137. **Gluskoter, H. J.**, Trace Elements in Coal: Occurrence and Distribution, Circ. 499, Illinois State Geological Survey, Urbana, p. 154, 1977.
138. **Zubovic, P.**, Physiochemical properties of certain minor elements as controlling factors of their distribution in coal, *Adv. Chem. Ser.*, 55, 221, 1966.
139. **Zubovic, P.**, Geochemistry of trace elements in coal, in Symp. Proc., Environ. Aspects of Fuel Conversion Technol., II, EPA-600/2-76-149, U.S. Environmental Protection Agency, Washington, D.C., 1976, 47.
140. **Swaine, D. J.**, Trace elements in coal, *Trace Subst. Environ. Health*, 11, 107, 1977.
141. **Horton, L. and Aubrey, V. K.**, The distribution of minor elements in vitrain: three vitrains from the Barnsley seam, *London J. Soc. Chem. Ind.*, 69, S41, 1950.
142. **Finkelman, R. B., Stanton, R. W., Cecil, C. B., and Minken, J. A.**, Modes of occurrence of selected trace elements in several Appalachion coals, *Am. Chem. Soc. Div. Fuel Chem. Prepr.*, 24, 236, 1979.
143. **Finkelman, R. B.**, Determination of trace element sites in the Waynesburg coal by SEM analysis of accessory minerals, *Scanning Electron Microsc.*, 1, 143, 1978.
144. **Finkelman, R. B. and Stanton, R. W.**, Identification and significance of accessory minerals from a bituminous coal, *Fuel*, 57, 763, 1978.
145. **Akers, D. J.**, Uranium in Coal Carbonaceous Rocks in the United States and Canada: an Annotated Bibliography, Rep., 157, Coal Research Bureau, West Virginia University, Morgantown, 1978, 5.
146. **Wewerka, E. M. and Williams, J. M.**, Trace element characterization of coal wastes, DO, EPA-600/7-78-028, Environmental Protection Agency, Washington, D.C., 1978, 52.
147. **Wewerka, E. M., Williams, J. M., Vanderborgh, N. E., Wagner, H., Wanek, P. L., and Olsen, J. D.**, Trace Element Characterization and Removal/Recovery from Coal and Coal Wastes, National Technical Information Service, Springfield, Va., 1978.
148. **Hausen, L. D., Phillips, L. R., Mangelson, N. F., and Lee, M. L.**, Analytical study of the effluents from a high temperature entrained flow gasifier, *Fuel*, 80, 323, 1980.

149. **Barret, E. P.**, The fusion flow and clinkering of coal ash; a survey of the chemical background, in *Chemistry of Coal Utilization,* Vol. 1, John Wiley & Sons, New York, 1945, 496.
150. **Bickelhaupt, R. E.**, Effects of Chemical Composition on Surface Resistivity of Fly Ash, Prepared for Environmental Protection Agency, Contract 68-02-1303, NTIS PB-244 885, National Technical Information Service, Springfield, Va., August 1975, 41.
151. **Borio, R. W. and Hensel, R. P.**, Coal-ash composition as related to high-temperature fireside corrosion and sulfur-oxide emission control, *J. Eng. Power,* 94, 142, 1972.
153. **Morstin, K. and Woznick, J.**, Hydrogen in coal, the feasibility of determination by neutron methods, IAEA-SM-216/33, International Atomic Energy Agency, Vienna, 1977, 119.
154. **Demeter, J. J. and Bienstock, D.**, Sulfur retention in anthracite ash, U.S. Bur. Mines, Rep. Invest., 7160, 12, 1968.
155. **Bryers, R. W. and Taylor, T. E.**, An examination of the relationship between ash chemistry and ash fusion temperature in various coal size and gravity fractions using polynomial regression analysis, Trans. ASME, *J. Eng. Power,* 98, 528, 1976.
156. **Thiessen, G., Ball, G. C., and Grotts, P. E.**, Coal ash and coal mineral matter, *Ind. Eng. Chem.,* 28, 355, 1936.
157. **Reiter, F. M.**, How sulfur content relates to ash fusion characteristics, *Power Eng.,* 59, 98, 1955.
158. **Rees, O. W. et al.**, Sulfur Retention in Bituminous Coal Ash, Circ. 396, Illinois State Geological Survey, Reston, Va., 1966, 10.
159. **Dutcher, R. R., White, E. W., and Spackman, W.**, Elemental ash distribution in coal components — use of the electron probe, Proceedings 22nd Iron-Making Conf. Iron and Steel Division, *Am. Inst. Min. Eng.,* 22, 463, 1963.
160. **Fischer, G. L. et al.**, Size-dependence of the physical and chemical properties of coal fly ash, *Am. Chem. Soc. Div. Fuel Chem. Prepr.,* 22, 149, 1977.
161. **Gronhovd, G. H., Wagner, R. J., and Wittmaier, A. J.**, Comparison of ash fouling tendencies of high- and low-sodium lignite from a North Dakota mine, *Proc. Am. Power Conf.,* 28, 632, 1966.
162. **Gronhovd, G. H., Hark, A. E., and Paulson, L. E.**, Ash fouling studies of North Dakota lignite, in Technology and Use of Lignite, Inf. Circ. 8376, U.S. Bureau of Mines, Washington, D.C., 1967, 76.
163. **Gronhovd, G. H., Beckering, W., and Tufte, P. H.**, Study of factors affecting ash deposition from lignite and other coals, presented at ASME Winter Annual Meeting, Los Angeles, Calif., 10, 1969.
164. **Ray, S. S. and Parker, G. F.**, Characterization of Ash From Coal-Fired Power Plants, Interagency Energy Environ. Res. Dev. Progr. Rep. PB-265-374, EPA-600/7-77-010, Environmental Protection Agency, Washington, D.C., 12, 1977.
165. **Hendrickson, T. A.**, U.S. Coal, *Synthetic Fuels Data Handbook,* Cameron Engineers Inc., 1975, 142.
166. **Hendrickson, T. A.**, Ash in coal, *Synthetic Fuels Data Handbook,* Cameron Engineers Inc., 1975, 147.
167. **Gluskoter, H. J. and Lindhal, P. C.**, Cadmium — mode of occurrence in Illinois coal, *Science,* 181, 264, 1973.
168. **Frost, J. K., Santoliquido, P. M., Camp, L. R., and Ruch, R. R.**, Trace elements in coal by neutron activation analysis with radiochemical separations, *Adv. Chem. Ser.,* 141, 84, 1975.
169. **Kuhn, J. K., Harfst, W. F., and Shimp, N. F.**, X-ray fluorescence analysis of whole coal, in *Advances in Chemistry,* Babu, S. P., Ed., Vol. 141, American Chemical Society, Washington, D.C., 1975, 66.
170. **Dreher, G. B. and Schleicher, J. A.**, Trace elements in coal by optical emission spectroscopy, *Adv. Chem. Ser.,* 141, 35, 1975.
171. **Debelak, K. A. and Schrodt, J. T.**, Comparison of pore structure in Kentucky coals by mercury penetration and carbon dioxide adsorption, *Fuel,* 58, 732, 1979.
172. **Schultz, H., Hattman, E. A., and Booher, W. B.**, Trace elements in coal, what happens to them? *Prepr. Pap. Natl. Div. Environ. Chem., Am. Chem. Soc.,* 15, 196, 1975.
173. **Zubovic, P., Hatch, J. R., and Medlin, J. H.**, Assessment of the chemical composition of coal resources, in Proc. U.N. Symp. World Coal Prospects, Katowice, 15-23.10, 1979, 68.
174. **Petrakis, L. and Grandy, D. W.**, Free radicals in coals and coal conversion. II. Effect of liquefaction processing conditions on the formation and quenching of coal free radicals, *Fuel,* 59, 227, 1980.
175. **Vypirakhina, S. S. and Aronov, S. G.**, Isotope method of determination active hydrogen in coal, *Solid Fuel Chem.,* 10, 13, 1976.
176. **Alemany, L. B., King, S. R., and Stock, L. M.**, Proton and carbon n.m.r. spectra of butylated coal, *Fuel,* 57, 738, 1978.
177. **Gerstein, B. C. and Pembleton, R. G.**, Pulsed nuclear magnetic resonance spectrometry for nondestructive determination of hydrogen in coal, *Anal. Chem.,* 49, 75, 1977.
178. **Abernethy, R. F. and Gibson, F. H.**, Rare Elements in Coal, Rep. BM-IC-8163, U.S. Bureau of Mines, Washington, D.C., 1963.

179. **Butler, J. A.**, Utah's new uranium mill, *Eng. Min. J.*, p. 152, March 1951.
180. **DeBrito, De, A. C.**, Spectrographic study of the ash of portugese anthracites, *Estudos. Notas Trabalhos Serv. Fomento Mineiros*, 10, 236, 1955.
181. **Hansen, L. D., Phillips, L. R., Mangelson, N. F., and Lee, M. L.**, Analytical study of the effluents from a high temperature entrained flow gasifier, *Fuel*, 80, 323, 1980.
182. **Wewerka, E. M., Williams, J. M., and Vanderborgh, N. E.**, Contaminants in coals and coal residues, 4th Natl. Conf. Energy and the Environment, LA-UR 76-2197, Los Alamos Scientific Laboratory, N.M., 1976, 23.
183. **Swanson, V. E., Medlin, J. H., Hatch, J. R., Coleman, S. L., Wood, G. H., Jr., Woodruff, S. D., and Hildebrand, R. T.**, Collection, Chemical Analysis, and Evaluation of Coal Samples in 1975, Open file Rep. 76-468, U.S. Geological Survey, Reston, Va., 1976, 503.
184. **Bertine, K. K. and Goldberg, E. D.**, Fossil fuel combustion and the major sedimentary cycle, *Science*, 173, 223, 1971.
185. **Gibson, F. H. and Selvig, W. A.**, Rare and Uncommon Chemical Elements in Coal, Tech. Pap. 669, U.S. Bureau of Mines, Washington, D.C., 1944, 23.
186. **Zilbermintz, V. A., Rusanov, A. K., and Kostrykin, V. M.**, The Question of the Distribution of Germanium in Fossil Coals, *Akad. V. I. Vernadskomu k Pyatidessyatiletiyu Nauch, Deyatelnosti*, 1, 169, 1936; Chem. Abstr., 33, 8384, 1939.
187. **Nazarenko, V. A.**, The occurrence of vanadium, beryllium, and boron in the ash of some coals, *Trav. Lab. Biogeochem. Acad. Sci. U.R.S.S.*, 4, 265, 1937; *Chem. Abstr.*, 32, 7770, 1938.
188. **Stadnichenko, T., Zubovic, P., and Sheffey, N. B.**, Beryllium Content of American Coal, Bull. 1084-K, U.S. Geological Survey, Reston, Va., 1961, 253.
189. **Goldschmidt, V. M. and Peters, C.**, The Geochemistry of Arsenic, *Nachr. Ges. Wiss. Göttingen Math Phys. Klasse Fachgruppe*, IV, 11, 1934; Chem. Abstr., 28, 5788, 1934.
190. **Grillot, R.**, Berlyllium in Coals, Centre d'Etudes et Recherches des Charbonnages de France, Publ. CEVCHAR No. 1195, 6, 1961.
191. **Abernethy, R. F. and Hattman, E. A.**, Colorimetric determination of beryllium in coal, U.S. Bur. Mines, Rep. Invest., 7452, 8, 1970.
192. **Nunn, R. C., Lovell, H. L., and Wright, C. C.**, Spectrographic analysis of trace elements in anthracite, in *Trans. Annu. Anthracite Conf. Lehigh University*, 11, 52, 1953.
193. **Deul, M. and Annell, C. S.**, The Occurrence of Minor Elements in Ash of Lower-Rank Coal from Texas, Colorado, North Dakota, and South Dakota, Bull. 1036-H, U.S. Geological Survey, Reston, Va., 1956, 155.
194. **Headlee, A. J. W. and Hunter, R. G.**, Elements in coal ash and their industrial significance, *Ind. Eng. Chem.*, 45, 548, 1953.
195. **Jedwab, J.**, Distribution of beryllium in some coal beds of Bonne-Esperance a Lambusart. (p. 69); Coal as a source of beryllium (p. 67); The presence of beryllium in Cerain Belgian coals (p. 77) *Bull. Soc. Belge Geol. Paleontol. Hydrol.*, 69, 227, 1960.
196. **Thilo, E.**, Results of analysis of two coal ashes, *Zeit. Anorg. Allgem. Chem.*, 218, 201, 1934.
197. **Leutwein, F. and Rosler, H. J.**, Geochemical investigations of paleozoic and mesozoic coals in Central and Eastern Germany, *Freiberg. Forsch.*, 19, 1, 1956.
198. **Hawley, J. E.**, Germanium content of some Nove Scotia coals, *Econ. Geology*, 50, 517, 1955.
199. **Wewerka, E. M., Williams, J. M., and Wanek, P. L.**, Assessment and Control of Environmental Contamination from Trace Elements in Coal Processing Wastes, ERDA Energy Res. Abstr. No. LA-UR-76-86, 1976; p. 7; **Wewerka, E. M., Williams, J. M., Wangen, L. E., Bertino, J. P., Wanek, P. L., Olsen, J. D., Thode, E. F., and Wagner, P.**, Trace Element Characterization of Coal Wastes, DOE LA-7831-PR, EPA-600/7-79, Environmental Protection Agency, Washington, D.C., 1979, 144.
200. **Hak, J. and Babcan, J.**, The geochemistry of germanium and berillyium in coals of the Sokolov basin, Geochem. in Czech., Trans. 1st Conf. Geochem., Ostrava, September 20 to 24, 1967, 163.
201. **Danchev, V. I., Strelyanov, N. P., Vasil'eva, G. L., and Nekrasova, L. P.**, Beryllium in Tertiary coal measures and siderite concentrations, *Litol. Polez, Iskopl.*, 5, 69, 1969.
202. **Drever, J. I., Murphy, J. W., and Surdam, R. C.**, The distribution of arsenic, beryllium, cadmium, copper, mercury, molybdenum, lead and uranium associated with the Wyodak coal seam, Powder River Basin, Wyoming, *Contrib. Geol.*, 15, 93, 1977.
203. **Odor, L.**, Beryllium content of Transdanubian eocene coals, *Magy. Allami Foldt. Intez. Evi. Jelentese*, p. 123, 1971.
204. **Phillips, M. A.**, Levels of both airborne beryllium and beryllium in coal at the Hayden Power Plant near Hayden, Colorado, *Environ. Lett.*, 5, 183, 1973.
205. **Losev, B. I., Nekrasova, Z. D., and Vasil'eva, I. V.**, Effect of treating coals with organic acids on the distribution of beryllium and gallium in pyrolysis products, *Khim. Topl.*, 2, 147, 1973.
206. **Eskenazi, G.**, Adsorption of beryllium on peat and coals, *Fuel*, 49, 61, 1970.
207. **Gladney, E. S. and Owens, J. W.**, Beryllium emissions from a coal-fired power plant, A, *J. Environ. Sci. Health*, A11, 297, 1976.

208. **Bencko, V., Vasilieva, E. V., Tichy, V., Konopikova, L., Horecka, J., and Symon, K.**, Immunological characterization of groups of people exposed to beryllium due to combustion of coal with increased contents of this harmful component, *Cesk. Hyg.*, 23, 11, 1978.
209. **Bekyarova, E. E. and Rushev, D. D.**, Forms of binding of germanium in solid fuels, *Fuel*, 50, 272, 1971.
210. **Kul'skaya, A. O. and Vdovenko, O. F.**, Spectrochemical determination of germanium, beryllium, and scandium in coal ash, *Khim. Fis. Khim. Spektral'n. Metody Issled. Rssd Redkikh Rasseyan. Elementov, Min. Geol. Okhrany Nedr. SSSR*, p. 135, 1961.
211. **Owens, J. and Gladney, E. S.**, Determination of beryllium in environmental materials by flameless atomic-absorption spectroscopy, *At. Absorp. Newsl.*, 14, 76, 1975.
212. **Gladney, E. S.**, Direct determination of beryllium in NBS SRM 1632 coal by flameless atomic absorption, *At. Absorp. Newsl.*, 16, 42, 1977.
213. **Tamura, N.**, Accuracy in the nondestructive neutron activation analysis of coal and beryllium for minor and trace elements using cobalt as a flux determinant, *Radiochem. Radioanal. Lett.*, 18, 135, 1974.
214. **Kear, D. and Ross, J. B.**, Boron in New Zealand coal ashes, *N.Z. J. Sci.*, 4, 360, 1961.
215. **Roga, B., Ihnatowicz, A., Weclewska, M., and Ihnatowicz, M.**, Boron contents of Polish coal, *Pr. Gl. Inst. Gorn. Ser. B*, No. 212, 1958, 7.
216. **Hutcheon, J. M.**, Fuel industries and atomic energy — some common interests, *J. Inst. Fuel*, 26, 306, 1953.
217. **Khan, I. A. and Sen, D.**, Separation of boron from coke, *J. Sci. Ind. Res. (India)*, 18B, 434, 1959.
218. **Bohor, B. F. and Gluskoter, J. H.**, Boron in illite as an indicator of paleosalinity of Illinois coals, *J. Sediment. Petrol.*, 43, 945, 1973.
219. **Duel, M. and Annell, C. S.**, The Occurrence of Minor Elements in Ash of Low-Rank Coal from Texas, Colorado, North Dakota, and South Dakota, Bull. 1036-H, U.S. Geological Survey, Reston, Va., 1956, 155.
220. **Rafter, T. A.**, Boron and strontium in New Zealand coal ashes, *Nature (London)*, 155, 332, 1945.
221. **Konieczynski, J.**, Occurrence of boron in hard coals, *Pr. Gl. Inst. Gorn.*, 482, 26, 1970.
222. **Kryukova, V. N., Shishlyannikova, E., and Koralis, O. I.**, Boron distribution in coals of the Azeisk deposit in the Irkutsk region, *Geokhimiya*, 7, 1115, 1975.
223. **Kunstmann, F. H., Harris, J. F., Bodenstein, L. B., and Van den Berg, A. M.**, The Occurrence of Boron and Fluorine in South African Coals and Their Behavior During Combustion, Bull. 63, Fuel Research Institute of South Africa, 45, 1963.
224. **Allan, F. J., Dundas, F. S., and Lambie, D. A.**, Determination of boron content of some Scottish coals and distribution of the boron between various density fractions, *J. Inst. Fuel*, 42, 29, 1969.
225. **Nakamura, K. and Kitamura, M.**, Manufacture of fertilizers containing boron, magnesia, and silica by burning with coal fuels, *Nenryo Kyokai-shi*, 49, 925, 1970.
226. **Lenz, U. and Köster, R.**, Boron compounds. XLVII. Quantitative determination of hydroxyl groups in lignites using activated triethylborone, *Fuel*, 57, 489, 1978.
227. **Millet, J.**, Determination of boron in industrial coals, in Congr. Groupe Avance. Methodes Anal. Spectrog. Prod. Met. 10e, Paris, 1957, 229.
228. **Skalska, S. and Held, S.**, Determination of trace contents of boron in graphite, coal, coke, and carbon black by the method of spectral emission analysis, *Chem. Anal. Warsaw*, 1, 294, 1956.
229. **Kunstmann, F. H. and Harris, J. F.**, Microanalytical method for the determination of boron in coal, *J. Chem. Metall. Min. Soc. S. Afr.*, 55, 12, 1954.
230. **Konieczynski, J.**, Spectrographic method for determination of trace amounts of boron in coals, coke, pitch, and tars, *Pr. Gl. Inst. Gorn. Komun. (Katowice) Stalinograd Ser. B*, 245, 1, 1960.
231. **Saxby, J. D.**, Atomic H/C ratios and the generation of oil from coals and kerogens, *Fuel*, 59, 305, 1980.
232. **Dobronravov, V. F.**, Influence of the petrographic composition on the nitrogen content of the Hard Coals of the Kuzbass, *Solid Fuel Chem.*, 10, 6, 1976.
233. **Karr, C., Jr., Estap, P. A., and Kovach, J. J.**, Spectroscopic evidence for the occurrence of nitrates in lignites, *Am. Chem. Soc. Div. Fuel Chem. Prepr.*, 12, 1, 1968.
234. **Ito, S.**, Separation of Nitrogen-Containing Constituents from Solid Fuel, German Patent 2504965, 1976, 16.
235. **Hamrin, C. E., Johannes, A. H., James, W. D., Sun, G. H., and Ehmann, W. D.**, Determination of oxygen and nitrogen in coal by instrumental neutron activation analysis, *Fuel*, 58, 48, 1979.
236. **Volborth, A., Miller, G. E., Garner, C. K., and Jerabek, P. A.**, Material balance in coal, II. Oxygen determination and stoichiometry of 33 coals, *Fuel*, 57, 49, 1978.
237. **Volborth, A., Miller, G. E., Garner, C. K., and Jarabek, P. A.**, Oxygen determination and stochiometry of some coals, *Am. Chem. Soc. Div. Fuel Chem. Prepr.*, 22, 1977; Symp. New Tech. in Coal Anal., presented at 174th Am. Chem. Soc. Natl. Mett., Chicago, Ill., 1977, 9.

238. Volborth, A., Miller, G. E., Garner, C. K., and Jerabek, P. A., Oxygen in coal ash, a simplified approach to the analysis of ash and mineral matter in coal, *Fuel*, 56, 123, 1976.
239. Kharitonov, G. V. and Zamai, A. A., Change of oxygen containing functional groups in coal during their oxidation by manganese tri- and tetracetate and lead tetracetate, *Zh. Khim.*, 1969; Abstr. 5P39, *Tr. Frunze. Politekh. Inst.*, 28, 61, 1968.
240. Wasilewski, P. and Kobel-Najzarek, E., Oxygen-containing functional groups in solid products of coal carbonization, *Zesz. Nauk Politech. Slask., Chem.*, 50, 332, 1970.
241. Schylyer, D. J., Ruth, T. J., and Wolf, A. P., Oxygen content of selected coals as determined by charged particle activation analysis, *Fuel*, 58, 208, 1979; Schlyer, D. J. and Wolf, A. P., A study of coal oxidation by charged particle activation analysis, in 4th Int. Conf. on Nuclear Methods in Environmental and Energy Research, BLN 27607, Columbia, Mo., 1980.
242. Churchill, H. V., Rowley, R. J., and Martin, L. N., Fluorine content of certain vegetation in western Pennsylvania area, *Anal. Chem.*, 20, 69, 1948.
243. Bradford, H. R., Fluorine in Western coals, *Min. Eng.*, 9, 78, 1957.
244. Lessing, R., Fluorine in coal, *Nature (London)*, 134, 699, 1934.
245. Kokubu, N., Fluorine in rocks, *Mem. Fac. Sci., Kyushu Univ., Ser. C*, 2, 95, 1956; *Chem. Abstr.*, 50, 11188, 1956.
246. Crossley, H. E., Fluorine in coal. II, III, IV, *J. Soc. Chem. Ind.*, 63, 284, 1944; Crossley, H. E., The inorganic constituents of coal; occurrence and industrial significance, *Chem. Age (London)*, 55, 629, 1946; *Inst. Fuel Bull.*, 67, 57, 1946.
247. McGowan, G. E., The determination of fluorine in coal, an adaptation of spectrometric methods, *Fuel*, 39, 245, 1960.
248. Abernethy, R. F. and Gibson, F. H., Method for determination of Fluorine in Coal, Rep. BM-RI-7054, Bureau of Mines, Pittsburgh, 1967.
249. Durie, R. A. and Schafer, H. N. S., The inorganic constituents in Australian coals. IV. Phosphorus and fluorine — their probable mode of occurrence, *Fuel*, 63, 31, 1964.
250. Abernethy, R. F., Gibson, F. H., and Frederic, W. H., Phosphorous, chlorine, sodium and potassium in U.S. coals, U. S. Bureau of Mines, Report of Investigations 6579, 34, 1965.
251. Sondreal, E. A., Elder, J. L., and Kube, W. R., Characteristics and Variability of Lignite Ash from the Northern Great Plains Province, Inf. Circ. 8304, U.S. Bureau of Mines, Washington, D.C., 1966, 39.
252. Anderson, C. H. and Beatty, C. D., Spectrographic determination of sodium and potassium in coal ashes, *Anal. Chem.*, 26, 1369, 1954.
253. Gluskoter, H. J. and Ruch, R. R., Chlorine and sodium in Illinois coals determined by neutron activation, *Fuel*, 50, 65, 1971.
254. Muter, R. B. and Cockrell, C. F., Analysis of sodium, potassium, calcium and magnesium in siliceous coal ash and related materials by atomic absorption spectroscopy, *Appl. Spectrosc.*, 23, 493, 1969.
255. Cooley, S. A. and Ellman, R. C., Analysis of coal and ash from lignite and subbituminous coals of Eastern Montana, Energy Resources of Montana, 22nd Annu. Publ. Montana Geological Society, 1975, 143.
256. Savranskaya, A. P. and Khanina, A. L., Rapid determination of calcium oxide and magnesium oxide in manganese ores, coal fines ash, and ground coke, *Zavod. Lab.*, 35, 559, 1969.
257. Kranz, M., Grala, M., and Krzymien, M., Gallium and aluminium in coal and gas dusts and ashes, *Poznan. Tow. Przyjaciol. Nauk Wydz. Mat.-Przyr. Pr. Kom. Mat. Przyr.*, 12, 3, 1968.
258. Panin, V. I. and Glushnev, S. V., Nature of aluminium oxide distribution in Moscow basin coals, *Khim. Tverd. Topl.*, 2, 87, 1974.
259. Panin, V. I. and Shpirt, M. Ya., Quantitative distribution of silicon, aluminium, iron, and calcium compounds in fly-ash of burned powdered coal, *Khim. Tverd. Topl.*, 3, 81, 1971.
260. Berg, W. A., Aluminium and manganese toxicities in acid coal mine wastes, *NATO Adv. Study Inst. Ser. E*, E7, 141, 1978.
261. Hitchen, A. and Zechanowitsch, G., Methods for the Analysis of Ilmenite, Titanium-Bearing Slags, and other Electric Furnace Slags. IVB. Determination of Aluminium in other Types of Ores and Slags, Can. Mines Br., Tech. Bull. 169, 1973, 37.
262. Navalikhin, L. V., Kireev, C. A., and Talinin, Yu. N., Rapid method for determining germanium, aluminium, and silicon contents in coal using a neutron generator, *Khim. Tverd. Topl.*, 4, 149, 1970.
263. Schultz, H. D., Vesely, C. J., and Langer, D. W., Electron binding energies for silicon minerals occurrings in a respirable coal dust, *Appl. Spectrosc.*, 28, 374, 1974.
264. Bernas, B., A new method for decomposition and comprehensive analysis of silicates by atomic absorption spectrometry, *Anal. Chem.*, 40, 1682, 1968.
265. Campbell, J. R., Distribution of phosphorus, *Mines Miner.*, p. 408, 1908.
266. Geer, M. R., Yancey, H. F., and Davis, F. T., Occurrence of phosphorous in Washington coal and its removal, *Trans. Am. Inst. Min. Metall. Eng.*, 157, 152, 1944.

267. **Selvig, W. A. and Seaman, H.**, Sulfur forms and ash-forming minerals in Pittsburgh coal, Mining and Metallurgic Investigations, U.S. Bur. Mines Cooperative Bulletin 43, Washington, D.C., 1929, 21.
268. **Van Krevelen, D. W.**, *Coal,* Elsevier, Amsterdam, 1961.
269. **Given, P. H. and Miller, R. N.**, Determination of forms of sulphur in coals, *Fuel,* 57, 380, 1978.
270. **Given, P. H.**, Problems in the chemistry and structure of coals as related to pollutants from conversion processes, in Symp. Proc. Environ. Aspects of Fuel Conversion Technol., EPA-650/2-74-118, St. Louis, Mo., 1974, 27.
271. **Ensminger, J. T.**, Coal: Origin, classification, and physical and chemical properties, in Environmental, Health and Control Aspects of Coal Conversion, Braunstein, H. M., Copenhover, E. D., and Pfuderer, H. A., Eds., Report Oak Ridge National Laboratories ELS-94, 1977.
272. **Casagrande, D. J. and Ng, L.**, Incorporation of elemental sulphur in coal as organic sulphur, *Nature (London),* 282, 598, 1979.
273. **Meyers, R. A.**, *Coal Desulfurization,* Marcel Dekker, New York, 1977.
274. **Casagrande, D. J., Idowu, G., Friedman, A., Rickert, P., Seifert, K., and Schlenz, D.**, H_2S incorporation in coal precursors: origins of organic sulphur in coal, *Nature (London),* 282, 599, 1979.
275. **Mukherjee, D. K. and Chowdhury, P. B.**, Catalytic effect of mineral matter constituents in a North Assam coal on hydrogenation, *Fuel,* 55, 4, 1976.
276. **Boateng, D. A. D. and Phillips, C. R.**, Examination of coal surfaces by microscopy and the electron microprobe, *Fuel,* 55, 318, 1976.
277. **Gluskoter, H. J. and Simon, J. A.**, Sulfur in Illinois Coals, Circ. 432, Illinois State Geological Survey, Urbana, 1968, 28.
278. **Gluskoter H. J. and Hopkins, M. E.**, Distribution of sulfur in Illinois coals, Depositional Environments in Parts of the Carbondale Formation; Western and Northern Illinois, Guidebook, Series 8, Illinois State Geological Survey, Urbana, 1970, 89.
279. **Gluskoter, H. J.**, Forms of sulfur, in *Coal Preparation,* American Institute of Mining, Metallurgical Petroleum Engineers, New York, 1968, 44.
280. **Gluskoter, H. J.**, Inorganic sulfur in coal, *Energy Sources,* 3, 125, 1977.
281. **Fowkes, W. W. and Hoeppner, J. J.**, Sulfur in lignite: forms and transformation on thermal treatment, U.S. Bur. Mines, Rep. Invest., 5626, 15, 1960.
282. **Hidalgo, R. V.**, Sulfur-clay mineral relations in coal, in Some Appalachian Coals and Carbonates, Models of Ancient Shallow; Water Deposition, Geological and Economical Survey, Morgantown, W.Va., 1969, 309.
283. **Casagrande, D. and Seifert, K.**, Origins of sulfur in coal; importance of the ester sulfate content of peat, *Science,* 195, 675, 1977.
284. **Greer, R. T.**, Nature and distribution of pyrite in Iowa coals, in *Proc. Electron Microsc. Soc. Am.,* Bailey, G. W., Ed., Claitor's Press, Baton Rouge, La., 1976, 620.
285. **Gray, R. J., Schapiro, N., and Coe, G. D.**, Distribution and forms of sulfur in a high volatile Pittsburgh seam coal, *Trans. Soc. Min. Eng. AIME,* 226, 113, 1963.
286. **Mansfield, S. P. and Spackman, W.**, Petrographic Composition and Sulfur Content of Selected Pennsylvania Bituminous Coal Seams, Special Res. Rep. SR-50, The Pennsylvania State University, University Park, 1965, 121.
287. **Neavel, R. C.**, Sulfur in Coal, Its Distribution in the Seam and in Mine Products, Ph.D. thesis, The Pennsylvania State University, University Park, 1966, 332.
288. **Reidenouer, D. R.**, The Relationship Between Sulfur Distribution and Paleotopography in Three Selected Coal Seams of Western Pennsylvania, M.S. thesis, The Pennsylvania State University, University Park, 1966, 196.
289. **Nazarova, L. N. and Berman, V. Yu.**, Interrelations between the sulfur content in a coal seam and components of the chemical composition of mine waters in the eastern Donets basin, *Gidrokhim. Mater.,* 50, 83, 1969.
290. **Cheek, R. B.**, Sulfur Facies of the Upper Freeport Coal of North-Western Preston County, West Virginia, M.S. thesis, West Virginia University, Morgantown, 1969, 66.
291. **Hopkins, M. E. and Nance, R. B.**, Sulfur content of the Colchester (No. 2), Coal member at the Banner Mine, Peoria and Fulton Counties, Illinois, in Depositional Environments in Part of the Carbondale Formation, Guidebook, Ser. 8, Illinois State Geological Survey, Urbana, 1970, 96.
292. **Eddy, G. E.**, Sulfur and Ash Distribution of Pittsburgh coal, Mathies Mine, Pennsylvania: Evaluation of Paleotopographic Control, Ph.D. dissertation, West Virginia University, Morgantown, 1971, 69.
293. **Schultz, H. D. and Proctor, W. G.**, Application of electron emission spectroscopy to characterize sulfur bonds in coal, *Appl. Spectrosc.,* 27, 347, 1973.
294. **Smith, J. W. and Batts, B. D.**, The distribution and isotopic composition of sulfur in coal, *Geochim. Cosmochim. Acta,* 38, 121, 1974.

295. McMillan, B. G., Sulfur Occurrence in Coal and Its Relationship to Acid Formation, Literature Review, Coal Research Bureau Technical Report No. 110, College of Mineral and Energy Resources, West Virginia, University, Morgantown, 1975, 18.
296. Von Demfange, W. C. and Warner, D. L., Vertical distribution of sulfur forms in surface coal mines spoils, *Pap. 3rd Symp. Surf. Min. Reclam.*, 1, 135, 1975.
297. Chadwick, R. A., Rice, R. C., Bennett, C. M., and Woodriff, R. A., Sulfur and trace elements in the Rosebud and McKay coal seams, Colstrip field, Montana, *Mont. Geol. Soc., Annu. Field Conf. Guidebk.*, 22, 167, 1975.
298. Chadwick, R. A., Woodriff, R. A., Stone, R. W., and Bennett, C. M., Literal and vertical variations in sulfur and trace elements in coal, in Fort Union Coal Field Symp., Eastern Missouri Coll., Billings, Mont., 1975.
299. Levene, H. D. and Hand, J. W., Sulfur staws in the ash when lignite burns, *Coal Min. Process.*, 12, 46, 1975.
300. Gladfelter, W. L. and Dickerhoof, D. W., Determination of sulphur forms in hydrodesulphurized coal, *Fuel*, 55, 355, 1976.
301. Lloyd, W. G. and Francis, H. E., Determination of sulfur in whole coal by X-ray fluorescence spectrometry, in Proc. of ERDA Symp. on X- and Gamma-Ray Sources and Appl., CONF-760539, University of Michigan, Ann Arbor, May 19 to 20, 1976.
302. Avgushevich, I. V., Yushchenko, E. V., Kulikova, E. S., and Kusenko, G. K., Determination of the sulfur content in solid fuels, *Solid Fuel Chem.*, 10, 60, 1976.
303. Shimp, N. F., Kuhn, J. K., and Helfinstine, R. J., Determination of forms sulfur in coal, *Energ. Sources*, 3, 93, 1977.
304. Soloman, P. R. and Manzione, A. V., New Method for sulphur concentration measurements in coal and char, *Fuel*, 56, 393, 1977.
305. Cooper, J. A., Sheeler, B. D., Wolfe, G. J., Bartell, D. M., and Schlafke, D. B., Determination of sulfur, ash, and trace element content of coal, coke, and fly ash using multielement tube-excited x-ray fluorescence analysis, *Adv. X-Ray Anal.*, 20, 431, 1977.
306. Attar, A. and Corcoran, W., Sulfur compounds in coal, *Ind. Eng. Chem. Prod. Res. Dev.*, 16, 168, 1977.
307. Paris, B., Direct determination of organic sulfur in raw coals, *Am. Chem. Soc. Div. Fuel Chem. Prepr.*, 22, 1977; Symp. on New Tech. in Coal Anal., presented at 174th Am. Chem. Soc. Natl. Meet., Chicago, Ill., 1977, 1.
308. Attar, A., Corcoran, Am. H., and Gibson, G. S., Transformation of sulfur functional groups during pyrolisis of coal, *Am. Chem. Soc. Div. Fuel Chem. Prepr.*, 21, 1976; Paper presented at 172nd Am. Chem. Soc. Natl. Meet., San Francisco, Calif., August 29 to September 3, 1976, 106.
309. Antonijević, V., Pravica, M., and Jovanović, M., Distribution of sulfur in ashes, *Hem. Ind.*, 33, 67, 1979.
310. Neavel, R. C. and Keller, J. E., Estimation of sulphur content in coal from titration of calorimeter bomb washings, *Fuel*, 58, 402, 1979.
311. Myers, G. A., Sulfur-Containing Coal or Lignite for Combustion, U.S. Patent 4148613, 1979, 4.
312. Thiessen, R., Occurrence and origin of finally disseminated sulfur compounds in coal, *Trans. Am. Inst. Min. Metall. Eng.*, 63, 913, 1920.
313. Thiessen, G., Forms of sulfur in coal, in *Chemistry of Coal Utilization*, Vol. 1, John Wiley & Sons, New York, 1945, 425; Thiessen, G., Composition and origin of the mineral matter in coal, in *Chemistry of Coal Utilization*, John Wiley & Sons, New York, 1945, 485.
314. Yancey, H. F. and Fraser, T., The Distribution of the Forms of Sulfur in the Coal Bed, Bull. No. 125, University of Illinois Engineering Experiment Station, Urbana, May 1921, 93.
315. Yancey, H. F. and Parr, S. W., Sulfur forms in coal: their distribution and control, *Ind. Eng. Chem.*, 16, 501, 1924.
316. Newhouse, W. H., Some forms of iron sulfide occurring in coal and other sedimentary rocks, *J. Geol.*, 35, 73, 1927.
317. Cady, G. H. and Leighton, H. M., The physical constitution of Illinois coal and its significance in regard to utilization, *Proc. Ill. Min. Inst.*, p. 93, 1933.
318. White, I. C., Geographic distribution of sulfur in West Virginia coal beds, *Am. Inst. of Min. Bull.*, 153, 932, 1919.
319. Parr, S. W. and Wheeler, W. F., Unit Coal and the Composition of Coal Ash, Bull. 37, University of Illinois Engineering Experiment Station, Urbana, 1908, 68.
320. De Waele, A., The occurrence of chlorine in coal, *Analyst*, 40, 146, 1915.
321. Selvig, W. A. and Gibson, F. H., Analysis of Ash From United States Coals, Bull. 567, U.S. Bureau of Mines, Washington, D.C., 1956, 33.
322. Daybell, G. N. and Pringle, W. J. S., The mode of occurrence of chlorine in coal, *Fuel*, 37, 283, 1958.

323. **Edgcombe, L. J.**, State of combination of chlorine in coal. I. Extraction of coal with water, *Fuel,* 35, 38, 1956.
324. **Kear, R. W. and Menzies, H. M.**, Chlorine in coal. Its occurrence and behaviour during combustion and carbonization, *B.C.U.R.A. Mon. Bull.,* 20, 53, 1956.
325. **Das Gupta, H. N. and Chakrabarti, J. N.**, Estimation of chlorine in organic combination in the coal substance, *J. Indian Chem. Soc.,* 28, 664, 1951.
326. **Ashley, G. H.**, Sulfur in coal, geological aspects, *Trans. Am. Inst. Min. Metall. Eng.,* 63, 732, 1919.
327. **Wandless, A. M.**, British coal seams, a review of their properties with suggestions for research, *J. Inst. Fuel,* 30, 541, 1957.
328. **Gluskoter, H. J. and Rees, O. W.**, Chlorine in Illinois Coal, Circ. 372, Illinois State Geological Survey, Urbana, 1964, 23.
329. **Gluskoter, H. J.**, Chlorine in coals of the Illinois Basin, *Trans. Soc. Min. Eng.,* 238, 373, 1967.
330. **Iapulacci, T. L., Demski, R. J., and Bienstock, D.**, Chlorine in coal combustion, U.S. Bur. Mines Rep. Invest., 7260, 11, 1969.
331. **Palmer, T. Y.**, Combustion sources of atmospheric chlorine, *Nature (London),* 263, 44, 1976.
332. **Vasyutinskii, N. A.**, Influence of chlorine and chlorides on the gasification of carbon, *Solid Fuel Chem.,* 10, 148, 1976.
333. MHD Power Generation Research, Development and Engineering, Rep. FE-1811-20, Montana Energy and MHD Research and Development Institute, Butte, 1976, 162.
334. **Kler, D. V.**, Genetic characteristics of accumulation of calcium in the Tunguska basin coals, *Reg. Geol. Nek. Raionov S.S.S.R. Mater. Nauchn. Stud. Konf.,* p. 91, 1976.
335. **Kler, D. V.**, Formation of the calcium mineralization of coals in the Tunguska and Taimyr basins, *Khim. Tverd. Topl. Moscow,* 6, 48, 1977.
336. **Balakhnin, M. V. and Merentsova, G. S.**, Neutralization of destructive process during the hydration of high-calcium ash of brown coal in the Kansk-Achinsk basin, *Izv. Vyssh. Ucheb. Zaved., Stroit. Arkhitekt.,* 17, 64, 1974.
337. **Corbaty, M. L. and Taunton, J. W.**, Treatment of Subbituminous Calcium Containing Coal and Inferior Grades of Coal, Brazilian Patent 78 01284, 1979, 36.
338. **Muralidhara, H. S. and Sears, J. T.**, Effect of calcium on gasification, *Coal Process. Technol.,* 4, 22, 1978.
339. **Otto, K., Bartosiewicz, L., and Shelef, M.**, Effects of calcium, strontium, and barium as catalysts and sulfur scavangers in the steam gasification of coal chars, *Fuel,* 58, 565, 1979.
340. **Martirosyan, G. G. and Safaryan, A. M.**, Production of calcium hydrosilicates and white portland cement during complex processing of Ekibastuz coal ashes, *Kompleksn. Ispol'z. Miner. Syr'ya,* 6, 70, 1978.
341. **Burek, R. and Palica, M.**, Effect of variations in the iron and calcium contents on ash determination by absorption methods using 80-keV x-rays, *Nukleonika,* 14, 1011, 1969.
342. **Masursky, H.**, Trace elements in coal in the red desert, Wyoming, Prof. Paper 300, U.S. Geological Survey, Reston, Va., 1956, 439.
343. **Komissarova, L. N., Shatskii, V. M., Guren, G. F., and Borisova, T. F.**, Distribution of scandium, gallium and germanium according to fractions of the gravitation separation of brown coal, *Izv. Vyssh. Uchebn. Zaved. Khim. Tekhnol.,* 18, 1256, 1975.
344. **Wait, C. E.**, The occurrence of titanium, *J. Am. Chem. Soc.,* 18, 402, 1894.
345. **Baskerville, C.**, The occurrence of vanadium, chromium and titanium in peats, *J. Am. Chem. Soc.,* 21, 706, 1899.
346. **Jones, J. H. and Miller, J. M.**, The occurrence of titanium and nickel in the ash from some special coals, *J. Soc. Chem. Ind.,* 58, 237, 1939.
347. **Reynolds, F. M.**, The occurrence of vanadium, chromium and other unusual elements in certain coals, *J. Soc. Chem. Ind.,* 67, 341, 1948.
348. **King, J. G. and Crossley, H. E.**, Methods for the Quantitative Analysis of Coal Ash, Dept. of Sci. and Ind. Res., Fuel Res., Phys. and Chem. Survey of the Natl. Coal Res., 28, 20, 1933.
349. **Kunstmann, F. H. and Gass, S. B.**, Occurrence of titanium in South African coals, *J. S. Afr. Inst. Min. Metall.,* 58, 165, 1957.
350. **Valeska, F. and Havlova, A.**, Spectrographic determination of titanium and manganese in coal, *Sb. Praci Hornickeho Ustavu Cesk. Akad. Ved.,* 1, 182, 1959.
351. **Yen, T. F.**, in *Trace Substances in Environmental Health,* Vol. 6, Hemphill, D. D., Ed., University of Missouri, Columbia, 1973, 347.
352. **Kyle, J. J. J.**, On a vanadiferous lignite found in the Argentine Republic with analysis of the ash, *Chem. News,* 66, 211, 1892.
353. **Mourlot, A.**, Analysis of a vanadiferous coal, *C. R.,* 117, 546, 1893.
354. **Almassy, G. and Szalay, S.**, Analytical studies on the vanadium and molybdenum content of Hungarian coals, *Magyar Tudomanyos Akad. Kem. Tudomanyok Osztalyanak Közlemenyei,* 8, 39, 1956; *Chem. Abstr.,* 52, 7656, 1958.

355. **Zilbermintz, V. A.**, Occurrence of vanadium in fossil coals, *C. R. Acad. Sci. U.R.S.S.*, 3, 117, 1935, *Chem. Abstr.*, 30, 1335, 1936.
356. **Zilbermintz, V. A. and Bezrukov, P. L.**, On the occurrence of vanadium in the Mesozoic coal measures of the South Urals, *Bull. Acad. Sci. U.S.S.R.*, 2, 417, 1936; *Fuel*, 16, 377, 1937.
357. **Jorissen, A.**, Presence of chromium and vanadium in coal from Liege, *Bull. Acad. Roy. Belg.*, p. 178, 1905, *J. Chem. Soc., Abstr.*, 88, 535, 1905.
358. **Jarkovsky, J. and Kupco, G.**, The geochemistry of trace elements especially vanadium, *Geol. Pr. Slovensk. Akak. Vied.*, 7, 101, 1956; *Chem. Abstr.*, 52, 6093, 1958.
359. **Headlee, A. J. W. and Hunter, R. G.**, Changes in the concentration of the inorganic elements during coal utilization, in Characteristics of Minable Coals of West Virginia, West Virginia Geological Survey, 1955, 150.
360. **Vorob'ev, A. L.**, Vanadium and nickel in coals of upper Silurian of the Alai and Turkestan Mountains, *C. R. Acad. Sci. U.S.S.R.*, 28, 250, 1940; *Chem. Abstr.*, 35, 3060, 1941.
361. **Abbolito, E.**, Presence of vanadium in the ash of some fossil fuels, *Ricerca Sci.*, 13, 563, 1942.
362. **Idzikowski, A. and Sachanbinski, M.**, Presence of rare element (germanium, gallium, vanadium, molybdenum) in the Belchatow brown coal deposit, *Pr. Nauk. Inst. Chem. Nieorg. Metal. Pierwiastkow Rzadkich. Politech. Wroclaw.*, 32, 145, 1975.
363. **Simpson, E. S.**, The Rare Metals and Their Distribution in Western Australia, Bull. 59, Misc. Rep. 35, Western Australia Geological Survey, 1914, 31.
364. **Butler, J. R.**, Geochemical affinities of some coals from Svalbard, *(Spitzbergen.) Kong. Ind. Handverk Norsk Polarinst.*, 96, 1, 1953.
365. **Yoshimura, J.**, Vanadium in carbonaceous minerals, *Bull. Inst. Phys. Chem. Res., (Tokyo)*, 9, 878, published with sci. papers, *Inst. Phys. Chem. Res. (Tokyo)*, 14, 271, 1930; *Chem. Abstr.*, 25, 1186, 1931.
366. **Zilbermintz, V. A. and Kostrykin, V. M.**, On the Distribution of Vanadium in Coals, Trans. All-Union Sci. Res. Inst. Econ. Mineral., Fascicle 87, 1936, 18.
367. **Vakhrushev, G. V.**, Exploration of Rare Elements in Bashkiriya — Southern Ural, Uchenye Zapiski Saratov. Gosuderst. Univ. N.G., *Chernyskevskogo*, 15, 124, 1940; *Chem. Abstr.*, 35, 6541, 1941.
368. **Iwasaki, I. and Ukimoto, I.**, Coal: rare elements in coal: vandium and chromium contents of coal, *J. Chem. Soc. Jpn.*, 63, 1678, 1942.
369. **Kessler, F. M. and Dockalova, L.**, The determination of manganese in coal ashes, *Paliva*, 35, 178, 1955.
370. **Lustigova, M.**, Determination of manganese content in coal ash, *Sb. Pr. U.V.P.*, 32, 149, 1976.
371. **Ueda, K., Tonogai, Y., and Iwaida, M.**, Determination of manganese in food coal-tar dyes, *Eisei Shikensho Hokoku*, 96, 71, 1978.
372. **Zaritskii, P. V.**, Mineralogy and geochemistry of manganese in terrigenous coal-bearing formations during diagenesis, *Dopov. Akad. Nauk. Ukr. S.S.R., Ser. B, Geol. Khim. Biol. Nauk*, 9, 786, 1978.
373. **Jones, D. J. and Buller, E. L.**, Analysis and softening temperatures of coal ash from coals in the Northern Anthracite field, *Ind. Eng. Chem., Anal. Ed.*, 8, 25, 1936.
374. **Brewer, R. E. and Ryerson, L. H.**, Production of high-hydrogen water gas from younger cokes, effects of catalysts, *Ind. Eng. Chem.*, 27, 1047, 1935.
375. **Estep, P. A., Kovach, J. J., Karr, C., Jr., Childers, E. E., and Hiser, A. L.**, Characterization of iron minerals in coal by low-frequency infrared spectroscopy, *Am. Chem. Soc., Div. Fuel Chem. Prepr.*, 13, 18, 1969.
376. **Katalymov, M. V.**, The manganese content of fertilizers, *Daklady Akad. Nauk SSSR*, 77, 447, 1951.
377. **Sukhov, V. A., Zamyslov, V. B., Shumeiko, V. P., Voitkovskii, Yu. B., and Lukovnikov, A. F.**, Behaviour of iron compounds in thermally treated coal production and their role in the oxidation process, *Khim. Tverd. Topl. Moscow*, 6, 30, 1978.
378. **Warne, S. St. J.**, The detection and elucidation of the iron components, present as carbonates, in coal, by variable atmosphere differential thermal analysis, *Therm. Anal.*, 5, 460, 1977.
379. **Gladfelter, W. L. and Dickerhoof, D. W.**, Use of atomic absorption spectrometry for iron determinations in coals, *Fuel*, 55, 360, 1976.
380. **Pak, Yu. N. and Starchik, L. P.**, Neutron-radiation determination of the iron content in coals, *Koks. Khim.*, 11, 10, 1974.
381. **Lefelhocz, J. F., Friedel, R. A., and Kohman, T. P.**, Mössbauer spectroscopy of iron in coal, *Geochim. Cosmochim. Acta*, 31, 2261, 1967.
382. **Liu, J. H.**, Moessbauer spectroscopic investigations of iron promoted Adam's platinum catalyst system and coal, *Diss. Abstr. Int.*, p. 4467, 1977.
383. **Aleksandrov, I. V., Patrushev, S. G., and Kamneva, A. I.**, Isolation of iron organomineral compounds from brown coal of the Kansk-Achinsk Basin, *Tr. Mosk. Khim. Tekhnol. Inst.*, 74, 74, 1973.
384. **Kamneva, A. I., Vlasova, S. G., and Aleksandrov, I. V.**, Content of various forms of iron in products of acid and alkaline hydrolysis of brown coal, *Tr. Mosk, Khim. Tekhnol. Inst.*, 70, 241, 1972.

385. **Kamneva, A. I., Vlasoma, S. G., and Aleksandrov, I. V.**, Detection of iron bound as complexes water-soluble organic mineral compounds of brown coal, *Khim. Tverd. Topl.*, 3, 47, 1973.
386. **Harry, N. S. and Schafer, K.**, Organically bound iron in brown coals, *Fuel*, 56, 45, 1977.
387. **Doughty, D. A. and Dwiggins, C. W., Jr.**, Characterization of the Valance State of Iron in coal dust, U.S. Bur. Mines, Rep. Invest., 7726, 15, 1973.
388. **Schafer, H. N. S.**, Organically bound iron in brown coals, *Fuel*, 56, 45, 1976.
389. **Ergun, S. and Bean, E. H.**, Magnetic separation of pyrite from coals, U.S. Bur. Mines, Rep. Invest., 7181, 25, 1968.
390. **Belly, R. T. and Brock, T. D.**, Ecology of iron-oxidizing bacteria in pyritic materials associated with coal, *J. Bacteriol.*, 117, 726, 1974.
391. **Dik, E. P., Soboleva, A. N., and Surovitskii, V. P.**, Mechanism of formation of ash deposits on semiradiation heating surfaces during combustion of coals with high iron content, *Vliyanie Miner. Chasti Energ. Topl. Usloviya Rab. Parogneratov, Mater. Vses. Konf.*, 2, 3, 1974.
392. **Childs, C. W., Ward, W. T., and Wells, N.**, Rattling iron concentrations from the Waikato coal measures, *N.Z. J. Geol. Geophys.*, 17, 93, 1974.
393. **Salmi, M.**, Trace elements in peat, Geol. Tutkimuslaitos, Geotek. Julkaisuja No. 51, 20, 1950; *Chem. Abstr.*, 45, 6977, 1951.
394. **Chasak, R.**, Spectrophotometric determination of lead, cobalt, tungsten, vanadium, nickel, and germanium, in *Coal and Accessory Rocks*, Vedeckovyzkum. Uhelny Ustav, Rozmn. Cstrava Radvanice, Czechoslovakia, 1973.
395. **Mott, R. A. and Wheeler, R. V.**, The inherent ash of coal, *Fuel*, 6, 416, 1927.
396. **Clarke, F. W.**, The Data of Geochemistry, Bull. 770, U.S. Geological Survey, Reston, Va., 1924, 841.
397. **Reynolds, F. M.**, Note on the occurrence of barium in coal, *J. Soc. Chem. Ind.*, 58, 64T, 1939.
398. **Shakhov, F. N. and Efendi, M. E.**, Geochemistry of the coals of the Kuznetsk Basin, *C. R. Acad. Sci. URSS*, 51, 139, 1946.
399. **Grosjean, A.**, Occurrence of millerite in the carboniferous of Belgium, *Bull. Soc. Belge Geol.*, 52, 34, 1943.
400. **Platz, B.**, Presence of copper in coal and coke, *Stahl Eisen*, 7, 258, 1887.
401. **Swaine, D. J.**, Occurrence of nickel in coal and dirt samples from the coal cliff colliery, *Aust. J. Sci.*, 23, 301, 1961.
402. **Dunn, J. T. and Bloxam, H. C. L.**, The presence of lead in the herbage and soil of lands adjoining coke ovens and the illness and poisoning of stock fed thereon, *J. Soc. Chem. Ind.*, 51, 100T, 1932.
403. **Savul, M. and Ababi, V.**, The copper, zinc, and lead content of several types of Romanian coal, *Acad. Rep. Populare Rominc, Filiala Iasi, Studi Cercetari Stiint. Chim.*, 2, 251, 1958.
404. **Fraser, D. C.**, Cupriferous peat: embryonic copper ore, *Can. Min. Metall. Bull.*, 54, 500, 1961.
405. **Ong, L. H. and Swanson, V. E.**, Adsorption of copper by peat, lignite and bituminous coal, *Econ. Geol.*, 61, 12414, 1966.
406. **Mikhailova, A. I. and Vlasov, N. A.**, Trace nutrient level in coals and carbon dioxide-humus fertilizers and their effect on the distribution of manganese, copper, and molybdenum in soil and plants, *Mikroelem. Biosfere Primen. Ikh. Sel. Khoz. Med. Sib. Dal'nego Vostoka, Dokl. Sib. Konf.*, 3, 79, 1971.
407. **Massey, H. F.**, pH and soluble copper, nickel, and zinc in eastern Kentucky coal mine spoil materials, *Soil Sci.*, 114, 217, 1972.
408. **Massey, H. F. and Barnhisel R. I.**, Copper, nickel, and zinc released from acid coal mine spoil materials of Eastern Kentucky , *Soil Sci.*, 113, 207, 1972.
409. **Sorenson, J. R., Kober, T. E., and Petering, H. G.**, Concentration of cadmium, copper, iron, nickel, lead and zinc in bituminous coals from mines with differing incidences of coal workers pneumoconiosis, *Am. Ind. Hyg. Assoc. J.*, 35, 93, 1974.
410. **Newmarch, C. B.**, Correlation of Kootenay coal seams, *Can. Min. Metall. Bull.*, 43, 141, 1950.
411. **Hinds, H.**, The deposits of Missouri, *Mo. Bur. Geol. Mines*, 11, 10 and 59, 1912.
412. **Jenney, W. P.**, The chemistry of ore-deposition, *Trans. Am. Inst. Min. Eng.*, 33, 445, 1903.
413. **Cannon, H. L.**, Geochemical Relations of Zinc-Bearing Peat to the Lockport Dolomite, Orleans County, New York, Bull. 1000-D, U.S. Geological Survey, 1955, 119.
414. **Dove, L. P.**, Sphalerite in coal pyrite, *J. Mineral. Soc. Am.*, 6, 61, 1921.
415. **Binns, G. J. and Harrow, G.**, On the occurrence of certain minerals at Netherseal Colliery, Leicestershire, *Trans. Inst. Min. Eng.*, 13, p. 252, 1896—1897.
416. **Green, S. J. and Thakur, B.**, The insoluble matter in coal tar. III. Some inorganic constituents of free carbon, *J. Soc. Chem. Ind.*, 67, 436, 1948.
417. **Grimmendahl, F.**, Occurrence of zinc compounds in coal and coal-carbonization products, *Tech. Mitt. Krupp. A Forschungsber.*, 5, 30, 1942; *Chem. Abstr.*, 37, 2541, 1943.
418. **Moss, K. N., Hirst, A. A., and Needham, L. W.**, The occurrence and estimation of lead and zinc compounds in coal, *Trans. Inst. Min. Eng.*, 79, 435, 1930.

419. Thurauf, W. and Assenmacher, H., Photometric zinc determination in coal and related products, *Brennst. Chem.*, 44, 21, 1963.
420. Weaver, C., The determination of zinc in coal, *Fuel*, 46, 407, 1967.
421. Morgan, G. and Davies, G. R., Germanium and gallium in coal ash and flue dust, *J. Soc. Chem. Ind.*, 56, 717, 1937.
422. Cooke, W. T., Occurrence of gallium and germanium in some local coal ashes, *Trans. R. Soc. S. Aust.*, 62, 318, 1938.
423. Inagaki, M., Gallium — a rare element in coal, *J. Coal Res. Inst. (Japan)*, 4, 1, 1953.
424. Inagaki, M., Moriya, S., and Yamagucki, T., Gallium as a by-product of germanium extraction from gas liquor, *Tanken*, 8, 1, 1957.
425. Inagaki, M. and Yamaguchi, T., Galium in Japanese coals, *Tanken*, 9, 161, 1958.
426. Kakihana, H., Qualitative spectrographic analysis of flue dusts, mainly produced in Japan. I. The flue dusts containing germanium, *J. Chem. Soc. Jpn., Pure Chem. Sec.*, 70, 226, 1949.
427. Volodarskii, K. Kh., Grekhov, I. T., Ratynskii, V. M., Shpirt, M. Ya., and Yurovskii, A. Z., Preparation of concentrations of rare and trace elements from fuel coals, *Tr. Inst. Goryuch. Iskop. Moscow*, 23, 64, 1966.
428. Bonnett, R. and Czechowski, F., Gallium porphyrins in bituminous coal, *Nature (London)*, 283, 465, 1980.
429. Asai, K. and Inagaki, M., Germanium and gallium in coal, *Kagaku no Ryoiki*, 6, 724, 1952.
430. Bertetti, I., Presence of gallium and germanium in some Italian coals, *Atti Accad. Ligure Sci Lett.*, 11, 53, 1954.
431. Takacs, P., Horvath, A., and Nadasy, M., Extraction of germanium and gallium from coal processing by-products, *Kohaszati Lapok*, 92, 327, 1959; *Chem. Abstr.*, 53, 19350, 1959.
432. Takacs, P. and Horvath, A., Possibilities of producing germanium and gallium from the by-products of the coal industry, *Kahaszati Lapok*, 7, 1959.
433. Zharov, Yu. N., Germanium and gallium in coals of north west part of the Zyryansk coal field, *Zh. Khim.*, 1976; Abstr. 10P31, *Tr. In-ta Goryuchikh Iskopaemykh M-va Ugol'n, Prom-sti S.S.S.R.*, 30, 29, 1975.
434. Gordon, S. A. and Saprykin, F. Ya., Distribution of Germanium and Gallium in Brown Coal, *Nauk Tr. Moskov. Gorm. Inst. Sbornik*, 27, 13—25, 1959.
435. Moody, A. H., The significance of iron in coal used for combustion, *Combustion*, 12, 22, 1941.
436. Manokhin, A. P. and Chernyak, A. S., Germanium and gallium behaviour during industrial semi-coking and gasification of long-flame coals, *Nauch. Tr. Irkutsk. Gos. Nauch. Issled. Inst. Redk. Tsvet. Metal.*, 19, 219, 1968.
437. Saprykin, F. Ya., Kulachkova, A. F., Lavrent'eva, M. M., and Pevzner, V. S., Gallium in coals and its associations with trace elements under different geochemical conditions of humid lithogenesis, *Tr. Vses. Nauchno-Issled. Geol. Inst.*, 295, 114, 1978.
438. Naser, M. I., Basily, A. B., and Raafat, A. M., Spectrographic estimation of gallium and germanium in coal and coke, *Indian J. Technol.*, 12, 359, 1974.
439. Goldschmidt, V. M., The presence of germanium in coal products, *Nachr. Ges. Wiss. Göttingen, Math. Physik. Klasse*, p. 1, 1930; *Chem. Abstr.*, 25, 3935, 1931.
440. Vistelius, A. B., New verification of the observations of goldschmidt on the role of germanium in mineral coal, *Doklady Akad. Nauk. S.S.S.R.*, 58, 1455, 1958.
441. Fleischer, M. and Harder, J. O., Geochemistry of germanium, The Chemistry and Metallurgy of Miscellaneous Materials, TID-5212, Quill, L. L., Ed., U.S. Atomic Energy Commission, Oak Ridge, Tenn., 1955, 93.
442. McCabe, L. C., Atmospheric Pollution, *Ind. Eng. Chem.*, 44, 113A, 1952.
443. Stadnichenko, T., Murata, K. J., Zubovic, P., and Hufschmidt, E. I., Concentration of Germanium in the Ash of American Coals — A Progress Report, Circ. 272, U.S. Geological Survey, Reston, Va., 1953, 34.
444. Machin, J. S. and Witters, J., Germanium in Flyash and its Spectrochemical Determination, Circ. 216, Illinois State Geological Survey, Urbana, 1956, 11.
445. Breger, I. A. and Schopf, J. M., Germanium and uranim in coalfield wood from upper devonian black shale, *Geochem. Cosmochem. Acta*, 7, 287, 1955.
446. Oka, M., Chang, H. C., and Gavalas, G. R., *Fuel*, 56, 3, 1977.
447. Oka, Y., Kanno, T., Ayusawa, S., and Haga, K., Rare elements in bituminous coal and lignite. I. Germanium in lignites from North Miyagi and South Iwate Prefectural Areas, *Bull. Res. Inst. Min. Dress. Metal.*, 11, 17, 1955.
448. Katchenkov, S. M., Concentration of germanium in coal, *Dokl. Akad. Nauk. S.S.S.R.*, 61, 857, 1948.
449. Santrucke, P., Germanium in the coal seams of the Cheb Basin, *Vestnik Ustred. Ustavu Geol.*, 33, 367, 1958.

450. **Santrucek, P.**, Germanium in brown-coal seams of the Sokolov District, *Vestnik Red. Ustavu Geol.*, 36, 75, 1961; *Chem. Abstr.*, 55, 12819, 1961.
451. **Schleicher, J. A.**, Germanium in Kansas Coal, Rep. of Studies, Geological Survey of Kansas, 1959, 163.
452. **Corey, R. C. and Myers, J. W.**, Germanium in coal ash from power plants, *Coal Utilization*, 12, 33, 1958.
453. **Corey, R. C. et al.**, Occurrence and Determination of Germanium in Coal Ash from Power Plants, Bull. 575, U.S. Bureau of Mines, Washington, D.C., 1959, 68.
454. **Howes, E. A. and Lees, B.**, The occurrence and recovery of germanium in large water-tube boilers, *J. Inst. Fuel*, 28, 298, 1955.
455. **Dundr, C. J.**, Recirculation of fly ash as a means of increasing the concentration of germanium, *Strojirenstvi*, 11, 435, 1961.
456. **Schleicher, J. A. and Hambleton, W. W.**, Preliminary Spectrographic Investigation of Germanium in Kansas Coal, Bull. 109, Rep. of Studies 8, State Geological Survey of Kansas, October 1954, 113.
457. **Gunduz, T. and Onabasioglu, I.**, Spectrophotometric determination of germanium in Turkish coals and modification of the method, *Commun. Fac. Sci. Univ. Ankara, Ser. B*, 16, 71, 1969.
458. **Pilkington, E. S.**, Survey of some australian sources of germanium, *Aust. J. Appl. Sci.*, 8, 98, 1957.
459. **Simek, B. G.**, The germanium content of coals of the Ostrau-Karwin Basin, *Chem. Listy*, 34, 181, 1942; *Chem. Zentr.*, II, 382, 1942.
460. **Simek, B. G., Coufalik, F., and Stadler, A.**, Germanium content of coals of the Ostrava-Karvina Basin, *Zapravy. Ustavu Vedecky Yvskum Uhli*, p. 167, 1948.
461. **Kunstmann, F. H. and Hammersma, J. C.**, The occurrence of germanium in South African coal and derived products, *J. Chem. Metall. Min. Soc. S. Afr.*, 56, 11, 1955.
462. **Belugou, P. and Dumoutet, P.**, The Extraction of Germanium from Coal, Notes Tech. R. 520, Centre d'Etudes et Recherches des Charbonnages de France, July 1953, 4.
463. **Wai, N. and Wang, J. S.**, Analysis of germanium in Taiwan coals, *Bull. Inst. Chem. Acad. Sinica*, 16, 59, 1969.
464. **Rouir, E. V.**, Germanium in the coals of Belgium, *Ann. Soc. Geol. Belg.*, 77, 284, 1954.
465. **Schreiter, W.**, The recovery of germanium, *Chem. Tech. (Berlin)*, 6, 141, 1954.
466. **Rishi, M. K.**, Germanium and other minor elements in the coals of Umaria, M. P., *J. Geol. Soc. India*, 11, 85, 1970.
467. **Subrahmanyan, S. and Madhavan Nair, A. P.**, Germanium in South Arcot lignite, *J. Sci. Ind. Res. (India)*, 14 B, 1955, 606.
468. **Banerjee, N. N., Rao, H. S., and Lahiri, A.**, Germanium in Indian coals, *Indian J. Technol.*, 12, 353, 1974.
469. **Mukherjee, B. and Dutta, R.**, Germanium in Indian coal ash, *Sci. Cult.*, 14, 538, 1949.
470. **Kranz, M. and Witkowska, A.**, Analysis for germanium in gasworks flue dust and coal ash, *Roczniki Chem.*, 35, 381, 1961.
471. **Gregorowicz, A.**, Germanium in Polish brown coals, *Prsemysl. Chem.*, 13, 700, 1957; *Chem. Abstr.*, 52, 8505, 1958.
472. **Mielecki, T. and Mraz, T.**, Germanium content of fly ash in certain boiler establishments of upper Silesia, Poland, *Przeglad. Gorniczy*, 13, 145, 1957.
473. **Mantea, S., Petrescu, N., and Peterescu, M.**, The presence of germanium in some coals and lignites from Romania, *Acad. Rep. Populare Romaine, Studdi Cercetari Met.*, 4, 548, 1959.
474. **Zilbermintz, V. A.**, Germanium in the coals of the Donetz Basin, *Miner. Suir'e*, 11, 16, 1936.
475. **Ershov, V. M.**, Relation of germanium to the organic matter in fossil coals, *Geokhimiya*, p. 605, 1958.
476. **Gordon, S. A., Volkov, K. Yu., and Menkovskii, M. A.**, Germanium content in coal, *Geokhimiya*, p. 384, 1958.
477. **Ratynskii, V. M.**, Accumulation of germanium in coals, *C. R. Acad. Sci. U.R.S.S.*, 40, 198, 1943.
478. **Ratynskii, V. M.**, Source of germanium in coal seams, *C. R. Acad. Sci. U.R.S.S.*, 49, 119, 1945.
479. **Ratynskii, V. M.**, Germanium in coal, *Tr. Biogeokhim. Lab., Akad. Nauk. S.S.S.R.*, 8, 183, 1946; *Chem. Abstr.*, 47, 7961, 1953.
480. **Ratynskii, V. M.**, Genesis of germanium mineralization and efficient methods for the combustion of germanium-containing coals, *Khim. Tverd. Topl., Moscow*, 2, 29, 1977.
481. **Ratynskii, V. M., Sendul'skaya, T. I., and Shpirt, M. Ya.**, Comparison of means for germanium accumulation in coal, *Tr. Inst. Goryuch. Iskop, Moscow*, 24, 3, 1968.
482. **Ratynskii, V. M., Sendul'skaya, T. I., and Shpirt, M. Y.**, Chief mode of entry of germanium into coal, in *Geochemistry of Germanium*, Weber, J. N. E., Ed., Dowden, Hutchinson, and Ross, Stroudsburg, Pa., 1973, 426.
483. **Travin, A. B.**, On ways of the accumulation of germanium in coals and further problems in research on the subject, *Izvest. Vostock, Filia, Akad. Nauk. S.S.S.R.*, 1, 44, 1957.

484. **Vinarov, I. V., Tselik, I. N., and Orlova, A. I.**, The problem of lixiviation of germanium by watei from coals, *Ukrainskiy Khimicheskiy Zh.*, 26, 383, 1960.
485. **Volkov, K. Yu.**, The regularities of distribution of germanium in coals of the Moscow-region deposits, *Mater. Geol. Polezn. Iskopaemym Tsental. Raionov Evrop. Chasti S.S.S.R., Moscow, Sbornik*, 1, 228, 1958.
486. **Sofiev, I. S., Semasheva, I. N., and Gorlenko, K. A.**, Conditions for the accumulation of germanium in lignite components, *Khim. Tverd. Topl.*, 6, 67, 1968.
487. **Diev, N. P. and Davydov, V. I.**, Germanium in tar by-products of the coking process, *Zh. Priklad. Khim.*, 30, 1685, 1957.
488. **Kostrikin, Y. M.**, Germanium in the tar by-products of the coking process, *J. Appl. Chem. (U.S.S.R.)*, 12, 1449, 1939.
489. **Inagaki, M.**, Germanium — a rare element in coal, *J. Coal Res. Inst. (Japan)*, 3, 72, 1952.
490. **Inagaki, M.**, Germanium and gallium in Japanese coals, *Nenryo Kyokai-shi*, 46, 684, 1968.
491. **Ono, K., Inada, Y., and Konno, I.**, Utilization of Lignite Containing Germanium, *Bull. Res. Inst. Min. Dress. Metall. Tohoku Univ.*, 11, 159, 1955.
492. **Beard, W. J.**, Germanium, Min. Res. I. C. No. 12, Canada, Mines Branch, Ottawa, 1955, 10.
493. **Gomez, M., Donaven, D. J., and Kent, B. H.**, Distribution of sulfur and ash in part of the Pittsburgh seam and probable mode of deposition, U.S. Bur. Mines, Rep. Invest., 7827, 44, 1974.
494. **Aubrey, K. V.**, The Germanium in Coal and in Some of Its Residual Products, Rev. Ind. Minerale, Spec. No., July 15, 1958, 51.
495. **Aubrey, K. V.**, Germanium in some of the waste products from coal, *Nature (London)*, 176, 128, 1955.
496. **Aubrey, K. V.**, Germanium in British coals, *Fuel*, 31, 429, 1952.
497. **Aubrey, K. V. and Payne, K. W.**, Volatilization of germanium during the ashing of coal, *Fuel*, 33, 1954.
498. **Hallam, A. and Payne, K. W.**, Germanium enrichment in lignites from the lower lias of Dorset, *Nature (London)*, 181, 1008, 1958.
499. **Ganguly, N. C. and Dutta, D. P.**, Spectrographic determination of germanium in coals, *J. Sci. Ind. Res., (India)*, 15B, 327, 1956.
500. **Fisher, F. L.**, Germanium, Mineral Facts and Problems, Bull. 585, U.S. Bureau of Mines Washington, D.C., 1960.
501. **Alekhina, V. I.**, Hydrolytability of the bond of germanium compounds to coal, *Protsessy Term. Prevrashch. Kamennykh Uglei, Pub: Nauka, Novosibirsk*, p. 421, 1968.
502. **Alekhina, V. I. and Adamenko, I. A.**, Germanium compounds volatilizing during the pyrolysis and combustion of coal, *Protsessy Term. Prevrashch. Kamennykh Uglei*, 408, 1968.
503. **Alekhina, V. I. and Lisin, D. M.**, Factors affecting the bond between germanium compounds and carbon compounds and hydrolyzability, in *Issled. Form Svyazi Germaniya Uglem Ego Provedeniya Pirolize Szhiganii*, Nauka, Novosibirsk, S.S.S.R., 1972, 36.
504. **Alekhina, V. I., Ryabchenko, S. N., and Lisin, D. M.**, Volatility of germanium during the thermal treatment of coals in relation to the degree of their metamorphiym, in *Protsessy Term. Prevrashch. Kamennykh Uglei, Izd. Nauka Sib. Otd. Novosibirsk, S.S.S.R.*, 1968, 383.
505. **Bekyarova, B. and Rushev, D.**, Determination of the vanadium and beryllium contents in various types of solid fuels and their industrially processed products, *God. Vissh. Khimikotkhnol. Inst., Sofia*, 16, 327, 1971.
506. **Ryabchenko, S. N. and Lisin, D. M.**, Distribution of germanium in products of the hydrolytic splitting of coals, in *Protsessy Term. Prevrashch, Kamennykh Uglei*, Izd. Nauka Sib. Otd. Novosibirsk, S.S.S.R., 1968, 361.
507. **Ryabchenko, S. N. and Lisin, D. M.**, Forms of germanium-carbon bonds, in *Issled. Form Svyazi Germaniva Uglem Ego Povedeniya Pirolize Szhiganii*, Nauka, Novosibirsk, S.S.S.R., 1972, 4.
508. **Ryabchenko, S. N. and Shiryaeva, K. N.**, Germanium content in products of the low-temperature extraction of coals, in *Protsessy Term. Prevrashch. Kamennykh. Uglei*, Nauka, Novosibirsk, 1968, 351.
509. **Syabryai, V. T., Kornienki, T. G., and Kuz'minskaya, I. N.**, Forms of germanium occurrence in organic substances of solid caustobioliths, *Dopv. Akad. Nauk. Ukr, S.S.R., Ser. B*, 33, 23, 1971.
510. **Syabryai, V. T., Kornienko, T. G., and Kuz'minskaya, I. N.**, Use of electrophoresis and electrolysis to study the forms of germanium occurrence in organic substances of solid caustobioliths, *Dopov. Akad. Nauk Ukr. S.S.R., Ser. B*, 33, 216, 1971.
511. **Zhou, Y. P.**, Two types of germanium distribution in coal beds, *Ti Chih K'o Hsueh*, 2, 182, 1974.
512. **Zahradnik, L.**, Germanium in the products of direct coal combustion and its extractability by muriatic acid, *Chemicky Prumysl.*, 9, 1959.
513. **Zahradnik, L.**, Germanium distribution among combustion products in a health with traveling grate, *Chemicky Prumysl.*, 9, 1959.

514. **Vnukov, A. V. and Sirotenko, A. A.,** Relation of germanium to the petrographic composition of coals, *Zh. Geol.;* Abstr. 11K31, *Geol. Razved. Mestorozhd. Polez. Iskop. Zabaikal,* p. 96, 1968.
515. **Thomo, N., Veizi, D., and Prifti, I.,** Effect of sodium chloride (NaCl) on the volatilization of germanium from coals during coking, *Bull. Shkencave Nat., Univ. Sheteteror Tiranes,* 31, 57, 1977.
516. **Sobynyakova, N. M., Starostina, K. M., Aleksandrova, L. N., Balikhina, S. I., Vasil'chikova, E. I., Krainova, L. P., Shmanenkova, I. V., El'khones, N. M., Nogaev, Yu. B., and Tsadolob., Yu. A.,** Sorption extraction of germanium and molybdenum during the processing of energy producing coal benefication products, *Tr. Vses. Nauch. Issled. Inst. Miner. Syr'ya,* 23, 59, 1972.
517. **Paraeev, S. V.,** Characteristics of the assay of coal seams for germanium, *Zap. Leningr. Gorn. Inst.,* 68, 30, 1975.
518. **Maleshko, A. Ya., Morozov, V. N., Sambueva, A. L., Tatarinova, N. A., and Shipitsyn, S. A.,** Spectrographic method for studying the kinetics of the yield of germanium during the thermal treatment of coal, *Irkutsk. Gos. Univ. Zhdanova Irkutsk U.S.S.R., Zh. Prikl. Khim. Leningrad,* 43, 1027, 1970.
519. **Lexow, S. G. and Maneschi, E. P. P.,** Germanium in Rio Turbio Coal, *Anales Asoc. Quim. Argentina,* 38, 225, 1950.
520. **Hampel, A. C., Ed.,** *Germanium, Rare Metals Handbook,* Reinhold Publishing, New York, 1954.
521. **Han, S. K.,** Separation of germanium from ammonia water in coal coking, *Hwahak Kwa Kongop.,* 18, 276, 1975.
522. **Fletcher, M. F.,** The Concentration and Location of Germanium in Some British Coals, Rep. 1232, Great Britain, Nat. Coal Board. Sci. Dept., Central Res. Establishment, December 1954.
523. **Fokina, E. I. and Klitina, L. V.,** Distribution of germanium in a coal seam, *Tr. Vses. Nauch. Issled. Geol. Inst.,* 132, 280, 1968.
524. **Fortescue, J. A. V.,** Germanium and other trace elements in some Western Canadian coals, *Am. Mineral.,* 39, 510, 1954.
525. **Fisel, S. and Gonteanu, A.,** Content of germanium in the coal washed for (preparation of) semicoke (Lupeni) or coke (Anina) and in the soot from some small power stations (Petrosani), *Lucr. Stiint. Inst. Mine Petrosani,* 9, 31, 1977.
526. **Farnand, J. R. and Puddington, E. I.,** Oil-phase agglomeration of germanium-bearing vitrain coal in a shaly sandstone deposit, *Can. Inst. Mining Metall. Bull.,* 62, 267, 1969.
527. **Bunkina, N. A., Bentsainov, Yu. V., Dubova, I. S., Makarova, N. D., Makarov, G. N., and Syskov, K. I.,** Germanium separation long-flame coal, *Tr. Mosk. Khim. Tekhnol. Inst.,* 74, 81, 1973.
528. **Balikhina, S. I., Vasil'chikova, E. I., Krainova, L. P., Smanenkov, I. V., El'khones, N. M., Nogaev, Yu. B., and Tsabolov, Yu. A.,** Sorption extraction of germanium and molybdenum during treatment of powder plant coal enrichment products, *Miner. Syr'e,* 23, 59, 1972.
529. **Adamkin, L. A.,** Relation between ash content and germanium content in coals, and its genetic significance, *Dokl. Akad. Nauk S.S.S.R. (Geochem),* 192, 1353, 1970.
530. **Adamkin, L. A.,** Facies analysis of germanium-containing coal seams of some deposits, *Litol. Polez. Iskop.,* 5, 82, 1973.
531. **Adamkin, L. A.,** Presence of germanium in genetic types of coals in some Transbaikal deposits, *Dokl. Akad. Nauk. S.S.S.R. (Geochem),* 114, 1427, 1974.
532. **Adamkin, L. A.,** Two types of germanium concentrations in coals, *Dokl. Akad. Nauk S.S.S.R. (Geochem),* 224, 194, 1975.
533. **Kulinenko, O. R.,** Relation between germanium content and seam thickness in Paleozoic paralic coal basins of the Ukraine, *Izv. Akad. Nauk S.S.S.R., Ser. Geol.,* 10, 111, 1976.
534. **Kostin, Yu. P.,** Distribution patterns of germanium in the oxidation zones of coal seams, *Metallogen. Osad. Osad. Metamorf. Prod.,* p. 179, 1973.
535. **Kostin, Yu. P. and Meitov, E. S.,** Genesis of high-germanium coals and criteria for their prospecting, *Izv. Akad. Nauk. S.S.S.R., Ser. Geol.,* 1, 112, 1972.
536. **Crawley, R. H. A.,** Sources of germanium in Great Britain, *Nature (London),* 175, 291, 1955.
537. **Shpirt, M. Ya. and Sendul'skaya, T. I.,** Distribution of germanium and types of germanium compounds in solid fuel, *Khim. Tverd. Topl.,* 2, 3, 1969.
538. **Salikova, G. E. and Sevryukov, N. N.,** Determination of germanium in hard coals, *Zavod. Lab.,* 36, 25, 1970.
539. **Zagorodnyuk, A. V., Magunov, R. L., and Stasenko, I. V.,** Phase analysis of ashes for germanium compounds, *Zavod. Lab.,* 39, 1060, 1973.
540. **Selenkina, M. S., Sazanov, M. L., and Zhukhovitskii, A. A.,** Determination of germanium in coal, soils and rock by gas chromatography, *Zh. Khim.,* 1974; Abstr. 5G110, *Tr. Vses Nauchno Issled. Geologorazved. Neft. Inst.,* 112, 152, 1973.
541. **Patzek, T.,** Determination of germanium in liquid products of chemical processing of coal. I. Modification of the color reaction with phenylfluorene and absorption measurements, *Koks, Smola, Gaz,* 13, 357, 1968.

542. **Patzek, T.**, Determination of germanium in liquid by products of chemical coal utilization. II. Estimation of germanium content in gas waters and coal tars. I, *Koks. Smola, Gaz,* 14, 8, 1969.
543. **Patzek, T.**, Determination of germanium in liquid by-products from chemical coal utilization. II. Estimation of germanium content in gas waters and coal tars. II, *Koks, Smola, Gaz,* 14, 84, 1969.
544. **Perkova, R. I. and Pecherkin, L. A.**, Determination of germanium in coal ash, *Izv. Vyssh. Ucheb. Zaved., Gorn. Zh.,* 12, 9, 1969.
545. **Ruschev, D. and Bekyarova, E.**, Determination of the rate of extraction of germanium component groups of solid fuels during chemical analysis, *Chem. Anal.,* 53, 569, 1971.
546. **Menkovskii, M. A. and Aleksandrova, A. N.**, The selection of conditions for ashing coal for germanium determination, *Zavod. Lab.,* 29, 1963.
547. **Nurminskii, N. N.**, Quantitative chemical determination of germanium and molybdenum in coals, *Zh. Khim.,* 1970; Abstr. 23G144, *Tr. Irkustsk Politekh. Inst.,* 44, 148, 1970.
548. **Nurminskii, N. N.**, Use of the Kjeldahl method for determination of germanium in coals, *Khim. Tverd. Topl.,* 3, 128, 1971.
549. **Houzim, V. and Volf, J.**, Determination of germanium in coal and its processing products, *Sb. Geol. Ved., Technol., Geochem.,* 9, 185, 1969.
550. **Kekin, N. A. and Marincheva, V. E.**, Spectrometric determination of germanium in coals. *Zh. Khim.,* 1971; Abstr. 18G108, *Sb. Nauch. Tr. Ukr. Nauch. Issled. Uglekhim. Inst.,* p. 197, 1971.
551. **Khizhnyak, N. D., Tsebrii, L. S., and Lenkevich, Zh. K.**, Determination of germanium in coals by derivative polarography, *Ukr. Khim. Zh.,* 37, 359, 1971.
552. **Andrianov, A. M. and Koryukova, V. P.**, Determination of germanium in cooler waters of coal-tar chemical plants, *Koks Khim.,* 7, 44, 1971.
553. **Adamenko, A. I., Losev, B. I., and Yavorskii, I. A.**, Separation of germanium by an ion-exchange method, *Khim. Tverd. Topl.,* 3, 65, 1972.
554. **Abbolito, E.**, Investigation of germanium in lignites of Umbria by X-ray methods, *Period. Mineral.,* 29, 9, 1960.
555. **Ryabchenko, S. N.**, Behaviour of germanium during the oxidation of coals, in *Protsessy Term. Prevrashch. Kamennykh Uglei,* Nauka, Novosibirsk, 1968, 370.
556. **Kulinenko, O. R.**, Origin of germanium in coals, *Izv. Akad. Nauk. S.S.S.R. Ser. Geol.,* 7, 53, 1969.
557. **Ryabchenko, S. N., Alekhina, V. I., Lisin, D. M.**, Nature of germanium compounds in coals, in *Protsessy Term. Prevrashch. Kamennykh Uglei,* Nauka, Novosibirsk, 1968, 376.
558. **Shirokov, A. Z., Lazebnik, P. V., and Dolgopolov, V. M.**, Origin of germanium in coals, *Izv. Dnepropetrovsk. Gorn. Inst.,* 54, 50, 1971.
559. **Shirokov, A. Z., Lazebnik, P. V., and Sedenko, S. M.**, Geological studies in relation to germanium raw materials, *Izv. Dnepropetrovsk, Gorn. Inst.,* 51, 3, 1972.
560. **Vekhov, V. A., Kuznetsova, L. M., and Sedenko, S. M.**, Migration of germanium (IV) with natural waters and its accumulation in coals, *Khim. Tekhnol.,* 16, 86, 1970.
561. **Sedenko, S. M. and Kuznetsova, L. M.**, Forms of migration and conditions of deposition of germanium under surface condition, *Izv. Dnepropetrovsk. Gorn. Inst.,* 51, 95, 1972.
562. **Sharova, A. K., Gertman, E. M., and Semerneva, G. A.**, Nature of the bonding of germanium with acids of coals, *Khim. Tverd. Topl.,* 2, 58, 1973.
563. **Kuznetsova L. M. and Vekhov, V. A.**, Sorption of germanium by Aleksandria brown coal, *Vop. Khim. Tekhnol.,* 31, 171, 1973.
564. **Lazebnik, P. V., Grinval'd, M. A., and Dolgopolov, V. M.**, Correlation between germanium and sulfur contents in coals from various areas of the USSR, *Zh. Geol. K.,* 1968; *Geol. Gorn. Delo,* p. 59, 1967.
565. **Adamenko, I. A.**, Germanium compounds volatilizing during the combustion of coal semicokes studies by an ion-exchange method, in *Protsessy Term. Prevrashch. Kemennykh Uglei,* Nauka, Novosibirsk, 1968, 426.
566. **Adamenko, I. A. and Yavorskii, I. A.**, Basic factors effecting the volatility of germanium from semicokes of coals during their combustion and thermal treatment, in *Protsessy Term. Prevrashch. Kamennykh Uglei,* Nauka, Sib. Otd., Novosibirsk, 1968, 396.
567. **Adamenko, I. A. and Yavorskii, I. A.**, *Principles of Germanium Behaviour in Various Gaseous Media, Szhiganii,* Nauka, Novosibirsk, 1972, 60.
568. **Akulova, T. I., Bentsianov, Yu. V., Makarov, G. N., Makarova, N. D., Syskov, K. I., and Shcheblanova, T. N.**, Effect of final temperature and heating rate of western-Donets-Basin gas coal on the transition of germanium-containing compounds into gaseous products, *Tr. Mosk. Khim.-Tekhnol. Inst.,* 80, 147, 1974.
569. **Medvedev, K. P. and Akimova, L. M.**, Forms of germanium separation during the coking of coals, *Khim. Tverd. Topl.,* 4, 66, 1970.
570. **Medvedev, K. P., Khar'kina, L. M., Kolesnik, V. M., and Semenenko, L. E.**, Extracting germanium from aqueous solutions of its salts by deposition with tannin or oak extract, *Otkrytiya, Izobret., Prom. Obraztsy Tovarnye Znaki,* p. 256, 1977; U.S.S.R. Patent 389702.

571. **Medvedev, K. P., Khudokormova, N. P., and Akimova, L. M.**, Dependence of germanium yield from condensation products (of coal), *Proizv-vo Koksa*, 3, 158, 1974.
572. **Medvedev, K. P., Khudokormova, N. P., and Akimova, L. M.**, Determination of germanium volatility and yield during heating and coking of coals and charges, *Zh. Khim.*, 1971; Abstr. 17105, *Sb. Nauch. Tr. Ukr. Nauch. Issled. Uglekhim. Inst.*, 23, 200, 1971.
573. **Medvedev, K. P., Khudokormova, N. P., and Akimova, L. M.**, Passage of germanium into condensation products under different coking conditions, *Koks. Khim.*, 6, 45, 1973.
574. **Singh, N. and Mathur, S. B.**, Survey of indigenous coals fly ashes, and flue dusts as a potential source of germanium, *NML Tech. J.*, 15, 42, 1973.
575. **Steward, K.**, Germanium in coal ash, *Gas J.*, 274, 279, 1953; *Gas World*, 137, 794, 1953.
576. **Pogrebitskii, E. O.**, Regularity in distribution of Ge in the Donets Basin coals, Zapiski Leningrad, *Gron. Inst.*, 35, 87, 1959.
577. **Headlee, A. J. W.**, Germanium and other elements in coal and the possibility of their recovery, *Min. Eng.*, 5, 1011, 1953.
578. **Iyer, V. G. and Sundaram, N.**, Germanium in coal ash, *Indian Min. J.*, 3, 182, 1955.
579. **Thompson, A. P. and Musgrave, J. R.**, Germanium, produced as a by-product, has become of primary importance, *J. Metals*, 4, 1132, 1952.
580. **Williams, A. E.**, Germanium from coal, *Iron Coal Tr. Rev.*, 163, 29, 1951.
581. **Chirnside, R. C. and Cluley, H. J.**, Germanium from coal. A British source of germanium for use in crystal valves, *G.E.C. J.*, 19, 94, 1952.
582. **Egorov, A. I. and Kalinin, S. K.**, Distribution of germanium in coals of Kazakhstan, *C. R. Acad. Sci. U.R.S.S.*, 26, 925, 1940.
583. **Eidel'man, E. Ya., Pakter, M. K., Shmanenkov, I. V., El'khones, P. M., Egorova, R. A., Dubrovskaya, D. P., Pozhidaev, A. T., and Forer, E. A.**, Dynamics of the removal of volatile substances and germanium during the layer coking of coals, *Khim. Tverd. Topl.*, 5, 77, 1972.
584. **El'khones, N. M., Egorova, R. A., Shmanenkov, I. V. and Skvortsov, Yu. I.**, Increasing the yield of germanium during coking of coal, *I. P. Bardin Razvit. Metall. S.S.S.R.*, p. 233, 1976.
585. **Bylyna, E. A., Losev, B. I., and Troyanskaya, M. A.**, Extraction of germanium from coal with gamma-irradiated carbon tetrachloride, *Izv. Akad. Nauk. S.S.S.R. Otdel. Tekh. Nauk*, 4, 124, 1958; *Chem. Abstr.*, 52, 17004, 1958.
586. **Aleksandrova, L. N., Ruzinov, L. P., Starostina, K. M., El'khones, N. M., Slobodchikova, R. I., and Ryzhova, T. G.**, Optimizing germanium leaching from coal ash, *Izv. Vyssh. Ucheb. Zaved., Tsvet. Met.*, 11, 148, 1968.
587. **Aleksandrova, L. N., Ruzinov, L. P., and Starostina, K. M.**, Optimization of the leaching of germanium from ashes based on the layer combustion of coal by sodium hydroxide solutions, *Izv. Vyssh. Ucheb. Zaved., Tsvet. Met.*, 13, 146, 1970.
588. **Aleksandrova, L. N., Ruzinov, L. P., and Starostina, K. M.**, Optimization of the leaching of germanium from sublimates of cyclone melting, *Tr. Vses. Nauch. Issled. Inst. Miner. Syr'ya*, 23, 22, 1972.
589. **Crook, S. R. and Wald, S.**, The determination of arsenic in coal and coke, *Fuel*, 39, 313, 1960.
590. **Simmersbach, O.**, Arsenic in coal and coke, *Stahl Eisen*, 37, 502, 1917.
591. **Duck, N. W. and Himus, G. W.**, On arsenic in coal and its mode of occurrence, *Fuel*, 3, 267, 1951.
592. **Kunstmann, F. H. and Bodenstein, L. B.**, Arsenic content of South African coals, *J. S. Afr. Inst. Min. Metall.*, 62, 234, 1961.
593. **Kogan, E. A. and Evdokimov, D. Ya.**, Kinetics of the separate and combined sorption of germanium (IV) and arsenic (III) on coals saturated with citric acid and o-hydroxyquinoline, *Ukr. Khim. Zh.*, 38, 541, 1972.
594. **Edgcombe, L. J. and Gold, H. K.**, Determination of arsenic in coal, *Analyst*, 80, 155, 1955.
595. **Crawford, A., Palmer, J. G., and Wood, J. H.**, The determination of arsenic in microgram quantities in coal and coke, *Mikrochim. Acta*, p. 277, 1958.
596. **Ault, R. G.**, Determination of arsenic in malting coal, *J. Inst. Brewing*, 67, 14, 1961.
597. **Benenati, F. E.**, Zn, Pb, Cd, and As in Soil, Vegetation, and Water Resources, Ph.D. thesis, University of Oklahoma, *Dis. Abstr. Int. B*, 35(a), 4420, 1975.
598. **Kekin, N. A. and Marincheva, V. E.**, Spectrographic determination of arsenic in coal and coke, *Zavod. Lab.*, 36, 1061, 1970.
599. **Hall, R. H. and Lovell, H. L.**, Spectroscopic determination of arsenic in anthracite coal ashes, *Anal. Chem.*, 30, 1665, 1958.
600. **Jackwerth, E. and Kloppenburg, H. G.**, X-ray fluorescence spectroscopic determination of trace arsenic in conjunction with a modified gutzeit method, *Z. Anal. Chem.*, 186, 428, 1962.
601. **Guscavage, J. P.**, Determination of arsenic, antimony and selenium in coal by atomic absorption spectrometry with a graphite tube atomizer, *Res. U.S. Geol. Surv.*, 5, 405, 1977.
602. **Spielholtz, G. I., Toralballa, G. C., and Steinberg, R. J.**, Determination of arsenic in coal and in insecticides by atomic absorption spectroscopy, *Mikrochim. Acta*, 6, 918, 1971.

603. **Santoliquido, P. M.**, Use of inorganic ion exchangers in the neutron activation determination of arsenic in coal ash, *Radiochem. Radioanal. Lett.*, 15, 373, 1973.

604. **Orvini, E., Gills, T. E., and LaFleur, P. D.**, Method for determination of selenium, arsenic, zinc, cadmium, and mercury in environmental matrices by neutron activation analysis, *Anal. Chem.*, 46, 1294, 1974.

605. **Jorissen, A.**, Molybdenum, selenium, etc., in coal from Liege, *Ann. Soc. Geol. Belgique*, 23, 101, 1896; *J. Chem. Soc. Abstr.*, 72, 265, 1897.

606. **Andren, A. W., Klein, D. H., and Talmi, Y.**, Selenium in coal-fired steam plant emission, *Environ. Sci. Technol.*, 9, 856, 1975.

607. **Pillay, K. K. S., Thomas, C. C., Jr., and Kaminski, J. W.**, Neutron activation analysis of the selenium content of fossil fuels, in Trans. Annu. Meet., Am. Nucl. Soc., Seattle, June, 1969.

608. **Belopolskaya, T. L. and Serikov, I. V.**, Chemical and x-ray spectral determination of selenium in coals, lignites, and rocks rich in organic matter, *Khim. Tverd. Topl.*, 6, 73, 1969.

609. **Barnes, J. H. and Lapham, D. M.**, Selenium: Pennsylvania's rarest mineral, *Pa. Geol.*, 3, 8, 1972.

610. **Gutenmann, W. H., Bache, C. A., Youngs, W. D., and Lisk, D. J.**, Selenium in fly ash, *Science*, 191, 966, 1976.

611. **Tupper, W. M.**, On the relative abundance of rubidium and strontium in vitrain ashes from coals in Nova Scotia, *Geochim. Cosmochim. Acta*, p. 314, July 1961.

612. **Newmarch, C. B.**, Correlation of kootenay coal seams, *Canadian Min. Met. Bull.*, 43, 141, March, 1950.

613. **Abernethy, R. F. and Gibson, F. H.**, Colorimetric method for arsenic in coal, U.S. Bur. Mines, Rep. Invest. 7184, 10, 1968.

614. **Arnautov, N. V. and Shipilov, L. D.**, Spectrographic determination of yttrium in coal ash, *Geol. Geofiz.*, 1, 115, 1960; *Chem. Abstr.*, 55, 6622, 1961.

615. **Degenhardt, H.**, Geochemical distribution of zirconium in the lithosphere, *Geochim. Cosmochim. Acta*, 11, 279, 1957.

616. **Ter Meulen, H.**, Distribution of molybdenum, *Nature (London)*, 130, 966, 1932.

617. **Jorissen, A.**, Distribution of molybdenum in the Liege coal fields, *Bull. Soc. Chim. Belgium*, 27, 21, 1913.

618. **Fiskel, J. G. A., Mourkides, G. A., and Gammon, N., Jr.**, A study of the properties of molybdenum in Everglades Peat, *Proc. Soil Sci. Soc. Am.*, 20, 73, 1956.

619. **Korolev, D. G.**, Some pecularities in the distribution of molybdenum in rocks of the Bylymskii coal deposit, *Geokhimiya* p. 420, 1957, *Chem. Abstr.*, 52, 13565, 1958.

620. **Matsuda, H. T. and Abrao, A.**, Benefication of Mo and U in sulfuric lixivation of ores. Separation of the Mo(VI) -U(VI) pairs by sorption of Mo on weakly anionic resin and fixation of U on strongly anionic resin, Separation of Mo(VI)-U(VI) pairs by absorption of Mo on coal and sorption of U on strongly anionic resins, Report No. IEA-291, *Nucl. Sci.Abstr.*, 1974, 29, 19130, 40, 1973.

621. **Dyatel, S. G. and Vasserman, L. I.**, Determination of molybdenum, Patent 658090, *Otkrytiya, Izobret., Prom. Obraztsy, Tovarnye Znaki*, 1979, 75.

622. **Briggs, H.**, Metals in coal, *Coll. Eng.*, 11, 303, 1934.

623. **Popović, A.**, Noble metals in the ash of some coals of the Timok Basin, *Bull. Soc. Chim. Belgrade*, 19, 305, 1954.

624. **Liebig, J. and Kopp, H.**, Jahresbericht über die fortschrite der reinen, pharmaceutischen, und technischen Chemie, Physic, Mineralogie, und Geologie für 1847—1848, Giessen, 1849, 1120.

625. **Klein, D. H.**, Mercury and other metals in urban spols, *Environ. Sci. Technol.*, 6, 560, 1972.

626. **Goldschmidt, V. M.**, Principles of distribution of chemical elements in minerals and rocks, *J. Chem. Soc.*, pt. 1, 655, 1937.

627. **Kessler, T., Sharkey, A. G., Jr., and Friedel, R. A.**, Analysis of Trace Elements in Coal by Spark-Source Mass Spectrometry, PB Rep., 214680/1, Govt. Rep. Announce, U.S. National Technical Information Service, 1973, 13.

628. **Hatch, J. R., Avcin, M. J., Wedge, W. K., and Brady, L. L.**, Sphalerite in Coals from Southeastern Iowa, Missouri, and Southeastern Kansas, Open file Rep. 76-796, U.S. Geological Survey, Reston, Va., 1976, 26.

629. **Borovik, S. A. and Ratinskii, V. M.**, Tin in the coals of the Kuznetsk Basin, *Dokl. Akad. Nauk. S.S.S.R.*, 45, 128, 1944.

630. **Pollock, E. N. and West, S. J.**, The determination of antimony at submicrogram levels by atomic absorption spectrophotometry, *At. Absorp. Newsl.*, 11, 104, 1972.

631. **Wilke-Dorfurt, E. and Romersperger, H.**, Iodine content of coal, *Zeit. Anorg. Allgem. Chem.*, 186, 159, 1930.

632. **Wache, R.**, Investigation of the occurrence of iodine in waters and coal and some experiments on the use of iodine as fertilizer, *Mitt. Lab. Preuss. Geol. Landesanstalt* No. 13, 43, 1931. *Chem. Abstr.*, 26, 4125, 1932.

633. **Stolper, W.**, Investigations of the iodine content of solid fuels and their products with reference to iodine extraction, Bergakademie, *Berlin*, 8, 587, 1956.
634. **Hinrichsen, F. W. and S. Taczak**, *Die Chemie der Kohle*, 3rd Ed. of Die Chemie der Steinkohle, Much, W. E., Leipzig, 1916, 523.
635. **Finn, C. P.**, An occurrence of barytes in the Parkgate (South Yorkshire), *Seam. Trans. Inst. Min. Eng.*, 80, 25, 1930 to 1931.
636. **Eskenazy, G.**, Rare earth elements in some coal basins of Bulgaria, *Geol. Balc.*, 8, 81, 1978.
637. **Kalinin, S. K., Ponomarev, P. M., and Fanin, E. E.**, Rhenium content in some oil shales in Kazakhstan, *Izv. Akad. Nauk. Kaz. S.S.R. Ser. Geol.*, 31, 48, 1974.
638. **Martin, A. and Garcia-Rossell, L.**, Uranium and rhenium in sedimentary rocks. II. Myocene basin of Granada, *Bol. Geol. Miner.*, 82, 65, 1971.
639. **Martin, A. and Garcia-Rossell, L.**, Uranium and rhenium in sedimentary rocks. III. Lignites of the Ebro depression, *Bol. Geol. Miner.*, 82, 78, 1971.
640. **Kirby, W.**, Mercury from coal tar, *J. Soc. Chem. Ind.*, 46, 422, 1927.
641. **Joensun, O. I.**, Fossil fuels as a source of mercury pollution, *Science*, 172, 1027, 1971.
642. **Poelstra, P., Frissel, M. J., Vander Klugt, N., and Tap, W.**, Behaviour of Mercury Compounds in Soils, Accumulation and Evaporation, in *Comparative Studies of Food and Environmental Contamination*, International Atomic Energy Agency, Vienna, 1974, 281.
643. **Crockett, A. B. and Kinnison, R. R.**, Mercury Distribution in Soil Around a Large Coal-Fired Power Plant, NTIS PB-269289, National Technical Information Service, Springfield, Va., 1977.
644. **Hall, J. H., Varga, G. M., and Magee, E. M.**, Trace elements and potential pollutant effects in fossil fuels, in Symp. Proc. Environ. Aspects of Fuel Conversion Technology, EPA-650/2-74-118, St. Louis, 1974, 35.
645. **Kalb, G. W.**, Total mercury mass balance at a coal-fired power plant, in *Trace Elements in Fuel*, Advances in Chemistry, Ser. 141, Babu, S., Ed., American Chemical Society, Washington, D.C., 1975, 154.
646. **Billings, C. E. and Matson, W. R.**, Mercury emissions from coal combustion, *Science*, 176, 1232, 1972.
647. **Anderson, W. L. and Smith, K. E.**, Dynamics of mercury at coal-fired power plant and adjacent coaling lake, *Environ. Sci. Technol.*, 11, 75, 1977.
648. **Ruch, R. R., Gluskoter, H. J., and Kennedy, J. E.**, Mercury Content of Illinois Coals, Environ. Geology Notes, 43, Illinois Geological Survey, Urbana, 1971, 14.
649. **Vasilevskaya, A. E. and Shcherbakov, V. P.**, Determination of mercury compounds in coal, *Dopovidi Akad. Nauk. Ukr. R.S.R.*, p. 1494, 1963.
650. **Vasilevskaya, A. E., Shcherbakov, V. P., and Klimenchuk, V. I.**, Determination of mercury in coal by dithizon, *Zavod. Lab.*, 28, 415, 1962.
651. **Rook, H. L., Gills, T. E., and LaFleur, P. D.**, Method for determination of mercury in biological materials by neutron activation analysis, *Anal. Chem.*, 44, 1114, 1971.
652. **Rook, H. L., LaFleur, P. D., and Gills, T. E.**, Mercury in coal: a new standard reference material, *Environ. Lett.*, 2, 195, 1972.
653. **Huffman, C., Jr., Rahill, R. L., Shaw, V. E., and Norton, D. R.**, Determination of Mercury in Geologic Materials by Flameless Atomic Absorption Spectrometry, Prof. Paper 800-C, U.S. Geological Survey, Reston, Va., 1972, C203.
654. **Heinrichs, H.**, Determination of mercury in water, rocks, coal, and petroleum with flameless atomic absorption spectrophotometry, *Z. Anal. Chem.*, 273, 197, 1975.
655. **O'Gorman, J. V., Suhr, N. H., and Walker, P. L., Jr.**, The determination of mercury in some American coals, *Appl. Spectros.*, 26, 44, 1972.
656. **Diehl, R. C., Hattman, E. A., Schultz, H., and Haren, R. J.**, Rate of Trace Mercury in the Combustion of Coal, Tech. Prog. Rep. 54, Pittsburgh Energy Research Center, 1972, 9.
657. **Billings, C. E., Sacco, A. M., Matson, W. R., Griffin, R. M., Coniglio, W. R., and Hanley, R. A.**, Mercury balance on a large pulverized coal-fired furnace, *J. Air Pollut. Control Assoc.*, 23, 733, 1973.
658. **Chow, T. J. and Earl, J. L.**, Lead isotopes in North American coals, *Science*, 176, 510, 1972.
659. **Block, D. and Dams, R.**, Lead contents of coal, coal ash, and fly ash, *Water Air Soil Pollut.*, 5, 207, 1975.
660. **Block, C.**, Determination of lead in coal and coal ashes by flameless atomic absorption spectrometry, *Anal. Chim. Acta*, 80, 369, 1975.
661. **Block, C. and Dams, R.**, Determination of trace elements in coal by instrumental neutron activation analysis, *Anal. Chim. Acta*, 68, 11, 1974.
662. **Wetherill, J. M.**, Field measurements of radon in the air of a coal mine in southwestern Alberta, *Can. Inst. Min. Metall. Bull.*, 61, 1335, 1968.
663. **Lloyd, S. J. and Cunningham, J.**, The radium content of some Alabama coals, *Am. Chem. J.*, 20, 47, 1913.

664. **Bayliss, R. J. and Whaite, H. M.**, A study of radium alpha activity of coal, coal ash, and particulate emission at a Sydney power station, *Air Water Pollut.*, 10, 813, 1966.
665. **Martin, A. and Garcia-Rossell, L.**, Uranium thorium ratio in carbonaceous sediments. II. Ebro Depression, *Bol. Real. Soc. Espan. Hist. Nat. Secc. Geol.*, 68, 65, 1970.
666. **Erdtmann, G.**, Determination of uranium and thorium in coal ash and power station precipitator ash and in bauxite and red sludge by activation analysis with epithermal neutrons, *Kerntechnik*, 18, 36, 1976.
667. **Evcimen, T. H. and Cetincelik, M.**, The importance of radioactive minerals (uranium and thorium) in Turkey for future nuclear energy production, in Dep. Radioact. Miner. Coal Miner. Res. Explor. Inst. Turkey Ankara World Energy Conf. Proc., 3, Ankara, 1978, 32.
668. **Berthoud, E. L.**, On the occurrence of uranium, silver, iron, etc., in the tertiary formation of Colorado territory, *Proc. Acad. Nat. Sci. (Philadelphia)*, 27, 363, 1875.
669. **Josa, J. M., Merino, J. L., and Villoria, Y. A.**, Lignitos radioactivos Espanoles, naturaleza y solubilizacion del urania, in *Proc. of Low-Grade Uranium Ores*, International Atomic Energy Agency, Vienna, 1967, 157.
670. **Alderman, J. K., Grady, W. C., Simcoe, E. J., and Simonyi, T.**, Uranium in Coal and Carbonaceous Rocks in the United States and Canada: an Annotated Bibliography, Rep. Coal Res. Bur. 160, West Virginia University, Morgantown, 1978, 21.
671. **Staatz, H. M. and Bauer, H. L.**, Uranium-Bearing Lignite Beds at the Gamma Property, Churchill County, Nevada, Trace Elements Memorandum, Rep. 226, U.S. Geological Survey, Reston, Va., June 1951, 4.
672. **Gott, G. B., Wyant, D. C., and Beroni, E. P.**, Uranium in Black Shales Lignites, and Limestones in the United States, Circ. 200, U.S. Geological Survey, Reston, Va., 1952, 31.
673. **Masursky, H., Pipiringos, N. G.**, Uranium-Bearing Coal in the Red Desert Area, Sweetwater County, Wyoming, Rep. TEM-341 A, U.S. Geological Survey, Reston, Va., 1952, 181.
674. **Masursky, H.**, Coal and Lignite — Red Desert Area, Sweetwater County, Wyoming, Geologic Investigations of Radioactive Deposits, Semiannual Progress Report, TEI 490, 1954, 155.
675. **Masursky, H.**, Uranium in Coals of the Red Desert Area, Rep. TEI-540, Wyoming, U.S. Geological Survey, Reston, Va., 1955, 162.
676. **Masursky, H.**, Uranium-Bearing Coal in the Eastern Part of the Red Desert Area, Bull. 1099-B, Wyoming, U.S. Geological Survey, Reston, Va., 1962, 152.
677. **Vine, J. D. and Moore, G. W.**, Uranium-Bearing Coal and Carbonaceous Rocks in the Fall Creek Area, Bonneville County, Idaho, Circ. 212, U.S. Geological Survey, 1952, 40.
678. **Vine, J. D.**, Geology of uranium in coaly carbonacerous rocks, in Uranium in Carbonaceous Rocks, Prof. Paper 365-D, U.S. Geological Survey, Reston, Va., 1962, 113.
679. **Vine, J. D.**, Uranium resources of eastern Montana, Mineral Potential of Eastern Montana — A Basis for Future Growth, 89th Congress, Ist, Session 12, 1965, 67.
680. **Schopf, J. M., Gray, R. J., and Felix, C. J.**, Coal Petrology, TEI 490, U.S. Geological Survey, Reston, Va., December 1954, 175.
681. **Schopf, J. M. and Gray, R. J.**, Microscopic Studies of Uraniferous Coal Deposits, Circ. 343, U.S. Geological Survey, Reston, Va., 1954, 10.
682. **Moore, G. W., Melin, R. E., and Kepferle, R. C.**, Uranium-Bearing Lignite in Southwestern North Dakota, Bull. 1055-E, U.S. Geological Survey, Reston, Va., 1959, 147.
683. **Breger, I. A. and Deul, M.**, The organic geochemistry of uranium. Contribution to the Geology of Uranium and Thorium by the United States Geological Survey and Atomic Energy Commission for the United Nations International Conference on Peaceful Uses of Atomic Energy, Geneva, Switzerland, Prof. Paper 300, U.S. Geological Survey, Reston, Va., 1955, 505.
684. **Breger, I. A., Deul, M., and Meyrowitz, R.**, Geochemistry and Mineralogy of a uraniferous subbituminous coal, *Econ. Geol.*, 50, 610, 1955.
685. **Breger, I. A., Meyrowitz, R., and Deul, M.**, Effects of destructive distillation on the uranium associated with selected naturally occurring carbonaceous substances, *Science*, 120, 310, 1954.
686. **Breger, I. A.**, Role of organic matter in the accumulation of uranium. Organic geochemistry of the coal-uranium association, in Form. Uranium Ore Deposits, Proc. Symp., 1974, 99.
687. **Zeller, H. D. and Schopf, J. M.**, Core Drilling for Uranium-Bearing Lignite in Garding and Perkins Counties, South Dakota and Bowman County, North Dakota, Bull. 1055-C, U.S. Geological Survey, Reston, Va., 1959, 59.
688. **King, J. W. and Young, H. B.**, High-Grade Uraniferous Lignites in Harding County South Dakota, Prof. Paper 300, U.S. Geological Survey, Reston, Va., 1955, 419.
689. **Denson, N. M.**, Summary of Uranium-Bearing Coal, Lignite and Carbonaceous Shale Investigations in the Rocky Mountain Region During 1951, TEM 341A, U.S. Geological Survey, Reston, Va., 1952, 44.

690. **Denson, N. M. and Gill, J. R.**, Uranium-Bearing Lignite and its Relation to Volcanic Tuffs in Eastern Montana and North and South Dakota, Prof. Paper 300, U.S. Geological Survey, Reston, Va., 1956, 413.
691. **Denson, N. M., Bauchman, G. O., and Zeller, H. D.**, Uranium-Bearing Lignite in Northwestern South Dakota and Adjacent States, Bull. 1055, U.S. Geological Survey, Reston, Va., 1959, 11.
692. **Denson, N. M.**, Uranium in Coal in the Western United States, Bull. 1055, U.S. Geological Survey, Reston, Va., 1959, 1.
693. **Denson, N. M., Ed.**, Uranium in Coal in the Western United States, Bull. 1055, U.S. Geological Survey, Reston, Va., 1969.
694. **Pipiringos, G. N.**, Uranium-Bearing Coal in the Central Part of the Great Divided Basin, Sweetwater County, Wyoming, Prof. Paper 300, U.S. Geological Survey, Reston, Va., 1956, 433.
695. **Gray, R. J.**, Laboratory Study of Uranium-Bearing Carbonaceous Shale and Impure Coal from Goose Creek District, Cassia County, Idaho, Trace Elements Investigations Rep. 669, U.S. Geological Survey, Reston, Va., 1957, 20.
696. **Porter, E. S. and Petrov, H. G.**, Recovery of uranium from lignites, *Min. Eng.*, 1004, 1957.
697. **White, E. W.**, Uranium Mineralization in Some North and South Dakota Lignites, M.S. thesis, Department of Mineralogy, The Pennsylvania State University, University Park, 1958, 79.
698. **Bachman, G. O. et al.**, Uranium-Bearing Coal and Carbonaceous Shale in the LaVentana Mesa Area, Sandovan Country, New Mexico, Bull. 1055-J, U.S. Geological Survey, Reston, Va., 1959, 295.
699. **Mapel, W. J. and Hail, W. J.**, Tertiary Geology of the Goose Greek District, Cassia County, Idaho, Box Elder County, Utah, and Elko County, Nev., Bull. 1055-H, U.S. Geological Survey, Reston, Va., 1959, 217.
700. **Gill, J. R.**, Reconnaissance for Uranium in the Ekalaka Lignite Field, Carter County, Montana, Bull. 1055, U.S. Geological Survey, Reston, Va., 1959, 167.
701. **Gill, J. R., Zeller, H. D., and Schopf, J. H.**, Core Drilling for Uranium-Bearing Lignite, Mendenhall Area, Harding County, South Dakota, Bull. 1055, U.S. Geological Survey, Reston, Va., 1959, 97.
702. **Ergun, S., Donaldson, W. F., and Berger, I. A.**, Some physical and chemical properties of vitrains associated with uranium, *Fuel*, 39, 71, 1960.
703. **Astheimer, L., Schenk, H. J., and Schwochau, K.**, Uranium enrichment from sea water by adsorption on brown coal, *Chem. Ztg.*, 101, 544, 1977.
704. **Hail, W. J. and Gill, J. R.**, Results of Reconnaissance for Uraniferous Coal, Lignite, and Carbonaceous Shale in Western Montana, Circ. 251, U.S. Geological Survey, Reston, Va., 1963, 9.
705. **Green, A.**, Recovery of Uranium from Lignite, U.S. Patent 3,092,445, 1963.
706. **Mitchell, R. J.**, Uranium-bearing lignite — North Dakota's newest industry, *Met. Min. Process.*, 2, 17, 1965.
707. **Steinberg, M.**, Value and prospects of recovery of uranium from U.S. coal deposits, Nucl. Sci. Abstr. 1969, 21829, U.S. at Energy Comm., 1969, 45.
708. **Cameron, A. R. and Birmingham, T. F.**, Radioactivity in Western Canadian coals, Paper 70-52, Geological Survey of Canada, 1970, 35.
709. **Chow, T. J. and Earl, J. L.**, Lead and uranium in Pennsylvania anthracite, *Geochem. Geol.*, 6, 43, 1970.
710. **Jeczalik, A.**, Geochemistry of the uranium in uranium-bearing pit coals in Poland, *Buil. Inst. Geol. Warshaw*, 224, 103, 1970.
711. **Little, H. W., Dirham, C. C.**, Uranium in Stream Sediments in Carboniferous Rocks of Nova Scotia, Paper,Geological Survey Canada, 1971, 17.
712. **Noble, E. A.**, Metalliferous lignite in North Dakota, University of North Dakota Guidebook No. 3, Misc. Ser. 50, North Dakota Geological Survey, 1972, 133.
713. **Danchev, V. I. and Kuznetsov, V. G.**, Distribution of clarke uranium in terrigenous strata of the lower carboniferous in the Orenburg district, *Izv. Vyssh. Uchebn. Zaved., Geol. Razved*, 19, 62, 1976.
714. **Danchev, V. I. and Strelyanov, N. P.**, Uranium-coal deposits and their main genetic types, *Geol. Rud. Mestorozhd.*, 15, 66, 1973.
715. **Danchev, V. I., Vinokurov, S. F., Nemyshev, M. V., Ol'kha, V. V., Ostrovskaya, G. Ya., Rasulova, S. D., and Strelyanov, N. P.**, Some features of the geochemistry of uranium in carbonaceous sediments, *Tr. Inst. Geol. Geofiz., Akad. Nauk. S.S.S.R., Sib. Otd.*, 286, 272, 1975.
716. **Ree, C. and Emmermann, K. H.**, Sedimentary occurrences of uranium in Rheinland-Pfalz, *Mainzer Geowiss. Mitt.*, 3, 81, 1974.
717. **Gentry, R. V., Christie, W. H., Smith, D. H., Emery, J. F., Reynolds, S. A., Walker, R., Cristy, S. S., and Gentry, P. A.**, Radiohalos in coalified wood: new evidence relating to the time of uranium introduction and coalification, *Science*, 194, 315, 1976.
718. **Logomerac, V.**, Waste materials of some industries as sources for obtaining uranium, *Hem. Ind.* 31, 349, 1977.

719. **Koglin, E., Schenk, H. J., and Schwochau, K.**, Spectroscopic studies on the binding of uranium by brown coal, *Appl. Spectros.*, 32, 486, 1978.
720. **Shin, H. S., Suh, J. H., and Hyun, B. K.**, A study on the induced polarization effects of the uranium bearing low grade anthracite and surrounding rocks in Goesan district, Korea. I, *Teahan Kwangsan Hakhoe Chi*, 15, 1, 1978.
721. **Nadal, P., Morales, G., and Gases, P.**, Beneficio de Lignites Uraniferous Espanoles, Combustion en Lecho Fluidizado, unpublished rep., 1980.
722. **Facer, J. F.**, Uranium in Coal, Rep. GJBX-56(79), U.S. Department of Energy, Grand Junction Office, Colo., May 1979.
723. **Moore, G. W. and Stephens, J. G.**, Reconnaissance for Uranium Bearing Carbonaceous Rocks in California and Adjacent Parts of Oregon and Nevada, Circ. 313, U.S. Geological Survey, 1954, 8.
724. **Patterson, E. D.**, Uranium in Carbonaceous Rocks, Geol. Invest. of Radioact. Depos., Semiannu. Prog. Rep. TEI-390, June 1 to November 30, 1953, 116.
725. **Snider, J. L.**, Radioactivity of Some Coal and Shale of Pennsylvanian Age in Ohio, TEI-404, U.S. Geological Survey, U.S. Atomic Energy Commission, Washington, D.C., 22, 1954.
726. **Lovering, T. G.**, Radioactive Deposits of Nevada, Bull. 1009-C, U.S. Geological Survey, Reston, Va., 1954, 63.
727. **Bergstrom, J. R.**, The General Geology of Uranium in Southwestern North Dakota, Rep. Inv. No. 23, North Dakota Geological Survey, Reston, Va., 1956, 1.
728. **Ferm, J. C.**, Radioactivity of Coals and Associated Rocks and Beaver, Clearfied, and Jefferson Counties, Pennsylvania, Rep. TEI-468, U.S. Geological Survey, U.S. Atomic Energy Commission, Washington, D.C., 52, 1955.
729. **Welch, S. W.**, Radioactivity of Coal and Associated Rocks in the Anthracite Fields of Eastern Pennsylvania, TEI-348, U.S. Geological Survey, U.S. Atomic Energy Commission, Washington, D.C., 31, 1953.
730. **Zeller, H. D.**, Reconnaissance for Uranium-Bearing Carbonaceous Materials in Southern Utah, Circ. 349, U.S. Geological Survey, Reston, Va., 1955, 9.
731. **Borovec, Z.**, Role of organic substances in the geochemistry of uranium, *Cas. Mineral. Geol.*, 19, 77, 1974.
732. **Kakimi, T., Hirayama, J., Sekine, S., and Ikeda, K.**, Possibility of syngenetic concentration of uranium in a coal field, *Chishatsu Chosasho Hokoku*, 232, 659, 1969.
733. **Uspenskii, V. A. and Pen'kov, V. F.**, Chief types of uranium containing organic substances in rocks, *Tr. Inst. Geol. Geofiz., Akad. Nauk. S.S.S.R., Sib. Otd.*, 286, 272, 1975.
734. **Agiorgitis, G. and Schermann, D.**, Uranium concentrations in carbonaceous matter in rocks of the copper deposit Mitterberg (Salzburg, Austria), *Tschermaks Mineral. Petrogr. Mitt.*, 19, 81, 1973.
735. **Horr, C. A., Myers, A. T., Dunton, P. J., and Heyden, H. J.**, Uranium and Other Metals in Crude Oils, Bull. 1100, U.S. Geological Survey, Reston, Va., 1961.
736. **Ujhelyi, C.**, Determination of small amounts of uranium in coal, *Magy. Kem. Foly*, 61, 437, 1955.
737. **Szonntagh, J., Farady, L., and Janosi, A.**, Chromatographic determination of uranium in the ash of Hungarian coals, *Magy. Kem. Foly*, 61, 312, 1955.
738. **Perricos, D. C. and Belkas, E. P.**, Determination of uranium in uraniferous coal, *Talanta*, 16, 745, 1969.
739. **Korkisch, J., Farag, A., and Hecht, F.**, A new method for the determination of uranium in phosphates, coal ash, bauxite by means of ion-exchange, *Z. Anal. Chem.*, 161, 92, 1958.
740. **Fujii, Y., Hirai, S., Okamoto, M., Aida, M., and Fukuda, H.**, Determination of trace amount of uranium in fossil fuels by neutron activation analysis. *Bull. Res. Lab. Nucl. React. (Tokyo Inst. Technol.,)*, 4, 23, 1979.
741. **Guttag, N. S. and Grimaldi, F. S.**, Fluorimetric determination of uranium in shales, lignites, and monazites after alkali carbonate separation. XV, in Collected Papers on Methods of Analysis for Uranium and Thorium, Bull. 1006, U.S. Geological Survey, Reston, Va., 1954, 111.
742. **Cronin, J. T. and Leyden, D. E.**, Preconcentration of uranium for x-ray fluorescence determination on chemically-modified filter, *Int. J. Environ. Anal. Chem.*, 6, 255, 1979.
743. **Weaver, J. N.**, Rapid, instrumental neutron activation analysis for the determination of uranium in environmental matrices, Reply to comments, *Anal. Chem.*, 47, 1207, 1975.
744. **Kovalov, A. A.**, Polygenetic uranium mineralization in coal-bearing formations, *Sov. Geol.*, 13, 59, 1970.
745. **Schmidt-Collerus, J. J.**, Investigation of the Relationship Between Organic Matter and Uranium Deposit, Report GJBX-130, US DOE, Washington, 1979.
746. **Zverev, V. L., Kravtsov, A. I., Voitov, G. I., Pryakhina, E. V., Titaeva, N. A., and Cheshko, A. L.**, Uranium and its isotopes in coal, *Dokl. Akad. S.S.S.R.*, 246, 1217, 1979.
747. **Hamersma, J. W., Kraft, M. L., Meyers, R. A.**, Applicability of the Meyers process for desulfurization of U.S. coal (A survey of 35 coals), Am. Chem. Soc. Div. Fuel Chem. Prepr. 22, 1977.

748. **Headlee, A. J. W. and Hoskins, H. A.,** Sulfur in the Coals, Characteristics of Minable Coals of West Virginia, 13, West Virginia Geological Survey, p. 25, 1955.
749. **Mielicki, T.,** The Chlorine Content of Polish Coals, Biul. Inst. Nauk. - Badawczego Przemyslu Weglowego Komun., 36, 5, 1958, *Chem. Abstr.*, 48, 977, 1954.
750. **Bentisianov, Yu. V., Kel'tsev, A. V., Makarov, G. N., Madarova, N. D., Syskov, K. I.,** Effect of some factors on the transfer of germanium-containing compounds into gas products from coking of brown coal, Ref. Zh., Khim. 1973, Abstr. 19P89, *Saint. Anal. Strnkt. Oreg. Soedin*, 4, 123, 1972.
751. **Savul, M. and Ababi,** The copper, zinc, and lead content of several types of Romanian coal, Acad. rep. populare Romine, Filials Iasi, *Studi Cercetari Stiint. Chim.*, 2, 251, 1958.

INDEX

A

B

C

T

U

V

W

X

Y

Z